INTRODUCTION TO
EXPERIMENTAL ECOLOGY

Intraspecific competition. Man *vs* Man
(Reproduced with permission from Radio Times Hulton Picture Library.)

Interspecific competition. Locust *vs* Man
(Reproduced with permission from the Anti-locust Research Centre, photographed by
D. G. Greathead.)

PLATE 1

Introduction to
Experimental Ecology

A Student Guide to
Fieldwork and Analysis

T. LEWIS M.A., Ph.D.

and

L. R. TAYLOR D.Sc.

Rothamsted Experimental Station,
Harpenden, Herts, England

ACADEMIC PRESS

LONDON · NEW YORK

ACADEMIC PRESS INC. (LONDON) LTD
24–28 Oval Road
London, N.W.1

U.S. Edition published by
ACADEMIC PRESS INC.
111 Fifth Avenue
New York, New York 10003

Library of Congress Catalog Card Number: 67–17591
ISBN Casebound edition: 0 12 447150 1
ISBN Paperback editon: 0 12 447156 0

Made and printed by offset in Great Britain by
William Clowes & Sons, Limited, London, Beccles and Colchester

Preface

Schoolteachers and students are increasingly aware that "biology" is more than the sum of anatomy, physiology and biochemistry. Changing syllabuses for university examinations, the Nuffield Scheme in the U.K., and the B.S.C.S. Green Book in the U.S., reflect this new awareness. To appreciate our rapidly changing surroundings, a better understanding is needed of how animals, including Man, respond to and are affected by their environment. The employment of ecologists in industry, agriculture and government is evidence of the recognition of this need. Yet properly trained ecologists are few, and non-biologists receive little or no ecological training even though some of them, town-planners, architects and engineers, largely create the environment we live in. This is partly because ecology has often been considered a pleasant, but valueless, country pastime, and partly because there is great difficulty in teaching such an all-embracing subject. It is relatively easy, with a little guidance, to collect ecological material, but much more difficult to extract from it a clear statement of the interdependence of organism and environment. This difficulty is increased by the long time scale of most ecological processes. Yet to be understood, ecology must be seen as evolution in action, and as applicable to individual people; each human life is an ecological experiment, an experiment in living in a society, in a rapidly changing man-made environment.

This book is an *introduction* to ecology because it concentrates on simple relationships. The subject of ecology extends far beyond the interdependence between species, and the dependence of the species' complex on habitat demonstrated by some exercises in this book. Some ecologists would say that true ecology begins where this book ends, although this is not our view. Nevertheless, until these basic relationships are clearly understood, more advanced ecology cannot be attempted and for the non-specialist, these simpler aspects of ecology demonstrate clearly the essential principles of variability of individuals and of populations, as well as Man's role in his environment. The aim of this book is to guide the teacher and student through the principles of ecology, to a realistic, quantitative understanding of the dependence of living animals on their changing environment.

Ecology is essentially a numerical subject. However, just as it is possible to be literate without being an expert grammarian, so we have attempted to lead the student to numeracy without being an expert mathematician. A similar argument applies to taxonomy, which is essential to ecology, and we have rejected the intractable problems of written, formal, keys in favour of simple,

v

partial, visual ones. The sections on Analysis (Chapter III) and Identification (Chapter VI) are tailored strictly to the requirements of the exercises; they are not texts in statistics or taxonomy and must not be judged as such. Both are presented diagrammatically and attempt to bridge the gap between elementary mathematics and biology and more specialized University courses. They are designed to give results quickly, easily, and accurately enough for our present purpose. Their use may be extended, with increasing facility, to the books listed as further reading. In the U.S. especially, the keys in this book will be only partially useful, but this applies to any key: if it is simple, it cannot be universal. Suitable visual keys exist in the U.S. and these are listed in Appendix N. The metric system and the centigrade temperature scale have been used wherever practicable. There are occasional exceptions as, for example, body temperature in Man, which is measured with medical thermometers, still calibrated in °F.

However effectively the ancillary subjects of taxonomy and mathematics are covered, the choice of exercises (Chapter IV) will determine whether the subject can be learned in practice. Plants and invertebrate animals are the most easily manipulated organisms and because moving animals illustrate more ecological principles than static plants, and because insects are the most common animals, these have been used as experimental material in most exercises.

Most of the exercises were devised during practical ecology courses or are parts of experiments. Typical results are included to show the type of information collected and how conclusions can be drawn from it. We do not suppose that all can be repeated in detail in all habitats, but with a little ingenuity most should be adaptable to local conditions. Repetition of exercises will suggest improvements and the accumulation of results from season to season will add greatly to their value and the interest to be derived from them. We have not given a long list of possible variants at the end of each exercise, but a few riders have been suggested where they seemed appropriate. The exercises are classified on p. 113 under headings listed in school and university Biology and Zoology syllabuses, and by requirements and facilities necessary to complete each one. In Chapter IV they are arranged to illustrate progressively more advanced concepts, and therefore generally become more difficult towards the end of the Chapter.

Our dual aim has been to provide practical examples of ecological principles and to show the relevance of ecology to Man, something that seems not previously to have been attempted in one book. In school the exercises can be done under the direction of teachers, or at higher levels the students can use the exercises as a guide in devising their own field experiments. Each exercise provides training in ecological methods and analysis, but in practice the experiments can be "open-ended" depending upon the amount of guidance given by the teacher. Naturalists on field courses and student teachers may find the outlook and ideas useful in developing local fauna surveys and training projects on a quantitative basis, and for undergraduates it offers a sound basis for more advanced studies.

We shall be pleased to hear of the application of exercises, successful or otherwise, and of any modifications adopted, as well as comments on errors and omissions.

Harpenden, Herts.   T. LEWIS
December, 1966   L. R. TAYLOR

Preface to the Fourth Printing

We would like to remind teachers that this is an *Introduction* to experimental ecology designed to give students a scientific approach to ecology by providing the practical and analytical methods necessary to enable them to formulate realistic problems and then collect appropriate evidence to answer them. The simple, direct, numerical method has been adopted to get away from the obscurantism to which some ecology has been prone. However, the essence of ecology is variability and students should not expect to obtain results identical with those presented here; the variability of the material will make it necessary to interpret results in terms of local, diurnal and seasonal environments. The long term experiments, which emphasize seasonal variability are now done in many secondary, and even primary schools. Their inclusion in this book does not imply that it is suitable only for university students.

Preface to the Students Edition

This book was first published just before the current wave of concern, especially amongst young people, for the environment. It was written in simple, non-technical, language to help teachers and students appreciate the ecological principles underlying the interdependence of living things. The critical approach presented is even more necessary in the present atmosphere when there is a danger of ecology becoming a facile, emotional, panacea for all social ills.

The book is now widely used as a text in universities, colleges and schools in many parts of the world and its translation into other languages attests its success in its main objectives. However, it has been used mainly by teachers to formulate courses and projects, and this was not our original intention. Perhaps its greater value lies in the ease with which students, naturalists and ecological enthusiasts can, from their own observations, use the exercises to frame meaningful questions capable of simple analysis.

For these reasons this soft-backed student edition has been produced: not as a reference book, but as a practical guide to help appraise, sample, analyse and understand the multitude of field situations that the student ecologist or interested naturalist will inevitably encounter.

May, 1974
T. Lewis
L. R. Taylor

Contents

CHAPTER I

Introduction to Ecology

DEFINITION OF ECOLOGY

When a farmer grows a crop of wheat, he decides how to treat the soil, when to sow the seeds, how to feed and look after the growing plants and when to harvest the grain. In doing these things he applies his own judgement and the experience of generations to help the wheat plant withstand frost in winter, grow quickly in competition with weeds in spring, to be strong enough to withstand pests in summer and ripen before the autumn mists and rain. He seeks to enable the plant to take every advantage of its environment by selecting seed that germinates and roots well in the local soil, grows at the right temperatures and flowers when the light is right and that resists the local pests and diseases. He hopes to produce a large crop of good food and make a profit and, as individuals, we may judge him to be a good farmer if he succeeds. However, from the point of view of his community, a local population of the species Man, he may be judged a better farmer if he improves the land as he farms it, for bad farming can ruin land by allowing soil to erode, to become poisonous or water-logged, or to decline in fertility and Man, as a species, is interested in the future as well as the present use of the land.

In all these aspects of growing his crop, the farmer is manipulating the ecology of wheat plants, which have been selected by plant breeders over a long period of time and many generations, to give the best results in a particular environment.

Over an even longer period of time, natural environmental pressures, such as heating and cooling of the earth and fluctuations in rainfall, acting upon living organisms have produced all the morphological, anatomical, physiological and behavioural differences by which all existing species, and individuals, are characterized. *These many and varied environmental pressures are continually changing, hour by hour, century by century and ecology is the study of the way in which individual organisms, populations of the same species and communites of populations respond to these changes.*

THE SCOPE OF ECOLOGY

Ecologists attempt to describe how organisms behave in nature and explain such fundamental questions as why certain organisms live in a particular place, what regulates their numbers and what differences occur within and between individuals and populations.

1

Ecological studies, therefore, embrace many disciplines and aspects of science. Neither the individuals in populations, nor the environments are uniform. Men, for example, are not all the same height; so to describe height adequately in a population of men, a statistical technique must be employed to define both the average height of the population and the limits of variability. As variability occurs not only between the heights of men, but in every physical, physiological and behavioural detail of living organisms, many features of populations must be described by statistical methods. Mathematical reasoning is required to define many responses of individuals to environments. Taxonomic techniques are needed to separate species and subspecies of organisms and to name them. The responses of plants and animals to rapid changes in the environment, such as cooling when the sun is temporarily covered by a cloud, are physiological and behavioural phenomena, whereas long-term fluctuations, such as a gradual change in annual rainfall, may introduce genetical and evolutionary aspects.

Studies of individual organisms and populations of organisms require different approaches. The behaviour of individuals is most often investigated in the laboratory and the behaviour of populations is mainly examined in natural habitats. Each is examined by different techniques and each obeys its own laws, just as human individuals and populations do in everyday life. People as individuals are born and die only once. Unlike breathing or sleeping, which are cyclical processes within each individual, there is no such thing as a cycle of birth and death for one person. Each of these two most important events in a man's life is a unique occasion and to study them would require constant observation, or the brief moment at which they occurred might be missed. In the population of a city, on the other hand, there is a continued sequence of births and deaths that can be recorded as rates, with seasonal and diurnal cycles that are peculiar to the population.

Another aspect of population ecology, interactions between species of plants and animals, is less familiar. This interdependence is vital to animal communities. For example, flowers supply food for bees; bees are eaten by mice; mice are hunted by owls, but owls also compete with foxes; foxes hunt rabbits and rabbits eat flowers, so affecting bees. Should one link be removed, the structure of the community may be seriously disrupted. Interactions within a species, e.g. between different populations of men (see Plate 1), are more familiar through historical records, and interactions between races are a basic human problem.

CONCEPTS OF ENVIRONMENT

Organisms must not be thought of as fighting against their harsh environment; they have evolved with it and their behaviour is adapted to it. Sea fish do not constantly resist life in salt water, though fresh-water fish, not adapted to live in it, would die. Each species takes advantage of the environment of which it has become a part. This is obvious when watching a seagull, or swift, soaring on invisible air currents. The environment they live in is not the same as Man's, for they can feel every slight movement of the air in which they are enveloped; not just a wind rushing past them as a man does. Because they are "part" of the air, they feel its twists and turns, spirals and slip-streams, velocity gradients and temperature gradients. They are as sensitive to these aerial patterns as a blind man's finger tips are to the texture of surfaces. Perhaps one of the most difficult and yet necessary disciplines in ecology is to learn to appreciate the environment, not as it is to Man, but as it is to another animal. For example, colours appear differently to men, who have colour vision, and dogs who are colour blind, because men and dogs have different visual perceptions.

Concepts of orientation may be even more difficult for men to imagine, for orientation is sensory perception plus the resulting behaviour. Different species use different senses and media with which to orientate. Human beings use mainly sight and depend on light to do so. Moths often use scent and depend upon the direction of the wind to contact each other; in human terms, this is as if people in the street suddenly became invisible when they pass going down wind.

Different sensory mechanisms are also employed by the same species to perceive the surroundings in different environments. For example, in the open air, honey bees orientate to a horizontal compass bearing, mediated by the plane of polarization of light, until they see a flower; then they fly towards it. In the different environment of the beehive, where there is no light, they communicate information about orientation to one another by scent and touch, while crawling on the vertical surface of the honeycombs. There, the plane of orientation is up and down, left and right, with gravity as the reference medium. In the transfer from one set of horizontal coordinates to another set of vertical ones, the behaviour of the bees is similar but the environment and sensory mechanisms employed have changed.

ECOLOGY AND MODERN MAN

Difficulties in imagining another animal's world may have been increased by Western culture which has regarded Man as a special creature and the individual as of primary importance. Other, more primitive, cultures, some-times had a more objective and, therefore, more ecologically sound, philo-sophy, which granted each species of animal a place in the system. Man has been successful in what is called "controlling his environment". These con-trols are impressive in their number and variety. Control of the media of sensory perception has extended his senses far beyond their unaided limits. For example, the control of light and electron flow in telescopes and micro-scopes has extended the range of our vision upwards to almost unimaginable distances and downwards to incredibly small dimensions; the environmental restrictions on our movement on and below the earth and sea and in the air have been removed by vehicles of all kinds; the soil, sea, sun and the winds are all made to work for Man, making his environment suit his needs.

Unfortunately, control of an environment has often led to its destruction.

Many rivers now flow polluted and lifeless through our great cities. The indiscriminate use of land for farming has sometimes turned forests into deserts and Man has destroyed many species of animals because their needs conflicted with his own. The things Man makes and the towns he lives in all become part of his environment and the problem of adjusting himself to changes in these is an ecological one. Groups of individuals of all species, including Man, have typical patterns of spatial distribution that are inadequately understood. Individuals of some species occur aggregated together, others are always solitary. Changing these spatial patterns suddenly may induce unexpected environmental tensions, hence the sudden growth of new cities presents ecological problems.

Many economic, social and political problems of today are the result of Man's ability in the past to change himself and his environment rapidly but with inadequate understanding of the ecological consequences of these actions. Improvements in the nutrition and health of crops have produced more food for men but the beneficial effects may be only temporary if they lead to increases in human populations that in turn require still more, and may start another cycle of food shortages, population instability and perhaps war. With an increased awareness of the interdependence of life and its environment, Man can more effectively manipulate his own surroundings for good or ill. An understanding of ecological principles is relevant to all human activities.

The need for more ecological knowledge and its wise application become increasingly important. During the life-time of most readers, the human population of the world will double. The problem is not only how to feed these people, but how to live with them in the new environment being so rapidly created. Most of the ecological principles demonstrated in this book apply to Man, and invertebrates are used as experimental material merely because they are easier to study than larger animals. Perhaps this book will encourage some of the next generation of scientists to accept the challenges ecology presents and to train themselves to meet it before it is too late.

CHAPTER II

Principles of Ecology

Individual organisms do not live in isolation but as breeding populations of the same species (Chapter IV, Ex. 25). Sometimes many populations of different species live together as "communities" occupying "habitats" in which food, climate and space are suitable. Different habitats support different populations of plants and animals (Chapter IV, Ex. 39) and these diverse populations and habitats together form ecosystems.

In the different ecosystems different animals may fulfil similar roles and so occupy similar "niches". For example, a rabbit in a wood and a grass-moth caterpillar, or webworm, on a river bank, although not closely related taxonomically, both eat grass and, therefore, occupy similar niches in the ecosystems they inhabit; a fox feeding on the rabbit occupies a comparable niche to a dragonfly preying on the grass-moth. The more similar the needs of different species, the more their niches overlap, but competition for food and space can be as intense between distantly related species as between closely related ones. For example, locusts usually eat grass and wild plants and do not then compete with man, but when large swarms migrate, they eat crops and so compete with man for food (see Plate 1).

THE ENVIRONMENT

An animal's environment has four basic components; weather, habitat, food and other organisms including other individuals of its own species. Together these affect birthrate, deathrate and dispersal and so cause populations to fluctuate in size and position.

WEATHER

Animals live almost everywhere on the earth, but the way different species are distributed is largely governed by climate. For instance, the polyphagus Cacao thrips *Selenothrips rubrocinctus* needs hot, humid weather throughout the year, and so is confined to equatorial regions; the Codling Moth *Cydia pomonella* which eats species of *Pyrus*, *Prunus* and walnut, needs a cooler climate where humidity fluctuates seasonally, and so it occurs in the temperate regions of northern and southern hemispheres (Fig. II:1). The ranges of the two species overlap only in South America where the Codling Moth occurs in equatorial latitudes at high, cooler altitudes and the thrips in the hot, humid plains because increases in altitude and latitude often have a similar effect on the flora and fauna.

5

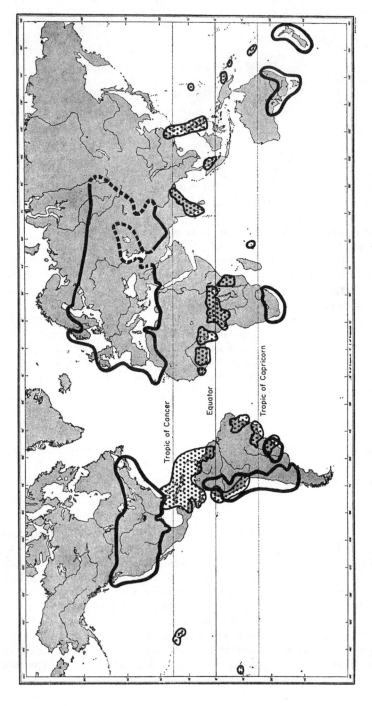

Fig. II:1. The limits of the geographical distribution of the Cacao thrips (*Selenothrips rubrocinctus*) (dotted zones) and the Codling moth (*Cydia pomonella*) (clear zones). (Commonwealth Institute of Entomology.)

Ecologists think of weather differently from most meteorologists because local differences, sometimes occurring within a few metres or even centimetres of an animal, can affect survival. Thus, within a broad climatic type there are local microclimates caused by differences in the geology, topography and vegetation (Chapter IV, Ex. 43). Different rocks near the surface affect the drainage and thus the temperature and moisture content of the soil. Small depressions, or even hedges planted across a valley, may create pools of cool air at night and plants and animals that are not frost-hardy may sometimes freeze and die there. The microclimate within woodland differs greatly from

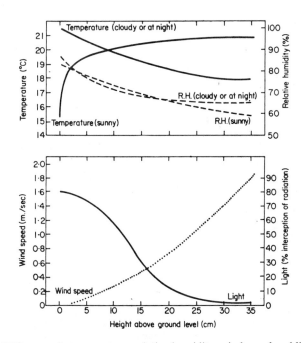

FIG. II:2. Differences in temperature, relative humidity, wind speed and light intensity occurring among grass between ground level and the tips of the plants, 35 cm tall. (Redrawn after Hughes.)

that in grassland and Fig. II:2 illustrates the variability in the microclimate of a grass sward. This variability directly affects the behaviour of animals living there (Chapter IV, Ex. 27). Sun-loving insects, like hover flies, are usually found flying above the vegetation or resting on flowers or the tips of plants. Others, like nabid bugs, live in the cooler, dark, moist bases of grasses during the daytime and climb upwards to search for food only at night when it is dark and humid. Similar microclimatic variations occur in many habitats such as in soil (Fig. II:3), on a cliff and among shingle (Fig. II:4).

Temperature, humidity and light are the most important weather factors and in nature the effects of one often predominate. Each may affect rates of growth, movement and the length of life of organisms, either directly or indirectly through their effect on other organisms, especially of plants that provide food or a place to live. Their separate effects can often be distinguished only by studying them independently.

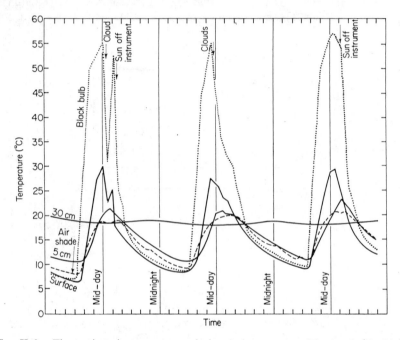

FIG. II:3.   Fluctuations in temperature during 3 days, measured in the air (shade), at the surface of the soil, and 5 and 30 cm below. The black bulb trace provides an estimate of radiation from the sun. The greatest fluctuations occur in positions exposed to the sun, and the least, deep in the soil. (Redrawn after Williams.)

Temperature

Normal life persists within narrow temperature limits of about -10 to $+50°C$. Individual species survive in a smaller range (Chapter IV, Ex. 10) and are active within even narrower limits. If too cold, cell proteins may be destroyed as ice forms, or as water is lost and electrolytes become concentrated in the cells; heat coagulates protein.

A small proportion of organisms, mostly vertebrates, can prevent their internal temperature from following changes of external temperature. Many animals are physiologically adapted to survive heat or cold and become inactive as they aestivate or hibernate.

Warm blooded animals, or homoiotherms, maintain fairly constant temperature by physiological mechanisms, except when their metabolism is upset during illness, but like all other characteristics, this "constant" body temperature differs slightly between individuals (Chapter IV, Ex. 1). In "coldblooded" animals, or poikilotherms, body temperature is similar to the temperature of the surroundings (Fig. II:5 and Chapter IV, Ex. 12). Because an animal's rate of growth largely depends on body temperature (Chapter

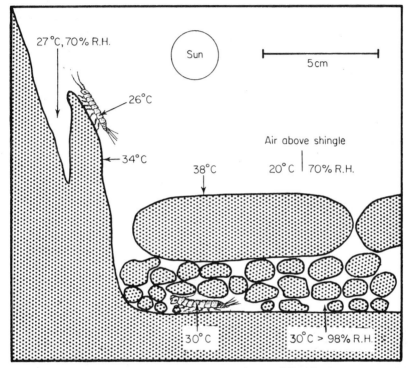

FIG. II:4.   Diagrammatic vertical section of a sandstone cliff and shingle inhabited by the Sea Slater (*Ligia oceanica*), showing microclimatic conditions and internal temperatures of the animals on a typical summer afternoon. (Redrawn after Edney *J. exp. Biol. 30* (1953).)

IV, Ex. 13) immature homoiotherms develop at about the same rate in all normal temperatures, but poikilotherms grow faster when it is warm and slower when it is cold.

The range of temperature favourable for activity and development of aquatic animals, especially marine animals, is smaller than for terrestrial ones. Deep sea temperature fluctuates only a few degrees; it varies more in shallow and, therefore, in fresh water, but even there the temperature cannot fall far below freezing or rise far above air temperature. Temperatures vary much more on land according to the area, season, daily weather and time of day. Temperature fluctuates rhythmically within a day and season and often produces rhythmical activity in animals, such as insects flying when it is warm enough and resting when too cold (Chapter IV, Ex. 11). The ranges of temperature favourable for the development of various animals are given in Fig. II:6a. The upper limit of the favourable range for the Sea Slug *Dendronotus frondosus*, which lives in cool oceans, is below the lower limit for the crustacean *Lernaea elegans*. Comparison of these temperature ranges with

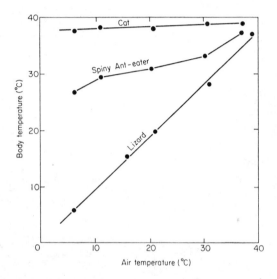

FIG. II:5. Animals in the evolutionary scale up to and including the reptiles (Lizard) are poikilotherms and their body temperature reflects their surroundings. Birds and the higher mammals (Cat) are homoiotherms, with constant body temperature in most air temperatures. In between are a few primitive mammals which have partial temperature control (Spiny Ant-eater). (After Martin.)

the annual temperature cycles at different latitudes (Fig. II:6b) indicates where and when the temperature is suitable for these animals to develop. Thus, the housefly can live out of doors in tropical latitudes throughout the year, but only for a few months in temperate regions where it spends much of the year inside warm buildings. The distribution of the Firebrat (*Thermobia domestica*) is even more affected by Man, for it lives at temperatures as high as 42°C and frequently inhabits kitchen hearths and the hobs of bakers' ovens. The related Silverfish (*Lepisma saccharina*) is a common household pest in temperate latitudes, but requires much lower temperatures than the Firebrat and develops best at 22°–27°C.

A few exceptional poikilotherms, especially insects, can regulate their temperature slightly. Hawk moths can raise the temperature of their flight muscles to 32°–36°C by vibrating the wings before take-off and gregarious butterfly larvae may raise their temperature 1½–2°C when clustered together. Locust hoppers may increase their temperature 10°C by basking sideways in the sun. Ants move their larvae to warm or cool places within the nest and bees maintain temperatures within their hives between 13° and 25°C by

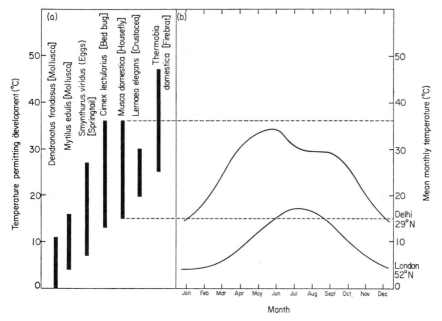

FIG. II:6. (a) The range of temperature suitable for the development of seven species of animals compared with (b) the annual temperature cycle at Delhi (29°N) and London (52°N), to show which species can survive in tropical and temperate latitudes. The dotted lines represent the temperature range within which the Housefly (*Musca domestica*) survives. (Data after Moore.)

fanning with their wings to evaporate water droplets when it is too hot, or releasing body heat through increased metabolic activity, when too cold.

However, for most poikilotherms, temperature fluctuations above or below tolerable limits kill all but resistant stages adapted to survive extremes. Thus, many species grow, mature and reproduce in less than half the year even in temperate latitudes, surviving inactive at other times.

Humidity

Too much or too little moisture may sometimes be harmful. In dry air too much water may be lost and in damp heat animals may be unable to cool themselves by evaporation.

Most terrestrial invertebrates are small and soon die when too much water is lost from their bodies during respiration, by excretion, or by transpiration through the skin or cuticle. The humidity of the atmosphere surrounding the

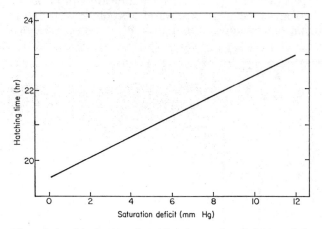

FIG. II:7.   The relationship between humidity (saturation deficit) and the time taken for eggs of the blowfly (*Lucilia sericata*) to hatch at 27°C. (After Evans.)

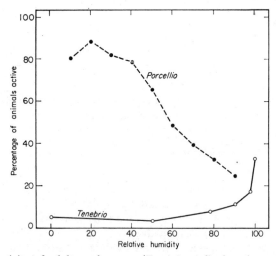

FIG. II:8.   Activity of adult meal worms (*Tenebrio molitor*) and wood lice (*Porcellio scaber*) in uniform humidities. (Redrawn after Pielou and Gunn.)

animal influences the water lost and often determines where the animal can live (Chapter IV, Ex. 8, 9) and its rate of development (Fig. II:7).
Activity also often depends on whether the atmosphere is dry or moist.

Adult mealworms (*Tenebrio molitor*) crawl most in moist air but the wood louse or sowbug (*Porcellio scaber*) crawls most in dry air (Fig. II:8).

Light

Light frequently determines which animals are in a habitat, because it controls the vegetation growing there. If the light intensities are unsuitable, the animals either move to more favourable illumination, or die. Fecundity too may be influenced by light intensity. The fruit fly (*Dacus tryoni*) lays most eggs in bright light and none are laid in darkness (Fig. II:9).

Many insects fly at sharply defined light intensities, different for different species. For example, *Drosophila subobscura* flies for half an hour at sunrise

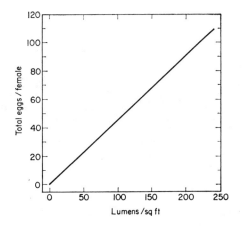

Fig. II:9. The relationship between light intensity and fecundity over a period of eight weeks in Queensland fruit flies (*Dacus tryoni*) kept at 26°C, 75% relative humidity and with paw paw mixture for food. (Redrawn after Barton-Browne.)

and sunset (33 f.c.) whereas the grass moth (*Crambus perlellus*) flies most at midnight (0·0008–0·02 f.c.). Day fliers, like hover flies and butterflies, respond only to bright light between about 2000 and 9000 f.c., whereas others like Tipulids are active over a much greater range of light intensities from dusk (10–100 f.c.) to starlight (0·00008 f.c.). The time of day at which dawn or dusk-flying insects appear, changes through the year with the changing time of sunset (Fig. II:10).

Wind

The most important effect of wind on small animals, especially invertebrates, is during dispersal (see p. 29). It affects the speed at which they move (Chapter IV, Ex. 21) and their ability to control their track in the air (Chapter IV, Ex. 22, 23). For vertebrates living in exposed places, like sheep on hills, it affects the rate at which their bodies lose heat. Also on exposed sites, especially on coasts, wind may inhibit the growth of many plants and trees, and make the fauna less diverse.

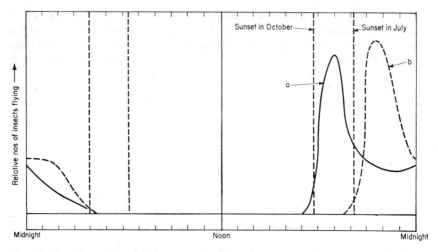

FIG. II:10. The relationship between the change in time of peak flight activity and the change in the time of sunset, for the fungus gnat *Mycetophila ocellus*. (a) The mean daily flight–periodicity curve for October and (b) for July. Sunset in October is about 2½ hr earlier than in July, so the time of maximum flight in October is similarly 2½ hr earlier.

HABITAT

In areas with similar climates, local differences in soil, vegetation and cultivation produce different habitats suitable for different species (Chapter IV, Ex. 35, 36, 37). Figure II:11 shows the relative abundance of various species of insects and spiders in environments created by different stages of plant succession. A grasshopper, an ant and a mirid bug were most abundant in the early stages of the succession, another mirid bug and a springtail in the climax forest.

In most parts of England, the climate is suitable for two snails, the large Roman snail (*Helix pomatia*) and *Zonitoides excavatus*, but *H. pomatia* occurs only on calcareous soil in or near beechwoods and *Z. excavatus* mainly on acid heaths. Similarly, crane flies and chafer beetles occur in all counties in the U.K., but leatherjackets, the larvae of crane flies, are most abundant in damp, clayey areas and chafer larvae in drier, sandy soil.

On a smaller scale, the suitability of habitats only a few metres or even centimetres apart may differ greatly. The large bark beetle, *Scolytus destructor*, which feeds on elm trees, prefers trunks and branches exceeding 15 cm diameter, but *S. multistriatus*, a smaller species, prefers slender branches of 5–8 cm in diameter (Chapter IV, Ex. 20). In a square centimetre of pine bark, as many as ten cereal thrips (*Limothrips cerealium*) may collect in tiny crevices to hibernate with their bodies pressed against the crevice walls, which must be between 0·1 and 0·5 mm apart. Wider or narrower crevices within a few centimetres of the crowded area may be empty and if there is a shortage of crevices of suitable size, population density is limited

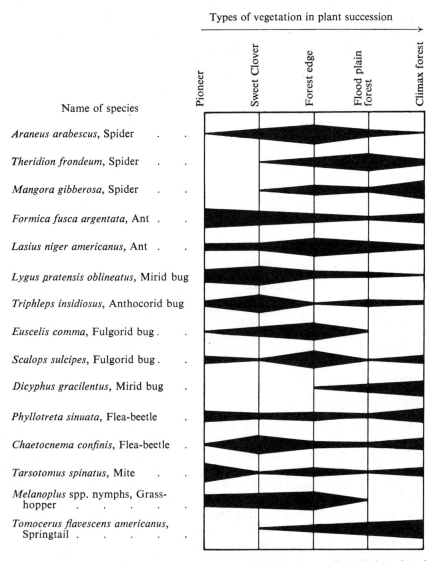

Fig. II:11. A graphical representation of the relative abundance of selected species of insects and spiders among different vegetation. The wider the line, the more abundant the species. The five different types of vegetation represent stages in the plant succession leading to red oak/maple forest in North America. (Redrawn after Smith.)

(Chapter IV, Ex. 40). Deep burrowing species of earthworms, like *Lumbricus terrestris*, live only where soil is deep but other species may inhabit shallow soil (Chapter VI, Ex. 38). Larvae of an Australian swift moth *Oncopera fasciculata*, in a field containing tufts of cocksfoot, may live in a burrow made in open ground or in one beneath a tussock. In dry weather, larvae in both habitats survive but if the ground is water-logged, only those that crawl up into tussocks escape drowning (Fig. II:12). Such minor variations within a habitat are one of the causes of aggregation in populations (Chapter IV, Ex. 25).

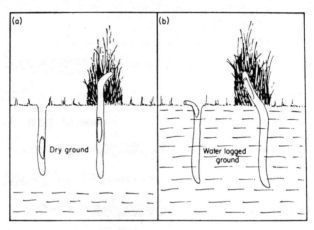

FIG. II:12.   In dry weather, larvae of an Australian swift moth, *Oncopera fasciculata*, survive in burrows in open ground or beneath tussocks of cocksfoot grass (a), but if the soil becomes water-logged, larvae in exposed burrows may drown, whereas those with burrows that penetrate tussocks probably survive (b). (Redrawn after Madge.)

FOOD

Availability of Food

For an animal to survive and reproduce it must have enough of the right kind of food, but the mere existence of enough food in the neighbourhood of the animal may not ensure its survival. Thus, in a pasture containing both sheep and the sheep tick, *Ixodes ricinus*, there is abundant food for the ticks, as each sheep has enough blood to provide very many tick meals. However, the tick has to feed at least three separate times in its life, once as a larva, once as a nymph and once as an adult, and on each occasion it must crawl on to a sheep. Hence, although the tick can survive considerable periods without food, whether it matures depends on it encountering a sheep on three critical occasions, which depends primarily on the frequency with which the pasture is grazed and on the number of sheep per unit area of pasture. Lack of food at a critical time, rather than an absolute shortage, can also affect hover flies. For their ovaries to mature, they must feed on pollen in spring.

Hence, when the weather is such that hedgerow plants are in flower when the flies emerge, the flies can mature quickly and lay many eggs, but when flowering is late they mature more slowly and lay fewer eggs.

Absolute shortages of food are more common than the kind already discussed. Absolute shortages may mean not only fewer animals survive, but whole populations may be destroyed. Thus, whether blowflies will develop on a piece of carrion in which eggs are laid will depend on the size of the carrion and the number of eggs. With too many eggs, no flies will emerge because no maggots get enough food to complete development.

Quality of Food

Not only quantity but quality of food can be important in determining the size of populations. A striking effect of quality on fecundity is shown by

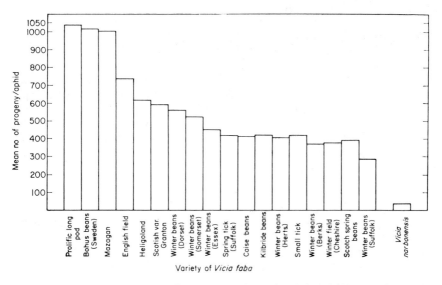

FIG. II:13. The mean number of progeny produced in 14 days by one parthogenetic female Bean Aphid (*Aphis fabae*) on different varieties of the field bean (*Vicia faba*) and on *Vicia narbonensis*.

the number of *Aphis fabae* produced when feeding on plants of different bean varieties. One parthenogenetic female was placed on plants of each variety and humidity and sunshine were the same for all. All aphids originated from the same egg (see p. 33) and so were genetically identical. The mean number produced on different varieties after 14 days ranged from 286 on a Suffolk type of winter bean to 1037 on Prolific Longpod beans. On *Vicia narbonensis*, the aphid-resistant wild bean from which the others have been developed, only 37 aphids were produced (Fig. II:13).

Food Chains

All food, carbohydrate, protein and fat, is synthesized by plants from simple nutrients obtained from the soil and air. Many animals do not receive these foodstuffs directly from plants, but second-, third- or sometimes fourth-hand, through a chain of predators after the plant tissue has been converted into animal tissue by a herbivore. Typical food chains usually have three or four links. For example, plant-feeding aphids provide food for lacewings, which are eaten by small birds which in turn are food for hawks or cats. Characteristically, the more links in the chain, the larger and fewer do animals become in each successive "link". Within such systems, the number of individuals of different species fluctuates, but the proportions of the species in different parts of the chain remain fairly constant. The effects of removing one link from such a system may be far-reaching. For instance, DDT is more toxic to mites than to springtails. After this insecticide is applied to the soil, the number of springtails soon exceeds the original number because predatory mites are fewer (Chapter IV, Ex. 34). Small quantities of chemicals, absorbed by animals early in the chain, may accumulate in those that eat them and become concentrated in the final predator, to harm its metabolism and reproduction. The desire for increased standards of living and food production must, therefore, be balanced against the potential danger to biological systems from the many thousands of tons of slowly destructible, toxic material spread annually by Man into his environment. These include the lead released from coal and "anti-knock" compounds in petrol, molybdenum and cadmium from industrial plant, the radioactive dust from nuclear explosions, and toxic pesticides and herbicides, which all eventually collect in the soil.

OTHER ORGANISMS

As well as feeding on each other, animals within a habitat are influenced in other ways by members of their own species, by other species, and also by visiting animals.

Intraspecific Effects

Scarcity or crowding in a single species may make survival difficult. Bisexual animals must be sufficiently abundant to meet a mate. The sheep tick is not found on some farms in northern England where the climate and terrain are suitable because the sheep are too scattered for ticks to find them and unless a male and female tick become attached to the wool of the same sheep and mate, they cannot reproduce. With few ticks the chance of a sheep collecting male and female and their finding each other is slender, so unless ticks are already common on a farm and average about one attached tick per sheep, they are unlikely to become established and increase.

The effect of crowding on the longevity of fruit flies is shown in Fig. II:14. There is an optimum density, at which they live longest; at greater or

lesser densities their lives are shorter. Their reproduction rate is also reduced
in dense populations (Fig. II:15).

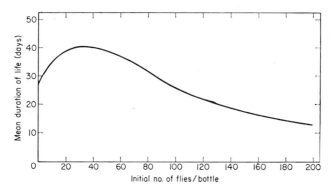

Fig. II:14.   The relationship between mean length of life and population density in
fruit flies (*Drosophila*). (Redrawn after Pearl, Miner and Parker.)

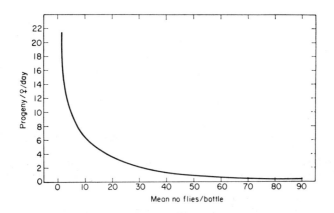

Fig. II:15.   The relationship between rate of reproduction in fruit flies (*Drosophila*) and
the density of the mated population. (Redrawn after Pearl.)

Interspecific Competition for Food and Space

Populations of different species living in the same environment may
influence each other in many ways; for example, they compete for food or
space, or disturb or kill each other. The growth of populations of *Paramecium
aurelia* and *P. caudatum* reared separately is shown in Fig. II:16a. Numbers
of *P. aurelia* increase the more rapidly but both populations attain a stable
size. When the species are reared together, all *P. caudatum* eventually die,
not because *P. aurelia* interferes with them directly, but because they eat the

same food more quickly (Fig. II:16b). Figures II:16c,d explain the effects of competition on each species separately. In contrast, different species of insects causing spangle galls on oak occupy different spatial zones on the same tree and on a single leaf. This lessens competition between the species and enhances survival of each (Chapter IV, Ex. 41, 42).

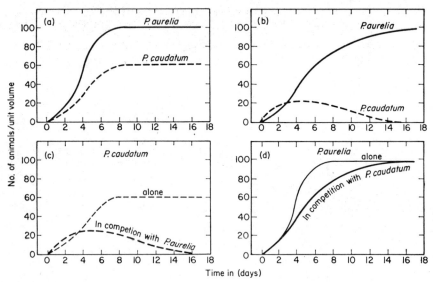

FIG. II:16.   Demonstration of interspecific competition between two species of
Paramecium—P. aurelia and P. caudatum.
(a) The increase in populations of each species when grown separately. P. aurelia grows most rapidly and the population becomes stable at a density of 100 per unit volume, compared with 60 per unit volume in the slower growing P. caudatum.
(b) The result of competition for food between the two species in a mixed culture.
(c) and (d) A comparison of population curves for each species grown alone and in competition.
(c) P. caudatum is eliminated by competition from P. aurelia.
(d) Competition slows the growth rate of the aurelia population, but this species eventually reaches its optimum density. (Redrawn after Gause.)

Predation and Parasitism

One species often preys on another (Chapter IV, Ex. 34). The density of a predatory population is usually greatest some time after the maximum density of its prey. This sequence may occur once, when the predators increase until all the prey are eaten; then the predatory population itself declines to extinction (Fig. II:17a). More often the sequence persists for many cycles, and the densities of prey and predators oscillate out of phase with each other (Fig. II:17b). Adding prey to a habitat renews food for the predator and produces a similar sequence of oscillations.

Predation of one animal by another often benefits Man. Up to 90% of the eggs laid by the first generation of Cabbage Root flies or Cabbage maggots (*Erioischia brassicae*) in the soil may be eaten by ground beetles, so lessening damage to cabbage roots. Ladybird (Lady Beetle) adults and larvae feed on aphids and scale insects and a ladybird *Rodalia cardinalis* introduced to California from Australia in 1889 has controlled the scale insect (*Icerya purchasi*) in orchards ever since. Many vertebrates, especially birds, prey on invertebrates such as snails and slugs and in a small area may greatly diminish numbers.

Most animals, including Man, are attacked by parasites (Chapter IV, Ex. 33). Figure II:18 summarizes the effect of parasites and other organisms on a population of Large White butterfly eggs (*Pieris brassicae*) laid on cabbages. Only 0·32% of the eggs laid became adults.

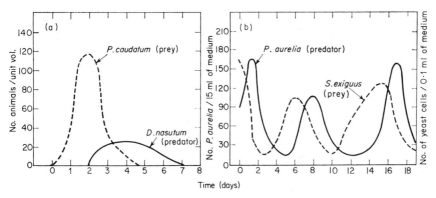

Time (days)

FIG. II:17. The principles of prey-predator relationships demonstrated with simple organisms, *Paramecium* spp., *Didinium nasutum* (Ciliophora) and *Saccharomyces exiguus* (Ascomycetes—Yeast).
(a) In the simplest example, the prey population (*Paramecium*) grows until the predatory population (*Didinium*) starts to multiply. The predatory population continues to increase and eventually exterminates the prey, but then has no food, so itself declines to extinction.
(b) Here the relationship between prey and predator is similar to (a) but when the decline of the prey population causes the predatory population to decrease, the prey recovers and begins to grow again. This enables the predator, with renewed resources, to grow also and such a system can persist for many cycles in this oscillating state. (Redrawn after Gause.)

Aggression

Individuals of the same or different species are often aggressive to one another especially during breeding and this causes a more even distribution of animals. A male stickleback (*Gasterosteus aculeatus*) guarding his nest, threatens and eventually chases intruding males from his territory. The further he moves from his own territory, the less aggressive he becomes and eventually he returns to his nest. Birds sing primarily to mark territorial boundaries and do so only so long as they hold a territory. Intruders of the same species are chased away unless they can establish dominance over part of the territory by more vigorous singing.

Similar intraspecific territorial behaviour occurs between many male dragonflies, so that they too become well-spaced within the habitat. When the mature males arrive at the mating site, each selects an area and defends it

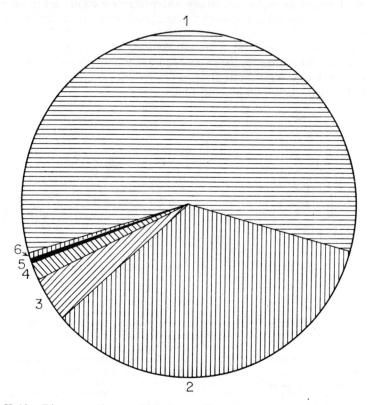

FIG. II:18. Diagrammatic representation of the relative importance of the mortality factors influencing a population of Large White butterflies (*Pieris brassicae*) in England, 1932. 1. Disease of caterpillars (59·17%). 2. Parasitized by larval parasite *Apanteles glomeratus* (34·38%). 3. Eaten by birds (4·25%). 4. Disease of pupae (2·7%). 5. Parasitized by pupal parasite (*Pteromalus puparum* (0·14%). 6. Emerged as healthy adults (0·32%). Assuming a sex ratio of 1:1, the number of individuals in a population fluctuates unless the percentage survival per generation is 2%. (Redrawn after Moss.)

against other males. The distance separating adjacent males is fairly constant for each species; for example, territories occupied by individual *Calopteryx virgo* males average 1·9 × 0·7 metres and for *C. splendens*, 2·6 × 0·9 metres. The smaller the species of dragonfly, the smaller the area occupied by each individual and the more males can live in a given habitat.

Intra and interspecific aggression, not solely associated with reproduction, occurs among ants and bees (Chapter IV, Ex. 26), which often need to defend their nests and surrounding areas. Ants near their own nests fight vigorously and win more often than those away from their nest area. *Formica sanguinea* often steals pupae from the nests of *F. fuscus*, a smaller species; many of these are eaten but adults emerge from some and become slaves in the *F. sanguinea* colony, to which they then give allegiance in preference to other colonies of *F. fuscus*.

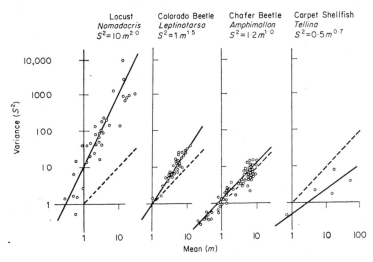

FIG. II:19.   Some animals aggregate into herds or swarms; others are solitary in their habits. This affects their disposition in space which is described by $S^2 = am^b$ where S measures the regularity of disposition and m is the population density. Locusts are highly aggregated with high S^2 values; Colorado beetles are slightly aggregated; Chafers are randomly distributed with points along the dotted line which represents randomness; Carpet Shellfish are solitary and the points fall below the dotted line.

Spatial Patterns

The maintenance of a territory by birds and dragonflies tends to distribute the individuals of a population fairly evenly within the region they inhabit. By contrast, ants and bees aggregate into dense colonies and the area occupied by a colony may depend upon its size. If the food of an animal occurs only in small areas, such as a clump of bullrushes, the animals tend to congregate there. The distribution of the bullrushes themselves depends on physical features such as patches of low-lying, wet ground. The environment has many components that act upon each individual in a population to affect its disposition. Each species responds differently to its environment and hence each species appears to have a characteristic spatial disposition (Fig. II:19, Chapter III, p. 85, Chapter IV, Ex. 23, 24).

In a similar way, the distribution of numbers of each of the many species within an area, or habitat, assumes a definite pattern or series (Plate 4). One particular species is always dominant, having a great number of individuals. Then follow two or three species with rather fewer but still many individuals, four or five species with even fewer individuals but still common, many species with a moderate number of individuals and a very large number of species with very few individuals (Chapter IV, Ex. 44 and Fig. IV:44:1).

For all living things, this pattern of community structure is similar with a few exceedingly common species and a vast number of species that are very rare; many of these are so rare that they are just differentiating into new species or just becoming extinct. Evolution is a constant, active process, often over-looked because the species with a short life span, and in which evolution is therefore most rapid, are small and insignificant.

The dominance of the most common species tends to be greater in uniform habitats than in variable ones and in colder climates than warm ones. Thus, there are millions of acres near the Arctic occupied by single species of conifers, whereas in a tropical rain forest there are many more tree species, and the most common is only slightly more plentiful than the next most common species. The tropical forest is more diverse than the coniferous forest.

Such patterns in the numbers and distribution of animals occur because the lives of all organisms are interdependent. If a rare species disappears from a community, its absence may pass unnoticed because its niche is quickly occupied by another species and the change in the whole pattern is slight. By contrast, when the most common species is wiped out, for example, by pesticide treatment, there is a reshuffling of the whole community until some other species dominates (Chapter IV, Ex. 34).

GENETIC ADAPTATIONS TO THE ENVIRONMENT

Evolutionary trends have been to develop resistance to unfavourable environments (Chapter IV, Ex. 19), and to facilitate survival, many animals have evolved specialized colouration, habits, morphological structures and physiological mechanisms.

ADAPTIVE COLOURATION

Adaptive colours and patterns, some of which aid concealment from predators, are common (Chapter IV, Ex. 15, 16, 17). More Small Tortoiseshell (*Aglais urticae*) pupae survive on grass banks, walls or among nettles than on fences, because they are more conspicuous to their predators when resting on a plain wooden background. The snail *Cepaea nemoralis* has several genetical variants that differ from one another in colour and pattern. On a green background, yellow snails are less likely to be eaten by thrushes than pink or brown ones. Thus, during a survey made in April when vegetation was sparse and bare earth visible, the percentage of yellow shells found on thrushes' anvils was 43% but at the end of May, when vegetation had grown and the background had become greener, only 14% of shells on anvils were yellow even though the proportions of different forms within the population remained fairly constant. Such selective predation may eventually affect the genetic composition of the population. The appearance of very dark (melanic) forms of moths in particular areas of Britain and Europe (see p. 159) has provided a striking example of this and of the speed with which adaptive colouration can spread through a population. Warning colouration is discussed on pp. 154 and 157.

CHANGES IN HABITS

Changes in feeding habits, and the development of physiological races which eventually become isolated genetically from the original population, enables species to extend their range. The capsid bug or plant bug *Plesiocoris rugicollis* is a common insect on willow in East Anglia, where before about 1914 it fed occasionally on currants but never on apple. By 1926 it had become a common pest on currants and also fed on apple thereby extending its range. In a similar way, presumably, the Apple Sucker (*Psylla mali*) extended its host range from hawthorn to apple and eventually evolved separate races. Now Apple Suckers collected from these two plants never breed together or feed on each other's host, though they are morphologically indistinguishable. The Stem and Bulb Eelworm (*Ditylenchus dipsaci*) has evolved many races, some confined to individual plant species like white clover, red clover, Narcissus or Tulip, and others like the "field strain" with a wide range of hosts including oats, onions, field beans, rhubarb and many weeds. Thus the ability to adapt feeding habits in a changing environment ensures the survival and often the successful spread of a species or group of animals. Among the bees (*Apidae*) especially, specialized feeding habits (Chapter IV, Ex. 28, 29) have evolved side by side with specialized floral structures (Chapter IV, Ex. 30) and bees can also adapt their behaviour to short term changes in the environment, such as daily changes in the time of nectar secretion or pollen production by plants (Chapter IV, Ex. 31).

PHYSIOLOGICAL ADAPTATIONS

Many insects and mites have developed resistance to insecticides and acaricides, by mechanisms that differ according to species and the pesticide. Thus, there are strains of houseflies resistant to DDT because they have a dehydrochlorination enzyme that makes DDT harmless. Resistant strains of *Drosophila melanogaster* have an oxidation enzyme that fulfils a similar function. Though detoxication is a physiological process, it has probably not evolved by mutation to cope with this particular unfavourable environment; it is more likely that a small proportion of insects always possessed these enzymes and thus they have been favoured by the changed environment and have multiplied without competition from the strain susceptible to DDT. This contest between insect and Man, the one producing resistant races while the other produces new insecticides will probably continue, as it has with bacteria resistant to antibiotics and with fungi able to attack hitherto resistant plants.

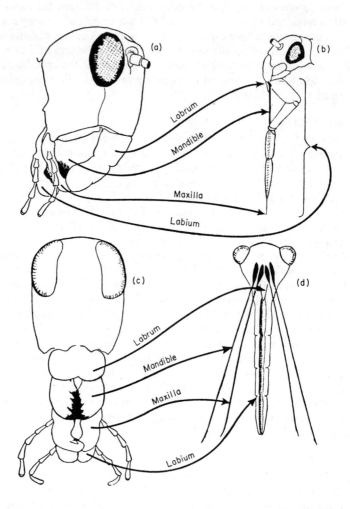

Fig. II:20.   Diagram illustrating the affinities of the piercing and sucking mouthparts
of plant bugs (Hemiptera) with the biting mouthparts of grasshoppers (Orthoptera).
(a) Head of grasshopper showing primitive biting mandibles which have a side to side
movement for biting and chewing plant tissue.
(b) Head of mirid bug showing sucking mouthparts with an up and down movement
for piercing plant tissue. The primitive mandibles and maxillae have evolved into thin,
piercing stylets, which lock together, after penetrating the plant, to form a sucking tube.
When not in use they are protected in the dorsal groove of a retractable sheath formed
from the lower lip (labium). The large upper lip of the grasshopper is represented in the
bug by a small flap (labrum).
(c) and (d) are exploded diagrams of grasshopper's and bug's mouthparts showing the
separate parts.

MORPHOLOGICAL ADAPTATIONS

There are many species which have evolved morphological adaptations to avoid predators, and mimicry of other animals and plant structures is common (Chapter IV, Ex. 15). Wings, permitting an aerial life, are more common in the most highly evolved invertebrates, the *Insecta*, than in more primitive *Arthropoda* (Chapter IV, Ex. 18). The mouthparts of many groups

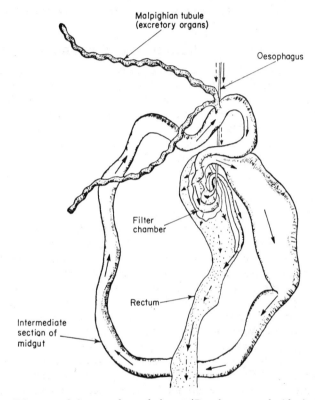

FIG. II:21.   Diagram of the gut of a scale insect (*Pseudococcus adonidum*) showing the filter chamber by which excess water is removed from the food. The chamber consists of the first and last sections of the mid-gut, closely coiled together and invaginated into the surface of the rectum. The solid arrows show the path of dissolved food through the gut, and the broken arrows the path of excess water. The intermediate section of the mid-gut, from which food is absorbed, lies free in the body cavity. (Redrawn after Pesson.)

of animals have been modified from the primitive biting forms to take maximum advantage of different types of food. In bugs (Heteroptera) and plant lice (Homoptera), the mandibles and maxillae have evolved long thin stylets enabling the insects to pierce leaves and feed on phloem sap (Fig. II:20). These juices are usually very dilute, so in homopterous insects, part of the mid gut has evolved into a filter chamber to enable excess fluid to pass directly from the first part to the last part of the mid gut, thereby short-circuiting the main digestive segment of the gut, which receives only the nutritious constituents of the plant juices (Fig. II:21). The evolution of such elaborate external structures and internal anatomy has probably taken much longer than the evolution of the colour differences (p. 159), but incessant change is an essential feature of living organisms and species with most genetic plasticity, like Man, are most likely to survive in a changing environment.

SURVIVAL IN UNFAVOURABLE CONDITIONS

Animals living in uniform or continually favourable environments with an abundance of food, like grain weevils infesting a large heap of grain or cockroaches in a warm kitchen, may continue to breed and develop there for many overlapping generations, because the environment is favourable for all stages. Such consistently suitable environments are rare outside the tropics and in most places different stages occur at different seasons (Chapter IV, Ex. 32), sometimes on different host plants or animals, and development is often sporadic.

HIBERNATION AND DIAPAUSE

Winter is passed in the most cold-resistant stage, which differs in different species. Among British Lepidoptera, the Purple Hairstreak butterfly and Vapourer moth overwinter as eggs, the Wall and Large Yellow Underwing as larvae, the Large White and Hebrew Character as pupae and the small Tortoiseshell and Herald as adults. During the winter some animals, such as bumble bees, are merely quiescent and become active on occasional warm days. Some long-lived species are quiescent for successive winters; larvae of the Wheat Blossom midge (*Sitodiplosis mosellana*) may remain in cocoons in the soil for up to 19 years before they pupate, but the environmental factors controlling such long quiescent periods are unknown.

To survive in temporarily unfavourable environments, many animals become dormant, their metabolism slows and growth and reproduction temporarily cease. This phenomenon, unlike quiescence, is obligatory and necessary before further development can proceed; it is called diapause and occurs in many different forms in some Crustacea, mites, snails and insects. It may take the form of arrested development in the egg, larva or pupa, or a failure to enlarge the reproductive organs in the adult. Its initiation and duration can be affected by weather, especially temperature and daily photoperiod, or by changes in the quality and quantity of food eaten, depending on the species. At the onset of diapause, mitosis usually ceases, the tissues become drier, respiration slows, and the fat body increases. For some species a period of arrested development in the cold is essential before normal growth can proceed. Codling moths overwinter as larvae in a dense silken sheath and rarely emerge from this to pupate before the end of February when it has been cold for several months. Larvae, removed to warmth before this, pupate erratically and the adults that emerge lay few eggs. Eggs of some British grasshoppers laid in late summer will not hatch at 25°C in September, but will do so in November, after cooler weather.

DISPERSAL AND MIGRATION

Unfavourable environments are avoided by many species and new breeding sites reached by dispersal (Chapter IV, Ex. 22, 23) which is an essential and regular event in the life cycle of many species like the Bean Aphid, *Aphis fabae* (Chapter IV, Ex. 24); movements may be over large ecological barriers like seas or mountains, or simply to neighbouring plants.

One particular stage in the life cycle of the species is usually particularly adapted to dispersal. Spiders often disperse in the air on silken threads as juveniles, and in a few insects like the Gypsy Moth (*Lymantria dispar*), whose adult ♀♀ do not fly, the small, hairy first stage larvae are carried by the wind. However, most insects disperse as adults. Their migration by flight is a form of dispersal especially associated with the need to find fresh breeding sites and usually occurs before they lay eggs. When migrating, insects fly more persistently and are less easily distracted than when moving between feeding and oviposition sites within a habitat. The length of their migratory flights may range from a few yards, as by ants, to many hundreds or even thousands of miles as by some aphids, moths and locusts. The whole of the journey in one direction, or the return journey, is not always made by the same individuals. The Monarch butterfly (*Danaus plexippus*) migrates from Canada and the northern states of the U.S.A. to hibernate in California and Mexico and, after hibernation, the same individuals may re-migrate north again. In contrast the Painted Lady butterflies, which breed in North Africa, migrate north to Britain in spring and their offspring migrate back to Africa in autumn.

The initial take-off into the air is always controlled by the insects themselves. Some migrations of large moths and butterflies consists of sustained flights in one direction, but the movement of many insects occurs in winds exceeding their flight speed, so their course almost entirely depends on the wind (Fig. II:22). Locusts can fly at about 15 m.p.h. but when migrating in swarms, individuals constantly change flight direction, so that the whole swarm drifts downwind at a slower speed than the individual locusts fly. They thus tend to collect in zones where winds from both sides of the equator converge and because the winds often bring rain, the arrival of the locusts in an area often coincides with heavy rainfall, which is essential for their eggs to develop. These areas of convergence move north and south with the seasons, so part of the locust population is always on the move.

Most species have the ability and tendency to disperse but some species do it more than others. For example, winged bean aphids (*Aphis fabae*) seem to have an irresistible urge to fly (Chap. IV, Ex. 24) and often leave apparently suitable hosts, but *Aulacorthum solani* is a relatively sedentary aphid and may even remain for a while on a non-host plant. Similarly, female cereal thrips fly in great numbers from wheat, oats and barley whereas winged females of the thrips *Amblythrips ericae* are abundant on large areas of *Erica* and

Calluna moorland in Britain, but few fly. Unfavourable aspects of the environment, such as poor food or crowding, often encourage dispersal and generally animals living in temporary habitats, like annual crops, disperse more rapidly than those in permanent habitats, like heath, moor or woodland. The dispersal of cereal thrips illustrates how large the losses can be when small animals are spread by wind. In vast numbers they are dispersed to places where they cannot survive and they die in ponds and water troughs, behind pictures in buildings and even in such improbable places as clocks and watches. However, enough always reach favourable habitats to hibernate and breed the following year when their enormous capacity to increase ensures the recovery of the population. The dispersive behaviour evolved by the

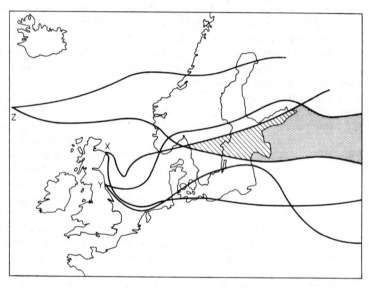

FIG. II:22. Mass migrations of Diamond-back moths (*Plutella maculipennis*) were observed on the east coast of Britain (X and Y) and by a weather ship at sea (Z) in June and July, 1958. The heavy lines mark the boundaries of the air masses that carried the migrating insects, which presumably originated in the shaded area where the three air masses overlap. The moths remained airborne by beating their wings, but their direction of movement was affected by the wind, and moths observed at Z must have been blown for 1000 miles over the Atlantic Ocean. (Redrawn after French and White.)

Common Corn thrips (*Limothrips cerealium*) and other species which disperse readily, ensures that a maximum number of habitats are colonized.

Many animals have a stage in which they withstand an unfavourable environment for long periods until dispersed by wind, water or by other animals, or, as with many parasites, until found by a fresh host. The body of each mature female nematode of the root-feeding genus *Heterodera* develops into a hard resistant cyst containing eggs, which may be spread from field to field by flooding or wind, or in soil moved by larger animals or by Man. Eggs of the large stomach worm of sheep (*Haemonchus contortus*) are deposited with the faeces in pasture and develop, in the soil, or especially in the equable environment of a cow pat, into infective larvae. These larvae survive on the herbage for a considerable time ready to infect any sheep that eat them (Fig. II:23).

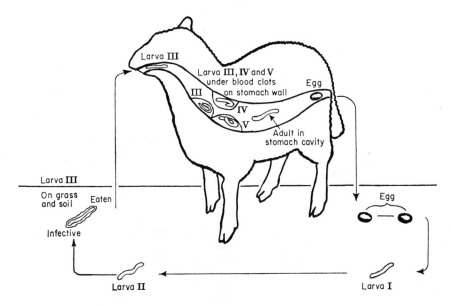

FIG. II:23. The life history of the Large Stomach Worm of sheep, *Haemonchus contortus*.

SURVIVAL AND FECUNDITY

In general, the more precarious the life cycle of a species and the smaller the chance of it finding a suitable environment, the greater is the fecundity of each female. The survival of the oil or blister beetle, *Sitaris muralis*, depends on the larvae being carried by bees to their nests and the chances of this are slender. Adult beetles lay up to 10,000 eggs near nests of wild bees (*Anthophora* sp.) and the minute, active, tough-skinned larvae hatch, then hibernate until spring, when a few successfully attach themselves to the hairy bodies of male bees, which appear earlier than females. When opportunity allows, the beetle larvae pass from a male bee to a female and are eventually carried to the nest where they slip off the bee and become sealed in a cell with a bee's egg and honey. In this new favourable environment the little, tough larva changes into a fleshy, legless grub, grows rapidly as it feeds on the egg and honey, the grub pupates, and emerges as an adult either in late summer or the following spring (Fig. II:24).

Many of the various effects the environment can have on the life cycle are seen in the annual cycle of the bean aphid (Fig. II:25). This species is well adapted to take maximum advantage of its environment. When food is plentiful and nutritious, it can reproduce quickly because it is parthenogenetic and viviparous; i.e. its females do not need to be fertilized by a male, and

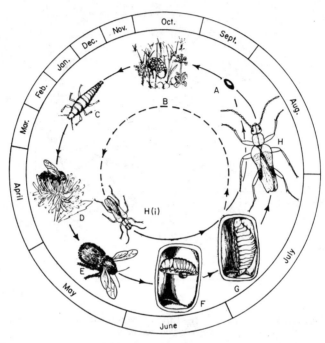

FIG. II:24.   The life history of an oil beetle, *Sitaris muralis*. A, egg of beetle; B, nest of bee, *Anthophora* sp. with its chimney-like opening near which each female beetle lays thousands of eggs; C, minute, active tough-skinned beetle larva (triungulin type), hatches and hibernates on vegetation until spring; D, male bee, on to whose body the beetle larva attaches itself; E, female bee, to which the beetle larva transfers itself and is carried to the bee's nest; F, cell in bee's nest into which beetle larva is sealed with honey and bee's egg, which it eats, and where it changes into a fleshy, ovoid, legless larva (eruciform type); G, last instar beetle larva in bee's cell; H and H(i) adult beetle which emerges from the bee's cell either in late summer or the following spring after over-wintering as stage G. The likelihood of an individual beetle larva surviving stages A–F is small. (Drawings of the individual stages are not to the same scale.)

they produce living larvae instead of laying eggs. When it is unnecessary to seek a fresh environment, all the animal's resources are devoted to rapid reproduction and energy is not even dissipated by growing wings.

The species survives winter as an egg on spindle trees (*Euonymus* sp.) or Guelder rose (*Viburnum opulus*). When temperatures increase in spring, wing-less females emerge from the eggs and feed on the young spindle leaves.

These females produce a second generation of wingless females, but the third generation of females is winged and in May and June they migrate from the woody host to annual plants, especially beans but also other quick-growing herbs that provide abundant nutritious food. In this favourable environment several wingless generations are produced, but when hosts deteriorate, or the aphids become overcrowded, winged forms are produced that fly to fresh

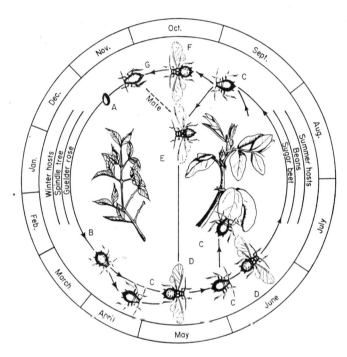

FIG. II:25. The life history of the Bean Aphid (*Aphis fabae*). The left-hand half of the circle represents the period spent on the winter host and the right-hand side, that spent on summer hosts. A, egg; B, wingless female (or stem mother); C, wingless viviparous female; D, winged viviparous female (migrant); E, male (migrant); F, female-producing autumn migrant; G, sexual, wingless egg-laying female. Throughout the year, reproduction is parthogenetic except in autumn when forms E and G mate, and the sexual female lays the overwintering eggs.

hosts and found new colonies. The aphids fly towards the light when they leave the plant, and are quickly carried upwards by turbulent winds, which may transport them for great distances. This sequence is repeated a number of times during a favourable summer until herbaceous hosts deteriorate finally in autumn. As days become shorter, winged males and winged, parthenogenetic females are produced, both of which fly and drift in the air until they find a spindle tree. There the parthenogenetic females produce another generation of wingless females that mate with the winged males and lay the eggs that overwinter.

MODIFICATION OF ENVIRONMENTS

MICROCLIMATES

Both solitary and social insects sometimes create specialized microclimates for their immature or adult stages. The yellow ant (*Lasius flavus*) builds its nest in the earth below a stone. The sun warms the stone faster than the

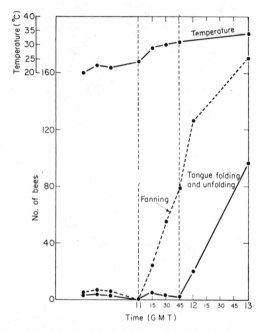

FIG. II:26. The relationship between overheating of the hive, and cooling behaviour (fanning and tongue-folding) in honey bees (*Apis mellifera*). When overheating (> 24°C) begins, many bees fan in the hot area. Further cooling is achieved when the temperature exceeds 30°C, by the evaporation of liquid droplets spread on the comb (tongue-folding and unfolding). (Redrawn after Lindauer.)

surrounding earth so the ants then bring their brood up from the nest to the underside of the stone, taking them down again at night, when the stone cools faster than the soil. Wood ants (*Formica rufa*) maintain the temperature of their nests up to 10°C above soil temperature by building their ant hills on slopes facing south. By day the ants make ventilation holes in the upper part of the hill, facing the sun, which they close at night. If the nest becomes too hot, wide openings are made on the northern, shaded side and connected by passages through the nest, which is cooled by the draught created. Field wasps (*Polistes gallica*) build nests in sunny places but prevent overheating by sprinkling water throughout the nest and fanning with their wings to increase evaporation and so cool the nest. Honey bees cool the hive in a similar manner and also by the evaporation of droplets of liquid from the mouth which they spread with their tongues to form a thin film over the comb (Fig. II:26). In winter the bees cluster to conserve the heat produced by their metabolism. In air conditioned and heated buildings Man's microclimate is almost completely controlled.

FOOD AND HABITAT

Some insects create their own specialized habitat in which delicate stages can develop. Many gall wasps (*Cynipidae*), gall midges (*Cecidomyiidae*) and aphids stimulate plants to produce galls, or proliferated masses of tissue, in which the animals can shelter and live amid abundant food and at a constant humidity. The galls are probably formed by the cambium tissues reacting to secretions from the larvae. Eighty-six per cent of all British gall wasps live on oak, and oak apples, marble galls and spangle galls formed by *Biorrhyza pallida*, *Cynips kollari* and *Neuroterus* spp. (Fig. IV:41:1) respectively, are well known. A few cynipids (7%) live on rose, where one of them, *Rhodites rosae*, produces the Robin's Pin Cushion or Mossy Rose Gall. Bean galls on willow are produced by sawflies (*Pontania* spp.) and the common purse galls found on poplar are produced by the overwintering generation of *Pemphigus bursarius*, an aphid that spends the summer on lettuce roots. The larvae of some wood-boring beetles feed on fungi that grow on a layer of wood chips and excreta in their burrows, the fungal spores being carried by the beetles from tree to tree. Honey bees manufacture the food they give to their larvae and feed future queens and workers differently, worker larvae receiving less food and of poorer quality than queen larvae.

However, such specialized modification of adult behaviour and the environment, for the benefit of the young, is less common in Invertebrates than in Vertebrates.

CHAPTER III

Analysis in Ecology

INTRODUCTION

Biological experiments yield less precise results than do physical experiments because only rarely do two animals respond to experimental procedures as identically as do two pieces of glass or two lumps of metal. This is evident in Ex. 1, p. 115, on the specific body temperature of Man. The errors of thermometry are small compared with the real differences in body temperature between different individuals of the same species. Nevertheless, body temperature is as specific in mammals as is specific heat in metals, provided it is measured in the appropriate way. This way is different from that used in physics; it involves taking many measurements instead of few, to allow for the difference between individuals and this in turn means additional stages of analysis using statistical, rather than mathematical, techniques to obtain the required values for the species. It also means that improvement in standards of measurement will improve the accuracy of the value for the species, but will not change the difference between individuals.

A second difference between physical and ecological experimentation is illustrated by Ex. 2, p. 117, in which the hypothesis states that, in developing mammals, stature depends upon age. It is not usually practicable to rear young animals under specially controlled conditions to eliminate those factors other than age, such as heredity, climate and nutrition, that might affect stature. In practice, it is necessary to accept the experimental material offered by nature, such as students in a school. In addition to the normal variability between two different individuals, there is also the variability introduced by the experiment itself. Because the experimental conditions in Ex. 2 could not be controlled, each different individual received a slightly different experimental treatment.

However, it is still possible to abstract from the resulting jumble of measurements, a precise statement about the relationship between, for example, age and stature (Ex. 2). The problem is to pick out the element of variability associated with the experimental treatment. In other words, when the experimental procedure cannot be manipulated, then the necessary manipulation must be done in the analysis.

The first step in this kind of analysis is to arrange the mass of primary measurements, or raw data, in comprehensive tables. Next, the relevant quantities must be examined in diagrams that express simply and graphically what can only be written at length. At a later stage, the relationships that are evident in the figures may be expressed statistically or mathematically; we shall not always progress so far in these exercises.

The complexity of the techniques presented differs. Results of many experiments in Chapter IV can be analysed without reference to Chapter III but there are some exercises for which reference to this chapter is essential. Cross-references are given where appropriate.

MAKING TABLES

Clear, concise tables are essential; they indicate the clarity of the thoughts behind them. It is difficult to give comprehensive rules for making tables, so much is common sense and depends upon the way the data are collected. A few suggestions are offered and there are many tables of varying degrees of complexity in the following exercises to use as a guide.

1. Give the table a title and legend so that other people can understand it without having to search through the description of the experiment.

2. Try to make the table self-explanatory by the logical arrangement of rows and columns. Remember that it is usually easier to extend the number of rows rather than the number of columns.

3. Make the first tables of "raw data", that is of the actual numbers counted. When percentages or other manipulated quantities are required, make secondary tables for these.

4. When using several subdivisions of categories, make the heading panels cover the appropriate columns or rows (Table III:1).

5. Do not put too much in one table. When summaries are needed they are usually best done in separate tables (Table III:2, 3, 4, 5).

6. Try not to leave blank spaces in tables. If a table is properly constructed, it is usually possible to fill every square by making a simple distinction between an observation giving a zero reading (0) and a missing observation (—).

7. Wherever possible, fill in the totals of rows and columns so that a grand total in the bottom right-hand corner can be used as a check for the final addition in both directions.

8. Distinguish between partial totals, accumulative totals and grand totals where several occur together.

9. Make clear how many observations are included; when average values are given, give also the sum and number of observations from which they are derived.

10. When numbers are given as percentages or converted in some other way, state clearly what the conversion is and what the original data are.

11. When giving measurements, give the units.

In primary data tables, the counts will usually be integers, whole numbers of animals, because animals do not occur in parts. In secondary tables, however, it is often necessary to decide how many digits to use. For example, if there are 100 insects of which 27 are beetles it is quite correct to state that "27% are beetles". If there are 101 insects of which 27 are beetles, it is

TABLE III:1

The distribution of *species A* and *species B* in 4 years' sampling by pitfall traps (6 traps per sample), operated 1 day and 1 night per week throughout the year

Date		No. of beetles per sample							Species total		Season total
		Woodland				Grassland					
		Day		Night		Day		Night			
Year	Season	A	B	A	B	A	B	A	B	A	B	A + B
1950	Spring	2	23	0	57	1	20	0	33	3	133	136
	Summer	38	13	0	19	14	7	1	12	53	51	104
	Autumn	21	12	0	7	8	9	0	18	29	46	75
	Winter	0	1	0	4	0	1	0	9	0	15	15
1951	Spring	1	6	0	19	0	6	0	27	1	58	59
	Summer	8	7	0	7	6	8	0	11	14	33	47
	Autumn	10	3	0	6	2	4	0	2	12	15	27
	Winter	1	1	0	0	1	1	0	1	2	3	5
1952	Spring	17	48	0	67	14	49	0	45	31	209	240
	Summer	72	34	3	57	50	11	1	55	126	157	283
	Autumn	39	28	0	29	24	14	0	10	63	81	144
	Winter	0	11	0	9	1	8	0	12	1	40	41
1953	Spring	12	31	0	79	1	18	0	57	13	185	198
	Summer	20	26	0	7	9	12	0	16	29	61	90
	Autumn	21	4	2	12	12	3	1	4	36	23	59
	Winter	0	2	0	1	0	5	0	0	0	8	8
	TOTAL	262	250	5	380	143	176	3	312	413	1118	1531

possible to write "26·7326732673% are beetles". This would be unreasonable; the problem is to know where to draw the line of reason. If there are 101 insects, each insect represents roughly 1% of the total meaning of the initial statement. It is not meaningful, therefore to be much more precise than 1%. As part of an insect cannot be caught, "27%" is a good answer for most purposes; "26·7%" is valid; "26·73%" is becoming pedantic and "26·732%" is meaningless. As we shall see later, the difference between 11 insects and 12 usually has more meaning than the difference between 14,711 and 14,712. There are occasions when the use of quantities that suggest a precision greater than reality cannot be avoided, but they should be avoided whenever possible. When using logarithms, only two places of decimals are necessary.

TABLE III:2

Seasonal occurrence of *species A* and *species B*

		Spring	Summer	Autumn	Winter	Full year
A	No.	48	222	140	3	413
	%	11	54	34	1	100
B	No.	585	302	165	66	1118
	%	52	27	15	6	100
Both spp.	No.	633	524	305	69	1531
	%	41	34	20	5	100

TABLE III:3

Time of activity of *species A* and *species B*

		Day	Night	Total
A	No.	405	8	413
	%	98	2	100
B	No.	426	692	1118
	%	38	62	100
Both spp.	No.	831	700	1531
	%	54	46	100

Averages are often difficult to define in tables. For example "the monthly average of the daily maximum temperature" is difficult to abbreviate without loss of meaning. It is sometimes better to label a column as the total, or average, of several other columns and to describe what this is in the legend, rather than to use incorrect or obscure abbreviations as column headings. When making meteorological observations and collecting samples of animals at regular intervals, be careful to record the correct date. For example, the maximum temperature read from a max.–min. thermometer at 9 a.m. refers to the previous day, but the minimum temperature to the current day because the minimum occurs about 3 or 4 a.m. Appendix J gives a table of standard weeks with the same dates every year. From this the seasons in Table III:1 can be defined accurately and are always the same.

TABLE III:4

Comparison of habitats of *species A* and *species B*

		Woodland	Grassland	Total
A	No.	267	146	413
	%	65	35	100
B	No.	630	488	1118
	%	56	44	100
Both spp.	No.	897	634	1531
	%	59	41	100

TABLE III:5

Annual fluctuations in populations of *species A* and *species B*

		1950	1951	1952	1953	Total
A	No.	85	29	221	78	413
	%	21	7	53	19	100
B	No.	245	109	487	277	1118
	%	22	10	43	25	100
Both spp.	No.	330	138	708	355	1531
	%	22	9	46	23	100

MAKING GRAPHS

Figures are used to express simply and graphically what can often only be written at great length. Thus the next step in an analysis is to express the collected data in the form of a figure. Once the characteristics requiring emphasis have been segregated in this way, they can be studied further without the distraction of the other innumerable features that constitute the whole living organism. The illustration of such characteristics in graphs of various kinds is therefore essential to reveal at a glance what information the basic observations contain. At a later stage, mathematical or statistical analysis may become necessary to ascertain the probability that any suspected relationship is valid. Making graphs is a simple art that can either help or hinder in emphasizing the required features. Adherence to a few basic rules will ensure that the graphs clarify and do not hinder analysis by distorting the truth.

SCATTER DIAGRAMS

Using as an example the data from Ex. 12, in which the basic observations are the temperatures of insects in different air temperatures (Table IV:12:1, p. 146), the following points should be noted:

1. The insect temperatures are in °C.
2. The air temperatures are in °C.
3. There is one measurement of insect temperature given for each air temperature.

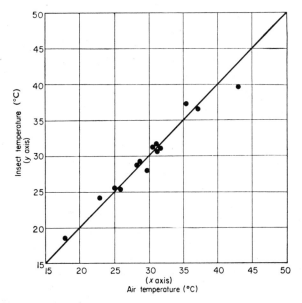

FIG. III:1. The effect of air temperature on the body temperature of caterpillars. Each point represents the mean temperature of six larvae measured by thermocouple in lateral folds of the cuticle. The "line of equality" is not fitted to the points.

4. There is every reason to suppose that if the air temperature changes, the insect temperature may change also: there is no reason to suppose that if the insect temperature changes, the air will follow suit.

The values from Table IV:12:1 are plotted in Fig. III:1. *Note these points:*

1. The quantities are marked on both axes and numbered at intervals (Fig. III:2:A).

2. Both axes are labelled, °C, and appropriately "air temperature" and "insect temperature" (Fig. III:2B).
3. The intervals for 1 degree are the same on both axes so that the figure is approximately square and not disproportionately exaggerated in one direction or the other (Fig. III:2C).

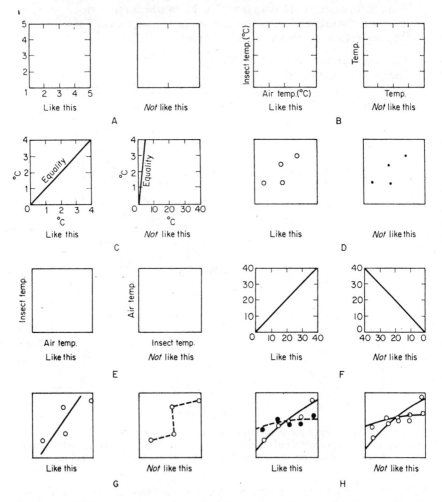

FIG. III:2. Making graphs; each pair of figures illustrates a point discussed in the text.

4. Each point is drawn clearly and boldly and it represents one pair of records (Fig. III:2D).
5. The air temperature occupies the horizontal x-axis. This is a convention, but a very valuable one, derived from statistical practice in which

the vertical y-axis is reserved for the character that is supposed to depend on the other measured value, plotted along the x-axis. In other words, if x changes, y will change too, but a change in y will not produce a change in x; y is the *dependent* variable and x is the *independent* variable. Both are called variables because they can have different values on different occasions. As a rule, y will be a biological characteristic and x is often a physical or environmental one, as in the present example (Fig. III:2E).

6. The figure has a legend, or description to make it intelligible to the reader.
7. The quantities increase as they progress away from the origin. It is best, where possible, to have the origin at the point (0, 0), i.e. where $x = 0$ and $y = 0$ (Fig. III:2F), but in the present example (Fig. III:1) this would waste too much space. Instead, both axes begin at the same temperature (15°C, 15°C).
8. A line is drawn to represent the *supposed* relationship between body and air temperatures (Fig. III:2G). It passes through all points where the air and body temperatures are equal, a *line of equality*, but it is near enough to the experimental values to illustrate, as a first approximation, that the body temperature is usually very close to air temperature. It does *not* say that air temperature is close to body temperature because of the accepted convention about the arrangement of the axes. If a line does not express an idea, omit it.
9. Distinguish clearly between different sets of observations by using different symbols or lines (Fig. III:2H).

Points representing pairs of observations taken simultaneously from two variables and plotted against a pair of axes in this way are said to form a "scatter diagram". In this kind of illustration, the points represent the factual observations and the line represents the hypothesis. Drawing a line through a scatter of points should be done with care. The line may be drawn according to some hypothesis derived quite independently of the points. In that case consider how well the points fit the line, for example the line of equality. Alternatively, the line may represent a hypothesis derived from the factual evidence represented by the points. Then the line should describe the general trend of the points; that is, the line may be fitted to the points. In the nature of biological variability it will never account for all the scatter.

A straight line is best drawn with the aid of a transparent ruler, preferably with parallel lines down its length, one of which is equidistant from both edges. With such a ruler, a line can usually be drawn that fulfils the following requirements:

1. It should pass very near to the average of all x's and all y's.
2. There should be approximately equal numbers and deviations of points on each side of the line.
3. It should be possible to draw two other lines, one on each side of and parallel to the first, which enclose very closely all the points. It is to fulfil this requirement that the parallel lines engraved on the ruler are most valuable. When a regression line is fitted (see p. 78), it may be found not to fall

exactly along the arbitrary line drawn as described. The reason will be referred to later (p. 81).

The points may fall so that both extremities are on one side of the line and the middle points on the other. If so, the points would be better described by a curve. To see this, look along the line, with the paper held almost perpendicular to, instead of parallel to, the plane of the eyes (see Figs. III:3, 4). This gives

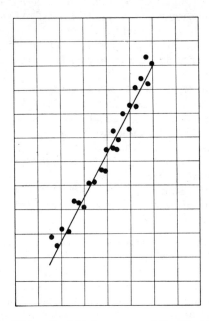

FIG. III:3.   Straight line or curve? See Fig. III:4.

a foreshortened view of the scatter and emphasizes any curvature. If the curvature can be clearly seen, a drawing instrument known as a French Curve will help to find an appropriate curve to fit the points.

There is very little scatter in Fig. III:1, because each observation actually represents the average of several readings (see Ex. 12, p. 144) and the insects are small, so body temperature quickly accommodates to air temperature. Some diagrams show much more scatter because the character to be measured varies between individuals; see for example Figs. VI:2:1, 2, p. 118, 119. This variability itself is a characteristic of living organisms, which differ from one another in so many ways that their variability may be so great as to almost obscure the features sought. Exercise 2 (p. 117) on body size and weight illustrates this point. Exercise 1 (p. 115) deals with body temperature, but this time in Man, whose temperature is usually regarded as constant and independent of the surrounding air. However, when temperatures of enough different people are measured carefully, even at the same time and place, differences are found (Table IV:1:1, p. 115).

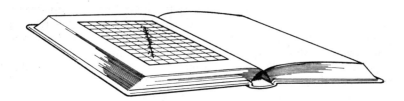

Fig. III:4. When the book is viewed from this angle, the scatter diagram on the opposite page will be seen to be best described by a curve.

FREQUENCY AND THE AVERAGE

If the temperatures of the 276 individuals in Ex. 2 are listed, the list is confused and without any sequence like the list of heights and weights in Table IV:2:1, p. 120. The list is therefore condensed into a table that gives the number of people with each of the temperatures found (Table IV:1:1, p. 115). This is called a "frequency distribution" and gives the frequency of occurrence of each temperature in the list. It shows that most temperatures fall within a narrow range, but that there are people both cooler and warmer than the majority. Ideally, any description of the temperature of the group should indicate the average temperature and the amount of variability. Note that by grouping the data as a frequency distribution, some individual detail has immediately been lost because a particular temperature can no longer be ascribed to a particular person. This is the first step in extracting from each separate individual the one feature to be considered as a characteristic of the species or population.

It should be noted that, whereas in biology a population consists of all the members of the same species that live interacting lives, in statistics a population means all those individuals behaving like the individuals in the sample under consideration. This may be part of a species population, for example the egg stage, or it may consist of a mixture of species such as all the animals in a pond. In the example above, the population, statistically speaking, is all students of the same age, sex and background as those in the class.

It would be quite feasible at this stage to take an average of all the temperatures and use this as a species characteristic. However, this would ignore entirely the differences between individuals. Although each temperature can no longer be ascribed to a named individual, it is still desirable to describe the extent of variability within the whole group and thus to define the limits beyond which normal temperature does not stray. Two species of animal may have the same average temperature, but the individuals of one species may differ more than of the other. In other words, *the variability itself is as characteristic of the species as is the average.*

To illustrate this, three sets of measurements are presented in Table III:6, supposed to represent the wingspan of three species of butterfly. The samples each consist of 100 individuals to simplify the procedure. It is assumed that measurements were accurate to 0·2 mm, so that the size categories listed, 30·0–30·9, etc., group the insects together slightly; there are, for example, 15 individuals of species *A* between 40·0 and 40·9 mm. Although it is not possible to state precisely what each individual measured, it is possible to describe collectively the characteristic wingspan of the species. Ignore the second column under each species for the present.

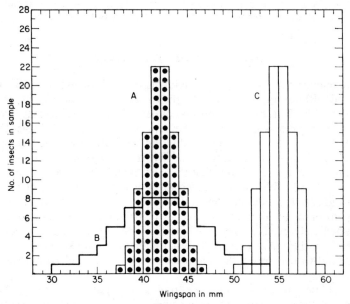

Fig. III:5. Frequency distributions for wingspan of three imaginary species of butterfly.

The data in Table III:6 are shown diagrammatically in Fig. III:5. Consider species *A*. A method of plotting similar to that used in scatter diagrams has been used, plotting separate points for each observation; but in this instance there is only one variable, wingspan, so the *y*-axis records only the number of insects or their frequency of occurrence at each wingspan. The diagram can be simplified by drawing in the outline of the columns occupied by the dots, so that the area enclosed represents the number of observations; it then becomes a histogram (species *C*). Finally all but the outline of the whole distribution may be eliminated (species *B*).

TABLE III:6

Wingspan of three imaginary species of butterfly

Wingspan mm	Species A No. of insects	Species A Cumul. Σ	Species B No. of insects	Species B Cumul. Σ	Species C No. of insects	Species C Cumul. Σ
30·0–30·9			1	1		
31·0–31·9			1	2		
32·0–32·9			1	3		
33·0–33·9			2	5		
34·0–34·9			2	7		
35·0–35·9			3	10		
36·0–36·9			5	15		
37·0–37·9	1	1	5	20		
38·0–38·9	3	4	7	27		
39·0–39·9	9	13	7	34		
40·0–40·9	15	28	8	42		
41·0–41·9	22	50	8	50		
42·0–42·9	22	72	8	58		
43·0–43·9	15	87	8	66		
44·0–44·9	9	96	7	73		
45·0–45·9	3	99	7	80		
46·0–46·9	1	100	5	85		
47·0–47·9			5	90		
48·0–48·9			3	93		
49·0–49·9			2	95		

TABLE III:6 *continued*

Wingspan mm	Species A		Species B		Species C	
	No. of insects	Cumul. Σ	No. of insects	Cumul. Σ	No. of insects	Cumul. Σ
50·0–50·9			2		1	
				97		1
51·0–51·9			1		3	
				98		4
52·0–52·9			1		9	
				99		13
53·0–53·9			1		15	
				100		28
54·0–54·9					22	
						50
55·0–55·9					22	
						72
56·0–56·9					15	
						87
57·0–57·9					9	
						96
58·0–58·9					3	
						99
59·0–59·9					1	
						100

The frequency distribution for species A should be compared with the other two hypothetical ones, species C having the same variability and a different average wingspan, and species B having the same average and a different variability.

Frequency distributions are the basic link between the behaviour of individuals and the reflection of this behaviour in whole populations and their description forms a complete section of statistics. They are not always symmetrical in shape as are examples in Fig. III:5. Figure III:6 shows two distributions that are asymmetrical in different ways. To make further use of this kind of diagram both the average of the group and the variability must be defined.

THE AVERAGE

There are several ways of measuring the average and in symmetrical distributions these different measures usually lead to the same result. Figure III:7 shows an asymmetrical, or skew, distribution from which three measures of the average can be obtained. The highest point on the distribution is at the temperature that occurs most often and is called the "mode." The temperature that separates the two equal areas A and B and thus has half the observations above it and half below it, is called the "median" or "50% point". Then there is the "mean" temperature which is found by adding up, or summing, all the temperatures obtained and dividing by the

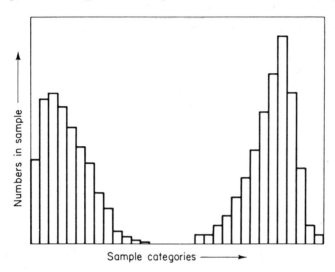

FIG. III:6.   Asymmetrical or skew frequency distributions.

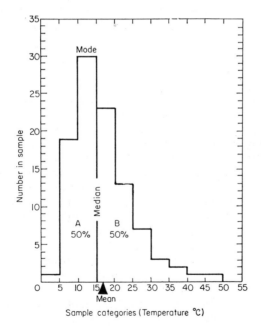

FIG. III:7.   Three different measures of the "average" in a skew distribution.

number of observations. In other words, the mean is the point of balance of the distribution. In the present example there is one observation between 0–5°; 19 between 5–10°; 30 between 10–15°; 23 between 15–20°; and so on. Thus the sum of the temperatures:

$$= 1 \times 2{\cdot}5 + 19 \times 7{\cdot}5 + 30 \times 12{\cdot}5 + 23 \times 17{\cdot}5 + 13 \times 22{\cdot}5 + 7 \times 27{\cdot}5$$
$$+ 3 \times 32{\cdot}5 + 2 \times 37{\cdot}5 + 1 \times 42{\cdot}5 + 1 \times 47{\cdot}5 = 1670$$

There are a total of 100 observations, hence the mean is:

$$= \frac{1670}{100} = 16{\cdot}7°C$$

This is usually written:

$$\bar{\theta} = \frac{\Sigma\theta}{N}$$

where $\bar{\theta}$ is the mean temperature (bar Theta),
$\Sigma\theta$ is read as "the sum of the temperatures",
N is the number of observations.

The mean is usually the best measure of the average value because it corresponds to the point of balance of the distribution; the turning moments about it are equal (see Fig. III:7). However, as the mean coincides with the median and the mode in the symmetrical distributions, a graphical method may be used to find the median and used as the average for many purposes. This graphical method also gives a measure of the variability and is therefore doubly useful.

VARIABILITY AND FREQUENCY

The variability contained in a frequency distribution is not easy to define without using statistical principles. The simplest definition would be to give the lowest and the highest temperatures and leave it at that. However, if a set of, say, 20 measurements were selected from a sample of 100, it is unlikely that the lowest and the highest values in the small sample would be the same as in the original larger sample. The chances of getting a very small or large value are greater if many rather than few temperatures are measured. It is therefore necessary to adopt a more uniform procedure to smooth out some of the errors caused by chance encounter with an exceptional value. It would help if a smooth curve could be drawn around the histogram and this is possible because the shape of frequency distributions have been studied and defined. There are, however, several steps in the process and some pitfalls.

If the sample categories in Fig. III:7 were narrower and the number of observations increased, it seems likely that the jagged outline of the distribution might become smoother. If this process were carried far enough, a smooth curve would result instead of a collection of pillars of different height. In other words, as the category size approaches zero, the histogram approaches a smooth distribution.

At the bottom of Fig. III:8 a symmetrical frequency distribution is fitted with such a smooth defining curve. In the upper part of the figure, the successive pillars of the frequency distribution below have been built up like steps, each step being raised until its base is level with the top of the preceding step. In this example, there are 100 measurements in the whole sample, so the height of the accumulated, or cumulative, total is 100 units.

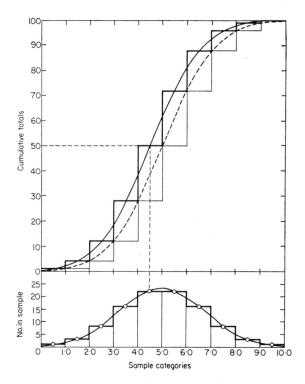

Fig. III:8.   Frequency curve, below, and cumulative frequency curve, above, of a symmetrical distribution.

The smooth curve required now passes through the centre of each step, as shown by the solid line. It is an S-shaped curve and the problem is to draw this curve accurately, remembering that this example is unrealistically regular in shape compared with what is likely to occur in nature. This problem is solved by the use of "probability paper" which converts the S-shaped curve into a straight line. In Fig. III:9, the same set of steps are drawn on probability paper and the S-shaped, or sigmoid, curve has been straightened. Much attention is given in biology to straightening curves because it is easier to fit a straight line to a set of points than to fit a curve. The straight solid line in Fig. III:9 passes through the middle of each step, just as do the curves in Fig. III:8.

Reversing the procedure, it is evident that by plotting the original set of observations on probability paper, a straight line could be drawn through them and then transferred back to the frequency distribution as a smooth

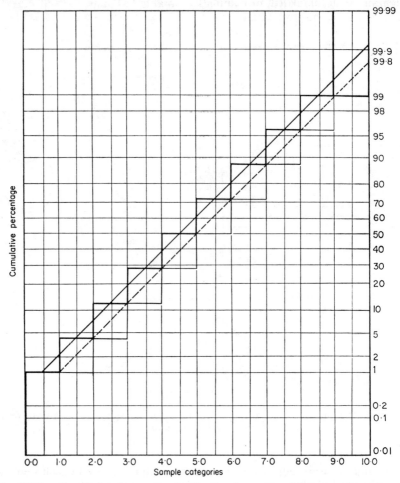

FIG. III:9. Cumulative frequency distribution from Fig. III:8, transformed to a straight line by plotting on probability paper.

curve. In practice, this is rarely necessary. Suppose that the smooth curve is the shape approached by the whole population of animals represented by a small sample. Knowing of its theoretical existence, its shape may be taken for granted and then from it may be derived only the quantities or statistics sought, namely the median and variability.

Unfortunately, these quantities are not obtained directly from the smooth curves shown; there is a flaw in the argument. To see this, look back at Fig. III:8. Because the cumulative total of 100 represents the whole sample, 50% or the median of the sample is represented by the point labelled 50 on the y-axis. Follow the horizontal dotted line until it cuts the solid sigmoid curve and then follow it down to the distribution below. It does not divide the distribution into two equal parts as required. There is a second sigmoid

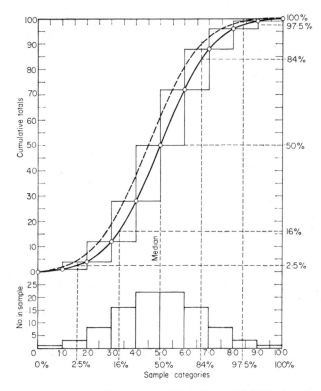

Fig. III:10.   The median, standard deviation and 95% limits of a symmetrical distribution.

curve, dotted in Fig. III:8, half an interval to the right of the solid curve. The 50% horizontal cuts this above the midpoint of the frequency distribution. This second curve is the one required to obtain the median and other statistics from the data.

Figure III:10 shows the same frequency distribution, but instead of the *outline*, the *area* of each column is emphasized. Consider the fifth column. It begins at $x = 4$ and ends at $x = 5$. Between those two limits there are 22 units, derived as in Fig. III:5. Progressing along the x-axis from 4 to 5, it is not possible to say precisely when those 22 units accumulate. All that is certain is that at $x = 4$ they are absent and by $x = 5$ all 22 are present. Look now, in the upper part of Fig. III:10, at the "step" from 4 to 5. The gain of those 22 units, from 28 at 4 to 50 at 5, is best represented by the diagonal solid line, not by the dotted line passing through the centres of the steps. Hence the solid line, not the dotted one, represents the rate of accumulation of units, or areas. Referring back to Fig. III:8, it is clear that the dotted and solid sigmoid curves have been inter-changed in making Fig. III:10. The area enclosed by the frequency distribution is therefore defined by the solid line in Fig. III:10 and this line passes, not through the middle of each step, but diagonally through the corners of the steps representing the accumulated columns. This solid sigmoid curve in Fig. III:10 is the integral of the smooth frequency distribution curve in Fig. III:8. It is shifted half an interval to the right, because it is concerned with areas, not points, and this must be kept in mind when transferring lines from probability paper to frequency distributions and *vice versa*.

95% LIMITS

The accumulated areas of the frequency distribution are represented correctly by the solid line drawn on probability paper in Fig. III:11. The y-axis is scaled as percentage; in the present example there are 100 units so this requires no conversion. The 50% point on the y-axis, or median, occurs above 5·0 on the x-axis, as Fig. III:10 indicated. The y-scale of probability paper is contracted in the middle and drawn out at the ends to convert the original sigmoid curve to a straight line. In doing this, the first and last 1% of the distribution becomes indefinitely extended. The straight line is lost at the ends because the first and last points given in Fig. III:10 cannot be plotted in Fig. III:11. This particular theoretical distribution curve evidently has no ends. This is why the outside limits of the distribution do not define its variability; such limits do not exist. An acceptable compromise is to define the position of the 2·5% and 97·5% values. Then, 95% of the distribution lies between these values. These "95% limits" often provide the most precise definition available from preliminary experiments. Other theoretical distribution curves have end points, but they are more difficult to use for other reasons.

A most difficult ecological concept to appreciate is that whereas the outcome of an experiment is never absolutely certain, the amount of variability in results is subject to precise definition. Once the appropriate system of observation is found, the accuracy achieved depends on the precision of measurement and persistence in replication, and should bear comparison with that in other branches of science.

STANDARD DEVIATIONS AND VARIANCE

In addition to the median and 95% limits, a third important statistic can be read off from the probability plot. If the 16% and 84% y-horizontals are drawn and perpendiculars dropped from their intersections with the distribution curve (Fig. III:11), the x-values of the sample categories will be found to lie half way between the median and the 95% limit. In this instance the median is at 5·0 and the 95% limits at 5·0 ± 3·4, i.e. at 1·6 and 8·4. The 68% limits, 16% and 84%, are 3·3 and 6·7, i.e. 5·0 ± 1·7. This quantity, ± 1·7,

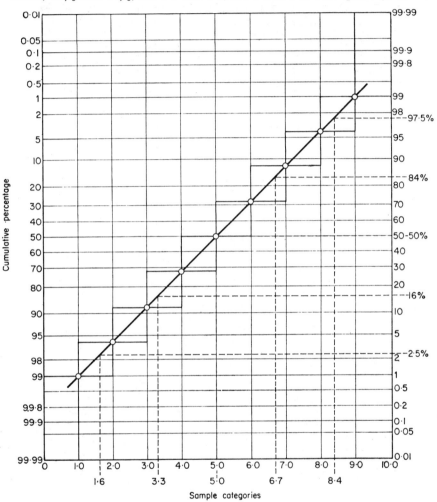

FIG. III:11.   Probability plot of Fig. III:10.   Note the vertical printed scales allow the paper to be used from top or bottom.

which encloses *two thirds* of the distribution even though it is only *half* as wide as the 95% limit, is called the *Standard Deviation*, usually abbreviated to S.D. It is statistically more fundamental than the 95% limits, appearing earlier in the calculations when distributions are computed instead of drawn. The S.D. can, of course, be positive, i.e. greater than the mean, or negative, i.e. less than the mean, and is written in the form Mean ± S.D. The square of each S.D. is the same, positive quantity and this is the *Variance*, the basic statistical measure of the population variability. From the variance

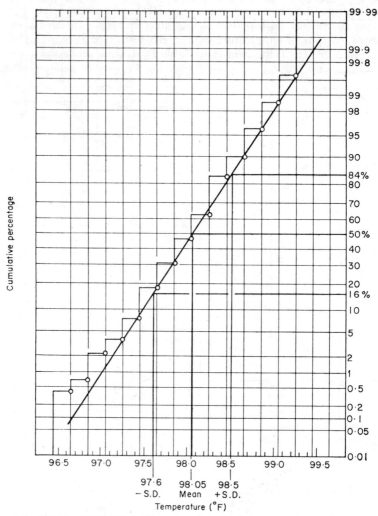

FIG. III:12. Probability plot of body temperature in students, from Ex. 1.

are derived S.D., 95% limits, and any other descriptions of the population variability. The idea of Variance is a central statistical principle with many valuable properties developed in more advanced studies. In Ex. 23 and 24, the Variance is used to define the spatial distribution of organisms; their physical disposition in space. The distinction between spatial "distributions" and the frequency "distributions" that describe them should be kept in mind. To avoid confusion, spatial arrangements are called "dispositions".

IRREGULARITY IN DISTRIBUTIONS

The frequency distributions shown in Figs. III: 5 and 8 are artificially regular in shape. In practice, such perfect shapes are rare even with large samples. The measurements of temperature used in Ex. 1 are typical of a good large sample. In Table III: 7, the data from Table IV: 1:1 are processed to obtain the frequency distribution and probability plot. In the probability plot, Fig.

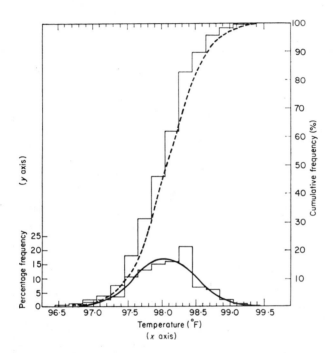

FIG. III: 13. Frequency distribution and cumulative frequency of body temperature with curves from the probability plot in Fig. III: 12.

III: 12, the straight line is fitted by eye and the mean (median) and S.D.s read off. The cumulative percentage distribution with its curve derived from the straight line in Fig. III: 12 and the frequency distribution with its smoothed curve, are drawn in Fig. III: 13. They show how difficult it would be to fit the curve directly to the histogram and how the irregularities are eliminated in the process. The calculated mean is 98·07, the fitted median 98·05, which is as close as one could expect with 276 units.

Because there were 276 units, there is an extra step in making the probability plot. The cumulative frequency distribution must be converted to a cumulative percentage distribution before it can be plotted. This step is shown in Table III:7, columns 5 and 6.

PROBABILITY AND VARIATION

The three frequency distributions in Fig. III:5, from Table III:6, are shown in Fig. III:14 as probability plots. The means may be compared; A

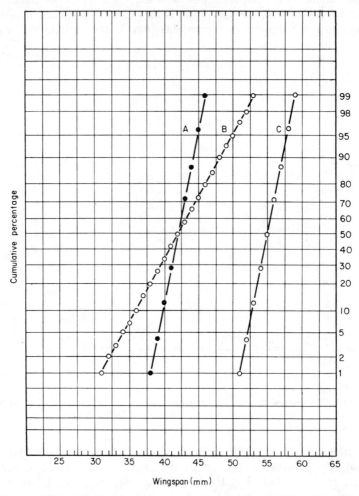

Fɪɢ. III:14.   Differences between means (or medians) and variabilities demonstrated on probability paper. Data from Fig. III:5; A and C have the same variability, so are parallel; A and B have the same median, so intersect at the 50% line.

and B intersect at the 50% line and are therefore the same, whilst C is different. However, A and C are parallel and hence have the same variability,

TABLE III:7

Frequency distribution and cumulative percentage distribution of the body temperature of 276 students between 0800 and 0900 hr

Temperature		Students		Cumulations	
Categories	Mean	Number	%	Nos.	%
96·5–96·6	96·55	1	0·4		
*96·65				1	0·4
96·7–96·8	96·75	1	0·3		
96·85				2	0·7
96·9–97·0	96·95	4	1·5		
97·05				6	2·2
97·1–97·2	97·15	5	1·8		
97·25				11	4·0
97·3–97·4	97·35	10	4·0		
97·45				21	7·6
97·5–97·6	97·55	28	10·0		
97·65				49	18·0
97·7–97·8	97·75	38	14·0		
97·85				87	31·0
97·9–98·0	97·95	41	15·0		
98·05				128	46·0
98·1–98·2	98·15	44	16·0		
98·25				172	62·0
98·3–98·4	98·35	57	21·0		
98·45				229	83·0
98·5–98·6	98·55	19	7·0		
98·65				248	90·0
98·7–98·8	98·75	17	6·0		
98·85				265	96·0
98·9–99·0	98·95	7	2·6		
99·05				272	98·6
99·1–99·2	99·15	3	1·0		
99·25				275	99·6
99·3–99·4	99·35	1	0·4		
99·45				276	100·0

$N = 276$, $\Sigma\theta = 27{,}067\cdot4$, $\bar{\theta} = 98\cdot07$

* Plot cumul. % at interval between categories, see p. 53 and Table III:8.

whilst B has a more shallow slope and is thus more variable. In this way two samples can be compared, even though one may be very large and the other small. In this example of the butterflies' wings it would be difficult to judge the variability in wing length from the original frequency distributions.

SKEW DISTRIBUTIONS

When frequency distributions are not symmetrical, statistical computation and interpretation of results is more difficult. Mode, median and mean are not the same and skill and experience may be needed even to decide which to use. In addition, it is often much more difficult to define the smooth distribution curve to which the sample approximates; much statistical theory and effort is directed towards this end and many different distributions have been

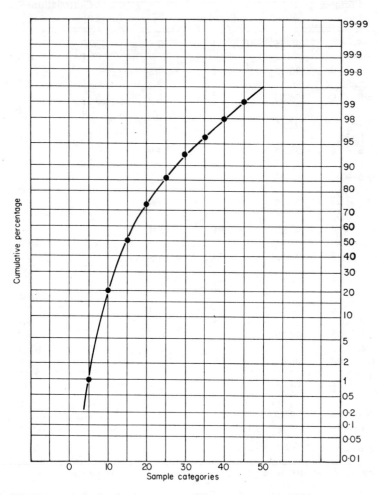

FIG. III:15. A skew distribution on probability paper is represented by a curve. Data from Fig. III:7.

mathematically defined. Skewness in biological distributions is commonly caused by the differential responses of animal populations to successive environmental changes.

Suppose heat is applied to a piece of iron. If the first 100 cal raises its temperature from 10° to 11°C, the second 100 cal will raise the temperature from 11° to 12° and the third 100 cal will raise it to 13°; the resulting rate of increase is arithmetic, each 100 units of heat raises the temperature by 1°

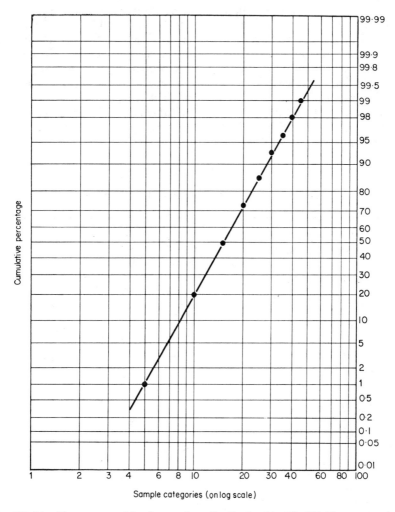

FIG. III:16. The curve resulting from a skew distribution (e.g. Fig. III:15) can sometimes be straightened by using log × probability paper. Note, the x axis increases from 1 to 10 and from 10 to 100 in equal intervals.

discounting losses to the atmosphere. Suppose now, the environmental temperature of a culture of yeast is raised. If the first rise of 5° causes the rate of reproduction to increase from 10 new cells per hr to 20, the next 5° is likely to cause an increase in reproductive rate from 20 to 40 and the third rise from 40 to 80. Each 5° rise multiplies the rate of increase of the population by 2. In other words, successive arithmetic increases in the environment, may produce geometric changes in the biological response.

Frequency distributions from this type of material are often symmetrical only when the x-axis is expressed on a geometric scale and become skew on an arithmetic scale. Figure III:7 is an example of such a skew distribution and Fig. III:15 shows its probability plot to be a curve. Figure III:16, however, shows the same material plotted on "log × probability" paper (see p. 73) as a straight line. Note that the x-axis of this plot is on a geometric scale such that the distance from 1 to 10 is the same as the distance from 10 to 100. The analysis of this distribution is given in Table III:8. The use of logarithms in ecology will be explained below, after considering what happens to the probability plot when two distributions overlap.

TABLE III:8

Stages in analysis of data for Figs. III:7, 15 and 16

1 Interval °C	2 Mean θ	3 No. in interval	4 Interval limits	5 $\sum$ to limit	6 Col. 2 × Col. 3	
			0	0		
0–5	2·5	1			2·5	
			5	1		
5–10	7·5	19			142·5	
			10	20		
10–15	12·5	30			375·0	
			15	50		Median
15–20	17·5	23			402·5	
			20	73		
20–25	22·5	13			292·5	
			25	86		
25–30	27·5	7			192·5	
			30	93		
30–35	32·5	3			97·5	
			35	96		
35–40	37·5	2			75·0	
			40	98		
40–45	42·5	1			42·5	
			45	99		
45–50	47·5	1			47·5	
			50	100		
					1670·0	

$n = 100$, $\Sigma\theta = 1670\cdot0$, $\bar{\theta} = 16\cdot7 =$ Mean (see p. 50)

BIMODAL DISTRIBUTIONS

Frequency distribution of measurements like wing length may sometimes give evidence of two elements in what was thought to be a single population. For example, males and females may be of different, but not widely nor obviously different, sizes. The resulting distribution shows this by having two maxima, or by being ungainly in form. In Table III:9, columns 2 and 3 give two separate, but overlapping, symmetrical distributions. Column 4 gives the sum of these, as would appear in a sample and this is shown in Fig. III:18. When the cumulative sum (column 5) is converted into percentage (column 6) and plotted as probabilities, a sigmoid (S) curve results (Fig. III:17, line S).

When this curve is clearly sigmoid, it can be separated graphically into its component parts quite simply and with sufficient accuracy for most purposes. The inflexion in curve S, the point at which a tangent to the curve crosses from one side to the other, occurs at 80% (Fig. III:17, Table III:9, column 6). This separates the probability curve into 2 parts, 80% and 20%, which correspond in proportion to the two separate distributions, D_1 and D_2, which together constitute the composite distribution. This can be checked in this artificially constructed example, by reference to the sums of columns 2 and 3. In practice, it is used to find the two component elements of the population. Dealing first with the larger, 80%, element from D_1, each point of the cumulative percentage, column 6, is scaled up by the proportion 100/80 since it is required to restore it to a full 100% distribution curve (column 8). A straight line is drawn through these new points, line 1, and extrapolated, that is extended beyond the last point, until it reaches a percentage about equivalent to where it began. In the present instance, the first point is at 1%, therefore the extrapolation reaches to near 99%.

Readings are taken from this line at each unit interval (column 9) and differences taken (column 10). These differences total 100, since they are read from the probability paper. The distribution should however contain 400 units, this being 80% of the original sum of 500 units. The distribution is correspondingly scaled up in column 11, plotted, and a curve drawn in Fig. III:18. One or two points are off the smooth distribution curve, because of errors in drawing or reading off, but the curve nevertheless is an excellent fit. Median and standard deviations can be read from line 1 in Fig. III:17 and used to describe the distribution.

The second element, with 20% of the total distribution, i.e. 100 units, requires an extra step in its extraction. In column 7, the values from column 6 have been subtracted from 100. They now increase from the bottom upwards and when scaled up to 100% ($\times \frac{100}{20}$, column 8) form another distribution. This time there are fewer points and because of the reversed direction of increase, the scale on the left-hand side of the probability paper must be used, instead of that on the right, as is more usual. The procedure is the same; draw a straight line (line 2); extrapolate to an equivalent value to the first point (99%) read off (column 12); take differences (column 13); and scale up (column 14). In this instance, the scale is unity, being 20% of 500,

TABLE III:9

Sequence of calculations for the separation of a bimodal distribution into two parts by a graphical method

1	2	3	4	5	6	7	8	9	10	11	12	13	14
Interval	D_1	D_2	$D_1 + D_2$	Cumul. Σ	Cumul. %	Subtract from 100	× $\frac{100}{80}$	from line 1	Diffs.	× $\frac{400}{100}$	from line 2	Diffs.	× $\frac{100}{100}$
1–2	4	0	4	4	0·8	—	1·0	1·0	1·0	4			
2–3	8	0	8	12	2·4	—	3·0	3·0	2·0	8			
3–4	20	0	20	32	6·4	—	8·0	8·0	5·0	20			
4–5	40	0	40	72	14·0	—	18·0	18·0	10·0	40			
5–6	56	0	56	128	26·0	—	33·0	33·0	15·0	60			
6–7	72	0	72	200	40·0	—	50·0	52·0	19·0	76			
7–8	72	0	72	272	54·0	—	69·0	69·0	17·0	68			
8–9	56	0	56	328	66·0	—	83·0	84·0	15·0	60			
9–10	40	1	41	369	74·0	—	92·0	92·0	8·0	32	99·4	0·6	1
10–11	20	12	32	401	80·0	—	100	97·0	5·0	20	90·0	9·4	9
11–12	8	37	45	446	89·0	11	× $\frac{100}{20}$ 55·0	99·1	2·1	8	53·0	37·0	37
12–13	4	37	41	487	97·4	2·6	13·0		0·9	4	13·0	40·0	40
13–14	0	12	12	499	99·8	0·2	1·0				1·0	12·0	12
14–15	0	1	1	500	100							1·0	1
Σ	400	100	500	500	100				100	400		100	100

and the column is used simply to eliminate decimal places. When plotted, the separation of the complex distribution into 2 smooth simple distributions is complete. The success of the separation may be assessed by comparing the two component distribution columns 11 and 14, with those from which they

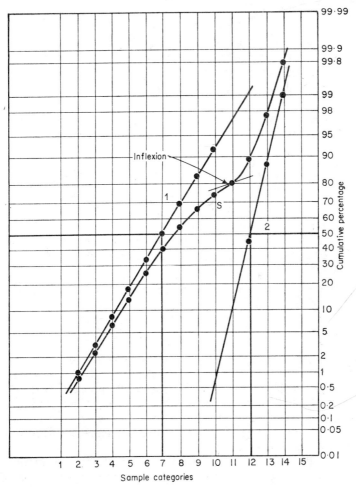

FIG. III:17. Overlapping distributions (S), as in Fig. III:18, can sometimes be separated into their two separate parts (1 and 2) by the graphical "inflexion" method" described in the text.

were originally derived, columns 2 and 3. This was a good analysis. Separations as distinct as this occur only occasionally. In practice the results may be improved by a process of trial and error once the major division has been made. Remember that, in plotting the reconstructed smooth distributions on

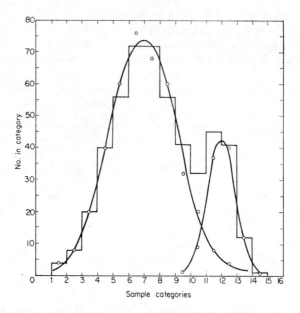

FIG. III:18. Overlapping distributions may appear as bimodal, or two-peaked curves, which can be separated using probability paper (see Fig. III:17).

top of the histogram, the points must be moved one half unit to the left as discussed earlier (p. 53).

LOGARITHMS

Logarithms are widely used in ecological work for several reasons. For example, Fig. III:19 shows daily records of the numbers of insects caught in a light trap. On the arithmetic scale it is hardly possible to see the small changes in mid-July and mid-August if the big changes in early and late July are to fit on the same page. The square root and logarithmic scales make the large catches appear relatively smaller and so make the large and small changes more easily comparable.

Skew distributions and mean values have already been mentioned (p. 59). Table III:10 shows daily catches of insects in a trap and the mean for the 10 days. The one very large catch on the sixth day causes the arithmetic mean (A.M.) to be bigger than all the other nine catches. This is evidently a poor definition for the average, which should indicate what one might expect to happen quite often. The second part of the same table gives the catches in logarithms and the mean logarithm. When this is converted back to numbers, it gives a geometric mean (G.M.) which falls near the median, five values above and five below, and is a much better description of what usually happens. Remember that the G.M. is always less than the A.M.

It has already been pointed out that frequency distributions need to be defined by their limits as well as their mean. This is equally true for scatter diagrams with two variables. Sometimes, as in Fig. III:1, there is so little scatter that this is hardly worth while. At other times, as in Fig. IV:2:1, 2,

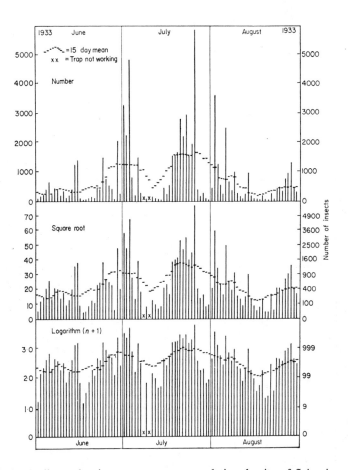

FIG. III:19. Daily catches in traps represent population density of flying insects and can change violently from day to day (top). Plotting the square root (middle) or logarithm (bottom) of the catch makes graphical presentation of results more manageable. (After Williams.)

there is a very wide scatter. When there are catches from two traps working in the same site, they provide two catches for each occasion and these can be plotted, one against the other (Fig. III:20, 21). On an arithmetic scale, as the catches get bigger, the scatter gets wider, so it is difficult to draw parallel

TABLE III:10

Arithmetic and geometric means

Catch (N)	Log ($N + 1$)
9	1·00
11	1·08
18	1·28
44	1·65
106	2·03
938	2·97
124	2·10
67	1·83
22	1·36
12	1.11
1351	16·41

A.M. = $\overline{N}$ = 135   Log ($N + 1$) = 1·64, G.M. = 44 − 1 = 43

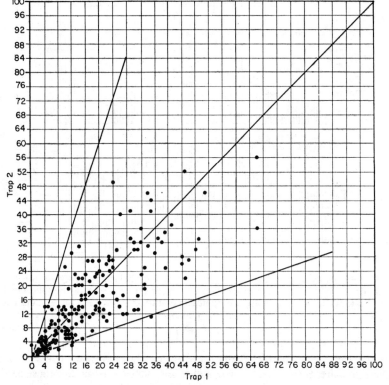

FIG. III:20. Arithmetic plot of daily catches of flies on one sticky trap against numbers caught on another (cf. Fig. III:21).

lines to enclose the points. On the logarithmic scale, however, the scatter remains much more even and is thus easier to define. The construction of Fig. III:21 will be explained later.

Logarithms are useful because ecology is often concerned with proportions, rather than actual numbers. For example, a severe frost may kill 20 plants in a flower bed. This statement has little meaning unless the total number of plants in the bed is also specified. If this was 200, interest is mainly

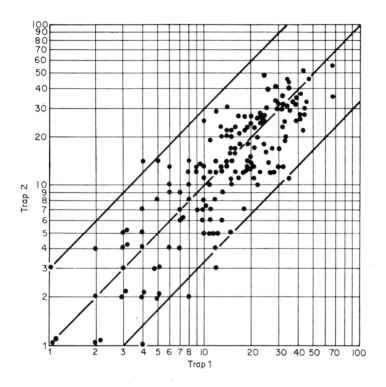

FIG. III: 21.   Data from Fig. III:20 plotted on double log graph paper (see p. 73).

in the proportion killed; 20 out of 200. In a larger bed of 2000 plants, we might reasonably expect the number dead to be 200, not 20. The proportions 20/200 or 200/2000 are easily defined as differences of logs, i.e. log 20 − log 200 is $1 \cdot 30 - 2 \cdot 30 = -1 \cdot 00$: log 200 − log 2000 is $2 \cdot 30 - 3 \cdot 30 = -1 \cdot 00$. The antilog of $-1 \cdot 00$ (or $\bar{1} \cdot 00$) is $0 \cdot 1$ equivalent to 10%. For convenience Fig. III:22 lists the equivalents of some negative and bar logs.

Exercise 43 (p. 236), on the effect of geology on insect fauna, demonstrates that the chalk downland and greensand heath affect the fauna and the structure of the populations living there (Table IV:43:2). But there is more information in this table than is covered by the simple statement that "geological types determine the fauna". It includes information on *how much* the fauna is affected by the different geological formations. Table III:11A is arranged taxonomically and gives the actual numbers in the samples. Table III:11B has been re-arranged according to the animals' percentage frequency

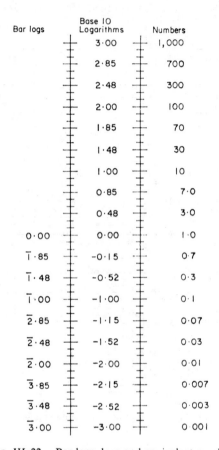

Bar logs	Base 10 Logarithms	Numbers
	3·00	1,000
	2·85	700
	2·48	300
	2·00	100
	1·85	70
	1·48	30
	1·00	10
	0·85	7·0
	0·48	3·0
0·00	0·00	1·0
$\bar{1}$·85	-0·15	0·7
$\bar{1}$·48	-0·52	0·3
$\bar{1}$·00	-1·00	0·1
$\bar{2}$·85	-1·15	0·07
$\bar{2}$·48	-1·52	0·03
$\bar{2}$·00	-2·00	0·01
$\bar{3}$·85	-2·15	0·007
$\bar{3}$·48	-2·52	0·003
$\bar{3}$·00	-3·00	0·001

FIG. III:22.   Bar logs, logs and equivalent numbers.

of occurrence at the two sites. This provides a list of orders in sequence, according to their relative commonness on chalk, for comparison with similar sequential lists for other characters (cf. Ex. 19, p. 162).

However, the percentage classification is crude, as seen from the groups rated as 100%; Orthoptera with 5 on chalk and 0 on heath is classified the same as Mollusca with 147/0. At the other end of the scale, Psocoptera with only 0/3 is placed last, whilst Hemiptera with 4/74, a much more reliable

TABLE III:11A

Numbers of invertebrates at two sites

	Chalk Downland (pH 7·4)	Greensand Heath (pH 5·4)
Annelida (earthworms)	46	0
Isopoda (woodlice)	218	8
Chilopoda (centipedes)	38	1
Diplopoda (millipedes)	37	0
Collembola (springtails)	25	3
Orthoptera (grasshoppers)	5	0
Hemiptera (bugs)	4	74
Dermaptera (earwigs)	12	0
Psocoptera (book lice)	0	3
Thysanoptera (thrips)	12	10
Lepidoptera (moths)	2	11
Coleoptera (beetles)	49	62
Diptera (flies)	21	15
Araneida (spiders)	73	104
Acari (mites)	9	17
Opiliones (harvestmen)	15	11
Mollusca (slugs and snails)	147	0
	713	319

TABLE III:11B

The percentage occurrence of invertebrate taxa on chalk as
compared with greensand

	% on Chalk
Annelida	100
Diplopoda	100
Orthoptera	100
Dermaptera	100
Mollusca	100
Chilopoda	97
Isopoda	96
Collembola	89
Diptera	58
Opiliones	58
Thysanoptera	55
Coleoptera	44
Araneida	41
Acari	35
Lepidoptera	15
Hemiptera	5
Psocoptera	0

The sample on chalk in Table III:11A is given here as a percentage of the total sample
from both sites, for each taxon.

result, is placed next to the last. It only needs a catch of three Psocoptera on chalk to move them to the middle of the list, but an unlikely 70 are needed to do the same thing to the Hemiptera.

TABLE III:12

Relative commonness of invertebrate taxa on Chalk and Greensand Heath

	Chalk Downland	Greensand Heath	% on Chalk Downs	Log (CD +1)	Log (GH + 1)	Diff. of logs (CD + 1) −(GH + 1)	Antilogs
Mollusca	147	0	100	2·17	0·00	2·17	148
Annelida	46	0	100	1·67	0·00	1·67	47
Diplopoda	37	0	100	1·58	0·00	1·58	38
Isopoda	218	8	96	2·34	0·95	1·39	25
Chilopoda	38	1	97	1·59	0·30	1·29	19
Dermaptera	12	0	100	1·11	0·00	1·11	13
Collembola	25	3	89	1·41	0·60	0·81	6·5
Orthoptera	5	0	100	0·78	0·00	0·78	6·0
Diptera	21	15	58	1·34	1·20	0·14	1·4
Opiliones	15	11	58	1·20	1·08	0·12	1·3
Thysanoptera	12	10	55	1·11	1·04	0·07	1·2
Coleoptera	49	62	44	1·70	1·80	−0·10	0·79
Araneida	73	104	41	1·87	2·02	−0·15	0·71
Acari	9	17	35	1·00	1·26	−0·26	0·55
Lepidoptera	2	11	15	0·48	1·08	−0·60	0·25
Psocoptera	0	3	0	0·00	0·60	−0·60	0·25
Hemiptera	4	74	5	0·70	1·88	−1·18	0·066

A more meaningful arrangement of the list is shown in Table III:12 in which the difference between the logs of the numbers, and the antilog of this, expresses their relative commonness as a proportion. Instead of having the first eight orders almost indistinguishable from each other, they are now graded according to the reliability of the sample, and similarly with the last three orders. No changes could be expected within the middle six orders, which are almost equally common at the two sites, because the numbers are all in or near the tens and hundreds and hence percentages already effectively placed them. To remove the main source of error, an effort should always be made to reach double figures in samples, but with "all or none" segregation, like those at each end of the list, this becomes impracticable.

Differences of logarithms are therefore usually preferable to percentages if the range of values is wide. Percentages can be used for simple demonstrations of more nearly similar quantities, especially when a whole is divided into several parts. An example is the division of a population of aphids into instars (Table III:13).

TABLE III:13

The division of a population of aphids into adults and instars

| | Adults | | Instars | | | Total |
		4th	3rd	2nd	1st	
Number counted	241	187	141	89	207	865
%	27·9	21·6	16·3	10·3	23·9	100

As there is no logarithm for 0, a complication arises when there are zero insects in the sample. In Table III:12, for instance, there were 0 Psocoptera on chalk downs and five orders with 0 on heathland. This difficulty is surmounted by using the logarithm of $(N + 1)$ instead of log (N), where N is the number of individuals in the sample, not to be confused with n meaning the number of units in the sample. Unity may be subtracted again after conversion back to numbers, but this is unnecessary where proportions are estimated; the proportion $(N_1 + 1)/(N_2 + 1)$ is used instead of the proportion N_1/N_2. This adjustment distorts the quantities, but mainly the small numbers, which were unreliable to begin with; the larger numbers are hardly affected. In return for this slight misrepresentation, much more useful information is obtained. For example, if one taxon does not occur on heathland, it would be rather difficult to present it in such a table of relative commonness without this adjustment. A table of $\log_{10} (N + 1)$ is given in Appendix D.

Another advantage of logs is to simplify the problem of significant figures. Only two places of decimals are necessary when taking logarithms of counts in ecology. This represents accuracy to the order of 1% and as mentioned earlier, improvements on this are rare. As the number in front of the decimal point in logs, the mantissa, is only 9 for one thousand million or $\bar{9}$ for one milli-micron, a total of three digits in logarithms represents any useful biological quantity.

LOGARITHMIC PAPER

The graphical use of logarithms is greatly facilitated by graph paper marked with logarithmic scales so that the numbers can be plotted directly, instead of reading the logarithms from tables and plotting the logs. Just as probability paper has the intervals contracted near the middle and extended near the ends to convert certain sigmoid curves to straight lines, logarithmic paper has the larger quantities successively more contracted, so that the interval from 1 to 10 is the same as the interval from 10 to 100 and from 100 to 1000, etc. Paper can be obtained with the logarithmic scale either along one axis, semi-log paper, or both axes, log × log or double-log paper.

The type of paper is known also by the number of cycles, i.e. orders of magnitude or multiples of 10, that it covers. Figure III:23 illustrates four different kinds of log paper and there are many more depending upon the range of numbers required. In Fig. III:23A the y-axis is in logarithms and

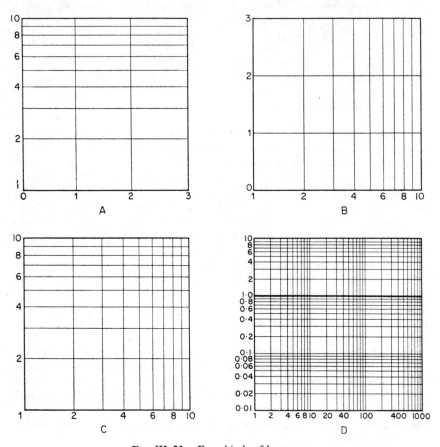

FIG. III:23.   Four kinds of log paper.
A. 1 cycle logs × numbers.
B. Numbers × 1 cycle logs.
C. 1 cycle "double log" or "log × log".
D. 3 cycle log × log.
Note: The origin on log × log paper is placed at (1, 1), which would be (0, 0) if the logarithms were inserted instead of the equivalent numbers.

the x-axis in ordinary numbers. In Fig. III:23B, the axes are reversed and in Fig. III:23C both axes are in logarithms. In each of these examples the log scale goes from 1 to 10, one log cycle. In Fig. III:23D there are three log cycles along each axis, y ranging from 0·01 to 10 and x from 1 to 1000. The origin is usually taken to be at $x = y = 1$ on log × log paper, i.e., at log x = log y = 0 if the logarithmic values were written instead of the equivalent numbers. The same resultant graph can, of course, be made by transforming all the values into logarithms and then plotting them on arithmetic coordinates. On multicycle log paper, the same units are printed for each cycle, running from 1 to 9. If the cycle is required to cover the range 100–900, the additional digits are superimposed by the user; see Fig. III:23D, in which the required units have been written in. Note the absence of zeros; with each diminishing cycle, the numbers get smaller but never reach 0. Note also that there are only nine major intervals to each cycle, not ten. Log × probability paper is also available.

It is instructive to see what happens to several simple equations when plotted on logarithmic paper. In Fig. III:24 expressions of the form $y = a + x$ are shown on arithmetic coordinates as parallel lines at 45° (A). In logs (B) these lines converge on the 45° line $y = x$ as x increases. Evidently the added constant is of no consequence when x exceeds about 100. In (C) the arithmetic expressions $y = bx$ are radiating lines through the origin. In logs (D) these lines are all at 45°, displaced by the *factor b*, as the arithmetic expressions $y = a + x$ were displaced by the *constant a* (compare D with A). This is because the lines represent log y = log b + log x; comparable with $y = a + x$. The four parts of Fig. III:25 all illustrate expressions of the form $y = x^a$. Parts A and B show these expressions on the arithmetic scale at different scale factors. Near the origin (A) the lines are curved and cross over at the point ($x = 1$, $y = 1$). As x increases (B) the lines diverge ever more rapidly so that x^2, x^4 and x^{16} become almost indistinguishable. A semi-log scale (C) begins to separate the lines a little, and on the log × log scale (D) separation is much improved and the lines are linear. For these kind of equations log × log paper is especially useful. Note that the "line of equality" $y = x$ is linear on both arithmetic and double-log scales, but on double-log scales, the line is always at 45° because of the accepted convention that log cycles always have the same scale on abscissa (x-axis) and ordinate (y-axis). On arithmetic scales the line of equality can have any apparent gradient, depending upon how the scales are chosen (see p. 42).

When asymmetrical data are transformed to logarithms to obtain linearity, for example by using log × probability paper (p. 61) for a distribution curve or log × log paper for a regression (p. 82), care must be taken with means and standard deviations. The mean of a scatter on log × log coordinates is $\overline{[\log (x + 1), \log (y + 1)]}$ and this is the mean of the [log (N + 1)]'s for each coordinate. It is not the log of the mean which is (log $\bar{x}$, log $\bar{y}$). Also, a linear log × probability transformation gives the mean *and standard deviation* in logarithms, thus; 2·00 ± 1·00. This is symmetrical *only* in logarithms; de-transformed it becomes 100 × or ÷ 10, i.e. upper S.D. = 100 × 10 = 1000; lower S.D. = 100 ÷ 10 = 10. Not only are the mean and S.D. anti-logged, but the + and − of the log scale become × and ÷ on the arithmetic scale.

Other transformations than logarithms are used in ecology; square roots (Appendix C, p. 357) sometimes are effective and there are other more

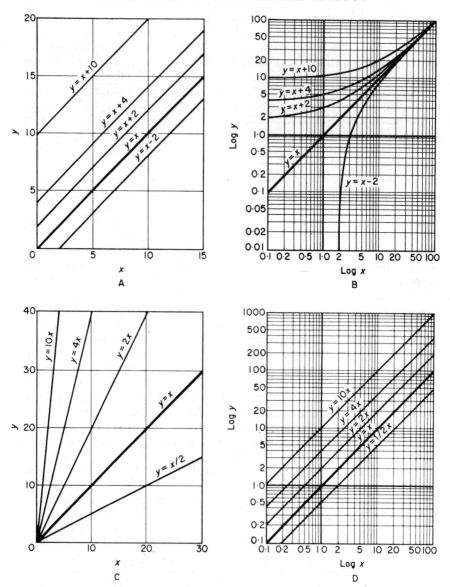

FIG. III:24. Mathematical expressions plotted on arithmetic and log paper.
  $y = x +$ constant
  A: on arithmetic coordinates,
  B: on log × log coordinates.
  $y = bx$
  C: on arithmetic coordinates,
  D: on log × log coordinates.

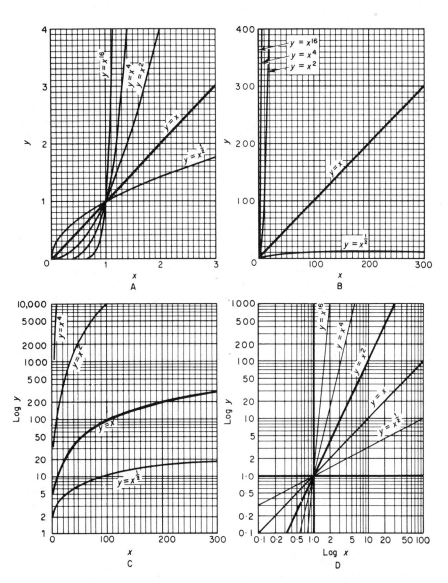

FIG. III:25.

$y = x^b$

A: on arithmetic coordinates,
B: on arithmetic coordinates but at a different scale,
C: on single log paper,
D: on log × log paper.

complex expressions, but their application is less common and, lacking appropriate graph paper, transformation must be done using tables.

REGRESSION ANALYSIS

When the scatter diagram of one dependent variable, such as plant height, on an independent variable, such as light intensity, shows an obvious tendency for the points to form a band sloping at an angle to the horizontal, it is usually possible to define the relationship by regression analysis. First it is necessary to decide which is the dependent and which the independent variable; in practice there is often quite a lot of interdependence. For example, insect body temperature is a variable, dependent on the independent variables of the physical environment. In contrast, environmental temperature in a bees' nest is partly dependent on numbers of bees, because bees generate heat.

The way the regression scatter is plotted determines the sequence of the analysis and its interpretation. In a sense, therefore, the making of a scatter diagram for regression analysis is stating the hypothesis that y is dependent, at least partly, on x. If a regression is successfully fitted and the relationship found to be "statistically significant" (p. 97) the hypothesis is provisionally vindicated. A significance test does not prove a hypothesis. It makes it more or less probably valid, but can never prove that another, better hypothesis does not exist. Regression analysis, therefore, is used most often, not to prove the existence of a relationship which is usually obvious, but to try to define that relationship so as to forecast the dependent variable from a given independent value. The object is to represent the relationship between the variables x and y by a line. Obviously the validity of the line will depend on the "goodness of fit" of the points to it. For example Fig. III:26A, B, C illustrates three stages in the progressive refinement of a cloud of points; the variance not belonging to the regression line but to other, extraneous factors is progressively stripped away. In (A), the extraneous variability is so great that the regression cannot be defined; it is not significant. In (B), the trend of the cloud is clearly visible, but the line accounts for only a part of the variance; a lot remains to be accounted for by other factors. In (C), every point is on the line and if the line is defined as $y = x$, every bit of evidence is accounted for; there is no variance left. This only happens in theory.

Fig. III:26D, E, F presents less obvious problems. (D) shows an "independent variable" that hardly varies. The dependent variable covers a range of values, but there is no regression. This may happen when the environment is experimentally uncooperative as, for example, when temperature remains constant on a dull day. (E) shows the reverse situation. The expected variation in the dependent variable did not occur; obviously the hypothesis used the wrong independent variable, and there is again no regression. In (F) there is a good regression, with extraneous variance, but although it looks as good as (B), its regression is less valid. Consider (D) and (E); both have no valid regression and they are the extreme positions for regression slope, $\tan 90° = \infty$ and $\tan 0° = 0$. (B) has the same shape but a highly valid regression, because it is half way between (D) and (E). Hence the validity of a regression depends both on the width of the scatter and also on the slope of the line. (F) is intermediate between (B) and (E) and hence its validity is also intermediate.

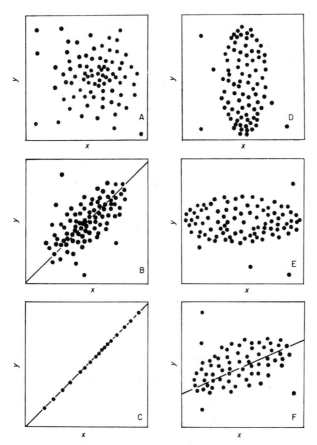

FIG. III:26. Different relationships between two variables, expressed as scatter diagrams, described in the text.

In considering this problem it must be remembered that, in the diagram, the apparent slope of the line can be changed by changing the scale of the axes. This, however, also changes the scatter of points about the line and so compensates for any change in the validity of the regression. The actual slope of the line can only be defined when 1 unit of y is made equal to 1 unit of x.

Before defining a relationship with a regression line, the validity of the relationship may be tested graphically, although such tests are not always very rigorous. To do this, first separate the scatter of points into quadrants by a *medial cross* (Fig. III:27, ungraduated lines) as follows. Draw a horizontal line dividing the points into equal numbers above and below; if there is an odd point this line will pass through it and it is subsequently disregarded. Draw a vertical line dividing the remaining points equally to left and right. Count the number of points in each quadrant and compare these numbers with Appendix I. If any quadrant contains as many or more than the "upper"

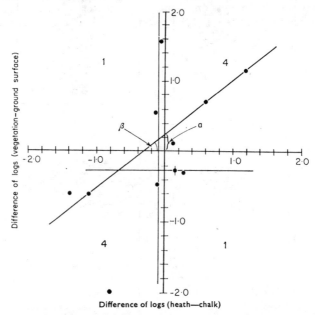

Difference of logs (heath—chalk)

FIG. III:27. Diagrammatic representation of the simple regression equation $y = a + bx$ where a (or $-a$) is the intercept on the y axis and b (or $-b$) is tan β (or tan $-\beta$), and a graphical test for a significant relationship.

quantity, or as many or less than the "lower" quantity, the two variables are significantly related. In the present example, the highest number is 4 but 5 is required for significance with $N = 10$ (one point is lost on the horizontal line). A computed significance test shows a significant relationship so, in this instance, the graphical test is the more exacting.

The regression line will be of the form:

$$y = a + bx$$

in which both a and b may be negative or positive and where a is the intercept on the y-axis (Fig. III:27), b is the tangent of the angle of elevation and is known as the regression coefficient: $b = \tan \beta$.

This angle can be found by graphical methods but it is much dependent on the variance of the y values. The total distribution of the ys comprises several smaller overlapping distributions along its length (Fig. III:28). Each of these is derived from the ys for a smaller range of xs. It is possible to visualize the regression line as that element in the composition of the total variance of y that is dependent on x. If this component could be defined and removed, the residual variance "about the regression line" would be considerably smaller than the original total variance. Evidently the nearer the slope is to 45°, ($b = \tan 45 = 1$), the bigger the component of the variance of y is associated with the regression, hence the greater validity of regression coefficients near 1 when the units of y and x are made equal in the diagram.

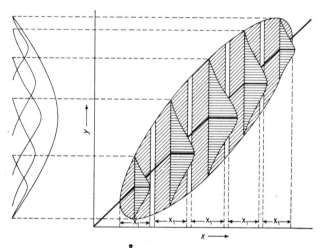

FiG. III:28. The scatter diagram of a regression shows a cloud of points which can be split up into a series of narrow strips, each on a unit of the x axis (x_1). Each strip then becomes a frequency distribution of the y values for a narrow range of values of x. Viewed from the y axis, these several frequency distributions can be seen separately. Below the y axis, they are drawn overlapping, and the total variability of the ys for the whole cloud of points, is their sum, shown as an enclosing envelope. Hence, fitting a regression can be thought of as splitting up the total variability into parts, and so isolating that element of variability associated with a change in the x values.

There is a further related problem if b is not 1. If the axes are reversed, and the regression of x on y is found by mistake, the resulting line will be different from the regression of y on x, because the variance of x is different from that of y; hence the component of variance accounted for by each regression line is different and the best line for y on x would not necessarily be the best for x on y. So an eye-fitted regression line is only likely to be completely valid if it is fitted to points with almost no scatter, like Fig. III:1. In Fig. III:28 the regression line does not pass down the axis of the scatter, although each distribution is symmetrical about it. It is possible, however, to approximate graphically to a regression line by reducing the variance of y. Consider first a linear regression.

If the x-axis is split up into regular intervals and means taken of the ys for each unit of x, the scatter due to the variance of the ys is reduced to a minimum. Carrying this process to its logical conclusion reduce the ys to two fixed points, with no variance, as in Fig. III: 29 from Ex. 21. Divide the abscissa by two vertical lines so that there are three nearly equal groups of points; the two end groups of the same size leaving any remainder in the middle group, e.g. in Fig. III:29 there are six points in each end group and eight in the middle. Find the medial points of the two end groups by drawing

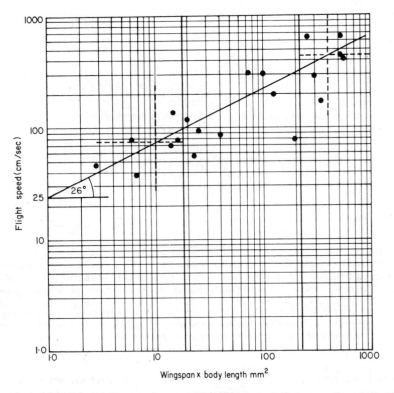

Fig. III:29. A regression line can be fitted graphically by splitting up the xs into three lots, and finding the median position of the ys in the end two groups.

crosses to divide the six points equally between upper and lower segments $(3 + 3)$ and left and right segments $(3 + 3)$. Join the two resulting medial points. Extrapolate the line back to the ordinate. Read off the intercept $(a = \log 25)$ and measure the angle of elevation (or depression) β $(+26°)$, making sure that the units of x and y are equal in the diagram. Find the regression coefficient, $b = \tan \beta = 0\cdot49$ and write the regression equation:

$$y = a + bx$$

If the units of x and y are not equal, b must be calculated as in the inset in Fig. IV:2, 2. In the present example this is unnecessary because the scales are the same on log × log paper. Hence:

$$\log y = \log 25 + 0\cdot49 \log x$$

De-transforming, because double-log coordinates were used (see Fig. III:25D):

$$y = 25x^{0\cdot49} \quad \text{or} \quad 25x^{0\cdot5} \text{ approx.}$$

which may be written:

$$y = 25\sqrt{x}$$

x is equal to wing span × body length, an area measure of overall size which makes some allowance for either big wings or a powerful body, both of which could contribute to flight speed. The square root of x therefore is a linear measure of size making some allowance for both factors in a way that neither would by itself. Using either wingspan or body size alone, the regression would have more scatter. Nevertheless the final expression may be stated simply; flight speed in insects is proportional to linear body dimensions. The true regression line passes through $(\bar{x}, \bar{y})$ and no provision was made for this. Hence the regression can be marginally improved by making it parallel to the present line, but passing through $(\bar{x}, \bar{y})$. This will change a but not b, and has not been done in Fig. III:29 for simplicity though it has in Fig. IV:45:2, p. 247. Like all graphical constructions, this regression will serve until a more stringent requirement makes a statistically defined one necessary. Then the appropriate numerical method will be found in a book such as Baileys' (1959) "Statistical Methods in Biology", and in quick reference form in Appendix A (p. 325), where the computed regression can be compared with the graphically derived one in Fig. III:29.

When the two medial points have been found on ordinary, arithmetic axes, it may appear that they lie on a curve and not on a straight line. As shown in Fig. III:3 (p. 44), this can often be seen in the original scatter by looking along it. Curvilinear regression equations, which are quadratics or polynomials, can be fitted, but for the present purposes transformation of the axes, as described on p. 66 *et. seq.*, is more appropriate. To do this, divide the data into five near-equal parts and so obtain five medial points. Plot these on single or double logarithmic paper to see if the points fall nearer to a straight line, referring to Figs. III:24, 25 (p. 76, 77) to find the appropriate relationships. If a suitable linear relationship results, treat the now linear scatter as described above. This had already been done in Fig. III:29. If a straight line can be obtained by this means, y can be read off the regression line at unit intervals of x and replotted as a curve in the original scatter diagram. Sometimes a square root transformation will succeed when the log fails. Unfortunately this means transforming each reading from a table (Appendix C, p. 357) because square root graph paper is not available.

Many lines representing ecological data can be straightened by such transformations but if a line cannot be straightened, the best alternative is to plot as many medial points as possible and draw a curve through them, with the help of a "French Curve". The relation of y to x obtained cannot be defined mathematically. Scatter diagrams do not always yield regression relationships. Some behavioural responses are "all-or-none", like fight or no-fight and flight or no-flight. These produce threshold relationships (p. 141).

In Fig. III:21 (and in the figures in Ex. 5, 6, 7) parallel lines have been drawn on either side of the regression line. These lines enclose most of the points and show that the scatter has limits similar to the 95% limits for a single distribution (p. 54). The limits have not, however, been fitted by any statistical process. They are merely indicators, showing that the limits remain fairly constant at different values of x. In contrast, the lines in Fig. III:20 diverge widely and hence limits change as x increases. When an appropriate transformation for the axes has been found, the limits should be parallel. In Ex. 5, 6 and 7, parallel lines have been drawn at the same values and help in judging by how much the points depart from the line of equality. True limits can be obtained by computation, but only when the regression line is also computed.

In order to obtain the equation $y = a + bx$ graphically in the example given above, the line was extended back to the y axis. This process, of extending a line beyond the limits of the data, is called extrapolation. In Fig. IV:2:1 it leads to a nonsensical value for height at birth, and is misleading because it implies that the relationship between height and age, existing between the ages of 12 and 17, exists at all other ages. The equation derived from the data applies only within the limits of the data, even though extrapolation is necessary to obtain it, for the line may be curved beyond these limits. Extrapolation is always open to this objection, although over short distances compared with the range covered by the data, e.g. 1 year in Fig. IV:2:1 it has a practical value provided the risk of error is remembered. The process of interpolation, reading off from the line a value for y at a given value of x within the limits of the data, is more dependable but even here, the resultant value of y has more validity the nearer it gets to the mean point $(\bar{x}, \bar{y})$. At that point it has all the validity of the whole data behind it and is the most accurate information obtainable from the observations.

Correlation analysis is another method of relating one variable to another but the statistical requirements for its application are more stringent and hence its use is more restricted. It will not be dealt with here and is mentioned only because the expression "A is correlated with B" is commonly used. This is often said when the relationship is a regression and it is not then strictly appropriate.

SPATIAL PATTERN

All the actions of an animal depend on its position in space; for example it cannot feed or reproduce unless near to food or a mate. Hence the spatial positions of animals are fundamental to a study of their ecology. Because the environmental requirements of different species differ, their resultant spatial arrangements differ. This is a familiar feature of populations of plants. Most grasses tend to spread evenly over a lawn, but there are often distinct patches of other plants such as buttercups, and daisies. This patchiness is easily demonstrated because it is extreme, compared with the even distribution of the grass plants. Sometimes the positions of animals in space, or their spatial disposition, is not so obviously patchy, although it is evidently less regular than tiles on a floor.

In Fig. III:30A, B, C, 200 entities, represented by dots, are arranged differently in the same two-dimensional surface, or space. In (A), the arrangement is regular, easily recognizable and rare in nature. In (C), there appear to be slight concentrations, or aggregations, in the bottom right corner and near the upper right corner. This degree of aggregation is very common in nature. It is not too obviously different from (B), the disposition of whose points in space is intermediate between (A) and (C); neither *regular* nor *aggregated* but *random*. Randomness is an important concept in statistics and in ecology because its spatial meaning can easily be defined and has a

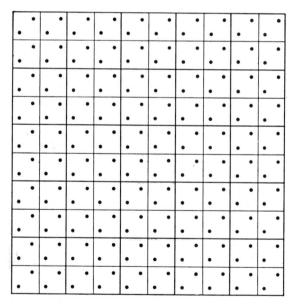

FIG. III:30.
A. A regular disposition of points in a plane, or space. The population represented here has a mean and no variance.

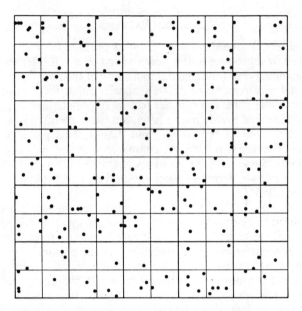

Fig. III: 30 B. Randomly spaced points with each point completely independent of its neighbours, have the mean density and variance equal.

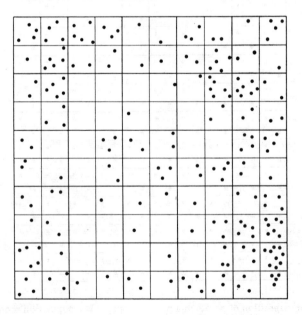

C. Aggregated points appear clustered, and have variance greater than mean density.

biological equivalent. Theoretically, randomness occurs when there is no interaction between individuals. Ideally therefore, *complete independence* between organisms and their environment would produce randomness. which is a useful concept because it has a well defined statistical equivalent. Although complete independence between individuals never occurs, the distribution of a few natural populations is indistinguishable from true randomness. Individuals more usually react together. If they repel each other, spatial regularity results; if they attract each other, aggregation results and this is the most usual natural condition. Very few regular dispositions occur in nature. One example is the regular shape of the comb in honey bees' nests which results from the necessity for the larvae to be as far apart as possible in a limited space.

At very low population densities, when a sampled area contains very few individuals, it is difficult, often impossible, to distinguish randomness from aggregation. Figure III:31A, B, C, D, E, F shows the disposition of from one to six random points in a given space. The points are the first six pairs from the table of random numbers in Appendix L (p. 373). The pairs of numbers were used as x and y coordinates to plot the points on a 1 to 100 scale. The first point happens, quite by chance, to fall almost at the edge of the square and similarly the second and so on. Only at the 53rd selection does a point fall within 10 units of the centre; then a second one occurs after 11 more selections and a third after only 3 more (see Appendix L.) There is, of course, more edge than middle available and another selection would give quite a different result. This is important in sampling practice and is dealt with in the following section.

Returning to Fig. III:30A, B, C, the mean density of the population in each instance is 2 per unit area, a total of 200 individuals in the complete population of 100 units. What differs between them is the number of individuals in each square, or unit. In Tables III:14 A, B, C, D, the numbers per square are reproduced and frequency distributions extracted from them. These are plotted in Figs. III:32, 33, excluding the regular distribution (A) which cannot be plotted. From the probability plots the standard deviation, positive and negative, is read off and the variance ($=$ S.D.^{2}) estimated. Note that the mean, which is known to be 2, falls at the median in Fig. III:32 which is linear as well as random. In Fig. III:33, the example of aggregation, the distribution is skew, the mean is not at the median and the S.D.s are asymmetrical, with the upper one bigger than the lower. As a result, the estimate of the variance made from this curve will be poor and the mean and the median are different, though close enough for many purposes. In Fig. III:30C, where the points are aggregated, the maximum number of organisms per unit is 9 and there are 23 empty units compared with a maximum of 6 organisms per unit and only 11 empty units when the points are random (Fig. III:30B and Table III:15).

Randomness and aggregation are more easily distinguished with larger mean values. The same procedure as above has therefore been repeated using the heavy squares in Fig. III:30 as units, each containing four of the small squares and giving a mean of 8 per unit. In the probability figures (Figs. III:32, 33), although the number of *points* increases, the accuracy of the estimate which is judged by the scatter of points about the line, does not change. This is because there are still only 200 *individuals* in the whole

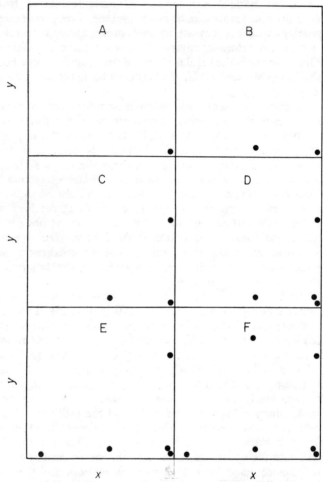

FIG. III:31.   Random points in space do not fall evenly over the area. The first five points in the random number table (Appendix L, p. 373) happen, by chance, to fall quite near the edge of the space available.

population. The means, S.D.s and the variances are compared in Table III:15 for the three kinds of disposition at two population densities. For perfectly regular dispositions we have a mean and no variance. A slight loss of regularity would introduce variance but it would be less than the mean.

Table III:14A

Fig. III:30B

Random 10 × 10 square, $N = 200$

Counts per unit area

6	4	3	2	0	3	2	2	1	2
2	1	2	3	0	3	0	5	0	1
2	4	1	3	1	2	2	1	3	2
1	1	4	4	3	0	0	2	1	3
1	2	1	0	1	4	2	3	4	2
3	2	1	3	2	0	4	2	2	2
1	2	3	3	1	4	4	1	3	1
4	3	2	3	4	0	1	2	2	0
1	2	1	1	1	0	1	1	2	1
3	1	2	2	2	2	4	4	1	1

$$\text{S.D.} = \frac{3{\cdot}4 - 0{\cdot}6}{2}$$
$$= 1{\cdot}4$$
$$s^2 = (1{\cdot}4)^2 = 1{\cdot}96$$

Frequency distribution of counts per unit area

Counts											Σ unit areas	Cumul. %	Σ counts
0	—	—	—	1	2	4	2	—	1	1	11		0
												11	
1	4	3	4	1	4	—	2	3	3	4	28		28
												39	
2	2	4	3	2	2	2	3	4	3	4	29		58
												68	
3	2	1	2	5	1	2	—	1	2	1	17		51
												85	
4	1	2	1	1	1	2	3	1	1	—	13		52
												98	
5	—	—	—	—	—	—	—	1	—	—	1		5
												99	
6	1	—	—	—	—	—	—	—	—	—	1		6
												100	
No. of units	10	10	10	10	10	10	10	10	10	10	100		200

Random distributions have variance and mean equal, within the limits of error of drawing and computing, whatever the population density, e.g. in Table III:14A, column 4, when the mean is 2, $s^2 = 1{\cdot}96$. Aggregated distributions have the variance greater than the mean and, as the population density increases, so the ratio of variance to mean increases (p. 23). This relationship between disposition and density in populations is expressed by a Power Law, which states that "variance is proportional to a fractional power of the mean in natural populations"

$$s^2 = am^b$$

where s^2 = variance, m = mean density } of the population.

a and b are population parameters, found as the intercept (a) and regression coefficient (b) on a log × log plot of s^2 against m (Ex. 23, 24, p. 178, 181).

TABLE III:14B

Fig. III:30B

Random 5 × 5 square, $N = 200$

Counts per unit area

13	10	6	9	4
8	12	6	5	9
8	5	7	11	10
10	11	9	8	6
7	6	5	10	5

$$\text{S.D.} = \frac{10{\cdot}7 - 5{\cdot}3}{2}$$
$$= 2{\cdot}7$$
$$s^2 = (2{\cdot}7)^2 = 7{\cdot}3$$

Frequency distribution of counts per unit area

						$\sum$ unit areas	Cumul. %	$\sum$ counts
0	—	—	—	—	—	0		0
1	—	—	—	—	—	0		0
2	—	—	—	—	—	0		0
3	—	—	—	—	—	0		0
4	—	—	—	—	1	1	4	4
5	—	1	1	1	1	4	20	20
6	—	1	2	—	1	4	36	24
7	1	—	1	—	—	2	44	14
8	2	—	—	1	—	3	56	24
9	—	—	1	1	1	3	68	27
10	1	1	—	1	1	4	84	40
11	—	1	—	1	—	2	92	22
12	—	1	—	—	—	1	96	12
13	1	—	—	—	—	1	100	13
No. of units	5	5	5	5	5	$\overline{25}$		$\overline{200}$

TABLE III:14C

Fig. III:30C

Aggregated 10 × 10 square, $N = 200$

Counts per unit area

4	4	5	3	1	1	3	2	1	4	$S.D. = \dfrac{3\cdot6 - 0\cdot1}{2}$
1	5	2	2	1	1	2	5	2	1	
2	5	0	0	0	1	1	6	7	2	$= 1\cdot75$
0	3	0	0	1	0	0	2	2	3	$s^2 = (1\cdot75)^2 = 3\cdot1$
2	0	0	3	2	2	0	0	4	4	
2	1	0	2	0	3	2	4	2	1	
2	2	0	1	1	1	1	0	2	4	
1	2	0	0	1	0	1	3	5	7	
5	1	0	0	0	1	0	3	3	9	
3	3	1	1	2	1	4	3	3	4	

Frequency distribution of counts per unit area

											Σ unit areas	Cumul. %	Σ counts
0	1	1	7	4	3	2	3	2	—	—	23		0
												23	
1	2	2	1	2	5	6	3	—	1	2	24		24
												47	
2	4	2	1	2	2	1	2	2	4	1	21		42
												68	
3	1	2	—	2	—	1	1	3	2	1	13		39
												81	
4	1	1	—	—	—	—	1	1	1	4	9		36
												90	
5	1	2	1	—	—	—	—	1	1	—	6		30
												96	
6	—	—	—	—	—	—	—	1	—	—	1		6
												97	
7	—	—	—	—	—	—	—	—	1	1	2		14
												99	
8	—	—	—	—	—	—	—	—	—	—	0		0
												99	
9	—	—	—	—	—	—	—	—	—	1	1		9
												100	
No. of units	10	10	10	10	10	10	10	10	10	10	100		200

TABLE III:14D
Fig. III:30C
Aggregated 5 × 5 square, $N = 200$

Counts per unit area

14	12	4	12	8
10	0	2	9	14
5	5	7	6	11
7	1	3	5	18
12	2	4	10	19

$$\text{S.D.} = \frac{13\cdot5 - 2\cdot9}{2}$$
$$= 5\cdot3$$
$$s^2 = (5\cdot3)^2 = 28\cdot1$$

Frequency distribution of counts per unit area

						Σ unit areas	Cumul. %	Σ counts
0	—	1	—	—	—	1		0
							4	
1	—	1	—	—	—	1		1
							8	
2	—	1	1	—	—	2		4
							16	
3	—	—	1	—	—	1		3
							20	
4	—	—	2	—	—	2		8
							28	
5	1	1	—	1	—	3		15
							40	
6	—	—	—	1	—	1		6
							44	
7	1	—	1	—	—	2		14
							52	
8	—	—	—	—	1	1		8
							56	
9	—	—	—	1	—	1		9
							60	
10	1	—	—	1	—	2		20
							68	
11	—	—	—	—	1	1		11
							72	
12	1	1	—	1	—	3		36
							84	
13	—	—	—	—	—	0		0
							84	
14	1	—	—	—	1	2		28
							92	
15	—	—	—	—	—	0		·0
							92	
16	—	—	—	—	—	0		0
							92	
17	—	—	—	—	—	0		0
							92	
18	—	—	—	—	1	1		18
							96	
19	—	—	—	—	1	1		19
							100	
No. of units	5	5	5	5	5	25		200

TABLE III:15 (Results to one decimal place)

Disposition type	Regular		Random		Aggregated	
Computed mean density	2	8	2	8	2	8
S.D. $(+s)$ ⎫	0	0	1·4	2·7	1·5	4·5
$(-s)$ ⎬ from Figs.	0	0	1·4	2·7	2·0	6·1
Variance (s^2) ⎬ III: 32, 33	0	0	2·0	7·3	3·1	28·1
s^2/m ⎭	0	0	1·0	0·9	2·0	3·8
Median density (from Figs.)	2	8	2	8	1·6	7·4
No. of zero units	0	0	11	0	23	1
Minimum counts per unit	2	8	0	4	0	0
Maximum counts per unit	2	8	6	13	9	19

The equation used to compute variance, when the graphical method is found inadequate, is given in Ex. 23 and Appendix A (p. 325).

Note in Table III:15 that when aggregated, the maximum number per unit is 19, compared with 13 when random. Also there are no random units with less than four individuals, but there is still one aggregated unit unoccupied. Hence the range of counts per unit is 0–19 when aggregated, 4–13 when random, and only 8 when regular. A much greater degree of aggregation than this is possible in nature.

SAMPLING

The object of taking a sample is to measure in a small population, the sample, relevant characteristics of the large population from which it came. The reasons for doing so are usually either that the whole population is too large to measure, or that the process necessitates destroying the animals so only a small part of the population can be examined. Whatever the reason, the sample must be truly representative of the whole population or the effort is wasted. If a sample is biased and represents only a part of the population, conclusions reached will be untrue when applied to the whole population. Any bias, which may easily occur when a subjective choice is made, must therefore be avoided. Such a choice may be exercised consciously and deliberately and a selection made of what is believed to be typical of the large population. If this belief is correct and the deliberately selected samples are typical, the average of the population may be correctly measured, although the variance will be seriously reduced. However, if the "typical" specimen can be so accurately chosen, the mean is already known and it is pointless to take such a sample. If the deliberately made choice turns out to be wrong,

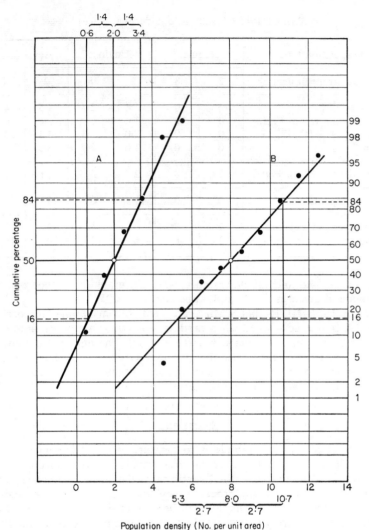

FIG. III:32. Changing the size of sample unit in Fig. III:30B from one square (A) to four squares (B) changes the mean density from 2 to 8 per unit and the variance increases with the mean (see Table III:15).

which is more likely, both the mean and variance will be affected. An unconscious bias (see Ex. 6, p. 128) may give the correct variance with a distorted mean. Even the behaviour of the animal may affect the sample (Ex. 5, 7, p. 126, 132). The best solution is to remove the personal element as far as possible and use objective processes in sampling.

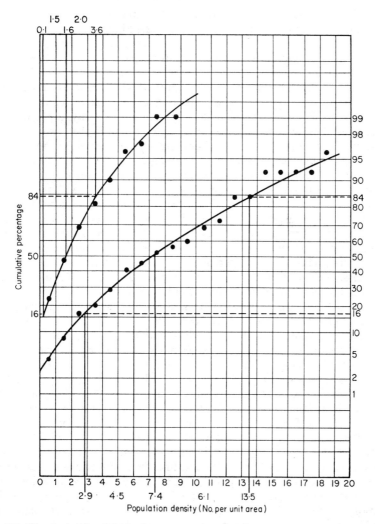

FIG. III:33. As in Fig. III:32, the mean and variance both increase with increase in sample unit size. However, with data from the aggregated disposition in Fig. III:30C, the ratio of variance to the mean is even greater at high densities (see Table III:15).

The whole sample usually consists of several to many "units", each of which is a standardized piece of the environment. For example, when sampling for caterpillars to discover which side of a tree they inhabit, each sample may consist of 20 separate leaves of similar size from different parts of the tree. If leaf size is thought to be an important environmental factor, or independent variable controlling the dependent variable "number of caterpillars per leaf", the sample would still be 20 units, each of 1 leaf, but in each sample the leaves would be a different size. Similarly a soil sample may consist of 20 units, each of a core of soil 6 cm diam × 8 cm deep. In each instance, the number of insects per unit is counted and the whole sample consists of all 20 counts.

The environmental units, leaves or cores of soil, are thus standardized objectively by size. The placing of each unit of the sample requires equally objective consideration. The placing of sample units in, say, a playing field, can be considered by analogy with what has been said in the previous section on the spatial pattern of organisms. Two practical aspects need consideration if the field population is to be truly represented. First, the area of the field should be covered as extensively as possible, for one end may grow clover and the other end grass. Second, it must be possible to make use of existing statistical techniques to extract the information needed, namely the mean and the variance, from the sample. The first requirement is clear. Each sample unit must represent as much ground as possible, so the units should not be aggregated but arranged in a regular or systematic pattern like Fig. III:30A. The second requirement is more difficult to fulfil. To avoid bias, the sample units must be independent of each other. Regular patterns of points are not independent; only random points are independent. Randomness avoids the persistent bias caused by aggregation in sampling and, given enough points, these will represent the area (Fig. III:30B). Random sampling is therefore ideal, but difficult to maintain. As Fig. III:31 shows, a very few randomly chosen units may seriously bias a single sample, although a second sample may not have the same bias, nor would a sample with more units. Hence random samples should have at least 5 units and preferably 10 or more. Also, choosing random sites is laborious. The following method simplifies it as much as possible. Draw 100 × 100 grid on graph paper. From Appendix L (p. 373) select as many pairs of numbers as needed. Using the pairs of random numbers as y- and x-coordinates, plot the points on the grid (this is how Fig. III:30B was made). Mark off 100 intervals along each side of the area to be sampled and find the intersections corresponding to the points on the grid. To apply this to an irregular plot requires ingenuity.

There are temptations to short-cut random sampling by replacing the rigid selection process with a haphazard one such as throwing an object and sampling at the point where it comes to rest. Such methods are not independent; they all depend to some extent on the observer, and have therefore none of the sound theoretical advantages of a random sample. Neither have they the effective lack of bias, and good area coverage, of a regular sample. If properly conducted random sampling is not feasible, use regular sampling and acknowledge this. Never describe a sample as random unless a full random procedure has been followed either from Appendix L or, if this is not suitable, from a text book of statistics such as Yates (1963) "Sampling Methods for Censuses and Surveys".

Regular samples are easy to devise and execute in natural surroundings. If the required population is artificially regular, such as plants grown in rows in a garden plot, the method is modified as follows. Count the rows (R) and measure their length (l). Compute the total length of row in the plot $(l \times R)$. Divide this by the required number of units (n) for the sample and so obtain the "unit row length" $[lR/n]$. Beginning in the middle of the plot, measure off a unit row length and take a sample unit such as the next stem or plant, or soil core, in a row. Repeat the process, working along the row to the end and doubling back along the next row. Carry over any partly completed unit row length into the next plant row each time. Upon reaching a corner of the plot, carry over the partly completed unit length to begin again at the diagonally opposite corner. The last sample unit should be taken near the starting point, in the middle of the plot.

Regular pattern sampling fails when such regularities as rows of plants are not taken into account as described above. Old pastures often have parallel ridges and hollows. A regular pattern of units in such terrain may sample only ridges or hollows, which may not be the object. This can be avoided by choosing the sampling pattern so that the intervals are not a sub-multiple of the rows. Alternatively, the grid may be set at a slight angle to the rows. Another alternative, perhaps the most refined type of sampling, is called "stratified". A straight-forward example of this is to divide the whole area into half as many sub-areas as the required sample units, and take 2 units at random from each sub-area. Stratified sampling has the advantages of both regular and random samples, but is still time-consuming.

The accuracy of any estimate of the mean population density obtained by sampling depends on finding some of the required organisms. It is not necessary, or even helpful, to have very large numbers in a sample, but every effort should be made to use units large enough to have at least one individual in each. A large number of zeros in the sample units makes any sample unreliable. If this means increasing the size of each unit to avoid having zeros, and hence decreasing the number of units to avoid using too much time, then a regular sampling pattern should be adopted.

When the sampling method consists of attracting very active animals from a relatively large area, a single unit sample may suffice. The spatial coverage is obtained by the movement of the animals. For example moths are often sampled by light trap, but remember that the sample represents only that part of the moth population susceptible to attraction by light (Exs. 22, 32, 44, 45).

SIGNIFICANCE

Because biological material is so variable, apparently simple decisions are sometimes difficult to make. For example, suppose a number of people or insects are offered a choice of, say, coconuts or peanuts, and 80% accept the first alternative, coconuts, thus leaving 20% who reject it in favour of peanuts. At first sight this may seem to be a fairly conclusive result justifying the hypothesis that most people, or insects, prefer coconuts. However, a vital piece of information is lacking; the number of people or insects observed.

If 800 out of 1000 people chose coconuts, and if the 1000 in the sample were representative of the 60,000,000 people in Britain, it might reasonably be inferred that most British people appear to prefer coconuts at the present time. If the experiment was repeated with different people, four times a year, and gave the same result, this hypothesis would be more valid and any seasonal cycle in taste could be discounted. If the experiment was continued in the same way for 10 years with the same consistent results, it would reflect a profound difference in the attitude of this generation to coconuts compared with peanuts which could confidently be accepted, and justifiably so, because such consistent results rarely happen in nature.

If, in contrast, the original experiment was done only once on 5 people, there would be no justification for maintaining the hypothesis. If only 1 person changed his mind, the 4 to 1 advantage would become 3 to 2, which is as near to equality as it is possible to get with a sample of 5 units. So, although the original percentages looked convincingly different they were misleading because of the inherent variability in the people. Note that this is not an assessment of the *biological* significance of the choice; there is no way of judging, from the evidence so far presented, if this preference has any relevance in the animals' ecology. All that is being considered is the significance, or otherwise, of the evidence in testing the hypothesis "people like coconuts more than peanuts".

The evidence based on 1000 observations is thus considered more significant than that based on 5. The next step is to define how much more significant it is. Thus some arbitrary, but objective level of significance is chosen, and the experimenter may decide to accept the evidence above this level and to reject evidence below it. When the evidence is accepted, it supports the hypothesis, not proving it, but making it worthwhile to investigate further. One possible arbitrary working level of acceptable significance is a 99 out of 100 chance of being right. The probability of a particular result occurring in an experiment by unbiased chance can always be estimated theoretically. If the actual result is compared with the theoretical chance result there is a method of stating how many times an experiment would have to be done before such a result occurred by chance. For example, in the choice experiment just described, if no preference for coconuts over peanuts existed, we should expect to get 50% choosing each. If there were 15 people involved, it could not be exactly 50% each time; sometimes it would be 8 to 7 for and sometimes 8 to 7 against, because people with no special preference would choose differently on different occasions. This would give a 50% result over many repeats. Similarly, even with 16 individuals, some would choose differently on different occasions and the result 8:8 would not occur every time; 9:7 would occur quite often; 10:6 would occur sometimes; 11:5 would be less frequent and 13:3 would be quite rare. In fact 13:3 would occur, according to probability theory, once in 20 experiments on the average. The result 13:3 may be considered significant evidence that the outcome was affected by some element in addition to pure chance; in this instance personal choice. If so, the level of significance is 1 in 20, or there is a 5% probability that the original hypothesis of "no choice only chance" is right or 95% certain to be wrong. Thus a 5% significance test gives the minimum difference between two groups necessary for the experimenter to be 95% confident of the result. When this test is not stringent enough a 1% or 0·1% level of significance may be used. At 0·1%, only once in a thousand times would such a result occur by chance. Such stringent tests are not given here, because results that would pass them are usually obvious. Statistical textbooks give the necessary tables. The difference necessary for a 5% significance test of a simple percentage separation into two fractions is given in Appendix F (p. 364). This table shows that as more people are examined, the smaller becomes the percentage difference acceptable as evidence.

DIFFERENCES BETWEEN TWO PROPORTIONS

Tests for significance differ according to the complexity of the experiment. If, instead of having one group of organisms that can be split into two opposing fractions, there are two groups each of which can respond in two different ways, the results can be expressed in a 2 × 2 table. For example the previous enquiry may be extended to ask if the insects and the people had different preferences for coconuts. The same choice could then be offered to a group of each, and the resulting percentages compared. The insects may divide into 60% for peanuts as against the people's choice of 80% for coconuts. Again the *actual numbers* of people and insects should be used for tabulation instead of percentages. The first step is to present the results in a simple 2 × 2 table of the numbers of people and insects responding in different ways, with totals:

Experiment	*Response*		*Totals*
	I	II	
	(choosing coconuts)	(choosing peanuts)	
I (people)	A	B	A + B
II (insects)	C	D	C + D
Totals	A + C	B + D	N

where N is $A + B + C + D$, i.e. the total number of people and insects observed. The problem is to decide if the proportion A/B is significantly different from the proportion C/D. Note that the question can be rephrased "is A/C different from B/D?" Hence the table can be inverted and the same tests applied. Using the same kind of reasoning as in the last section, a table can be made to give the 5% significance level as a quantity to be exceeded before accepting the evidence. The table in Appendix G (p. 365) is more complex because of the extra quantities, and applies only to small numbers: $N \leqslant 30$; $(A + B)$ and $(C + D) \leqslant 15$. To find the appropriate place in the table:

follow column 1 to find $(A + B)$,
follow column 2 to find B,
move across the rows to the correct column for $(C + D)$.

This has been done for the 2 × 2 table below and the resultant value for D marked in heavy type in Appendix G.

Experiment	Treatment		Totals
	I	II	
I	$A = 2$	$B = 12$	$(A + B) = 14$
II	$C = 8$	$D = 5$	$(C + D) = 13$
Totals	$(A + C) = 10$	$(B + D) = 17$	$N = 27$

The limit given in Appendix G (p. 365) for the position of D is 5, which coincides with the result obtained. The observed number in position D must be equal to, or less than, 5 for the ratios to be considered significantly different. Hence the difference between experiments, or treatments is just significant at the 5% level. If the observed value of B cannot be found in column 2 of the table, use A instead and the value in the body of the table then refers to C. In part 2 of this table, the value of D is also given at the 1% level. The present example does not pass this second test, for which a D of 3 or less would be required. The table gives two levels of significant differences directly from the original small counts, $A + B = C + D \leqslant 15$, but this table cannot be increased in size indefinitely; some index is required that can be computed from given quantities, to make tabulation simpler. This index is called χ^2 (chi squared) and χ^2 has been tabulated, for several levels of significance, against a quantity, "degrees of freedom" (d.f.), related to the number of separate elements in the analysis. For 2 × 2 tables, d.f. is always 1 and χ^2 for 1 d.f. is given in Appendix H (p. 369).

χ^2 IN 2 × 2 CONTINGENCY TABLES

As an example of the analysis of proportionate results from paired experiments, consider Table IV :22:1 (p. 172) of moths caught in a pair of light traps, one at 27 m and one near ground level. Most moths were sexed so it is possible to decide if the proportion of males to females remained the same at different heights. Using only those days when both traps worked and all insects were sexed, the corrected counts, which differ slightly from the raw data in Table VI: 22: 1 are:

	Females (A)	Males (B)	Totals (A + B)
Lower Trap	17	246	263
Upper Trap	(C) 32	(D) 458	(C + D) 490
Totals	(A + C) 49	(B + D) 704	(N) 753

PLATE 2. Rothamsted moth traps working at two heights in the quadrangle of Burleigh
Secondary Modern School, Hatfield, England. (Photo by courtesy of the Headmaster,
Mr. R. Braddock.)

Facing page 100.

From this compute the index:

$$\chi^2 = \frac{N[|AD - BC| - \frac{1}{2}N]^2}{(A + B)(C + D)(A + C)(B + D)} \tag{1}$$

The quantity $|AD - BC|$ must give a positive result; the vertical line brackets indicate this and it can be arranged by changing round the table thus:

	Males	Females	Totals
Lower Trap	(A)	(B)	(A + B)
	246	17	263
Upper Trap	(C)	(D)	(C + D)
	458	32	490
Totals	(A + C)	(B + D)	(N)
	704	49	753

Now computing

$$\chi^2 = \frac{753[|(246)(32) - (17)(458)| - \frac{753}{2}]^2}{(263)(490)(704)(49)}$$

$$= \frac{753(290 \cdot 5)^2}{(263)(490)(704)(49)}$$

$$= 0 \cdot 014 \qquad \text{d.f.} = 1$$

Reference to the χ^2 table (Appendix H, p. 369) shows that the hypothesis "there is no difference in the proportion of females at the two sites" is between 90 % and 95 % probable (with 1 d.f.).

The first species listed in Table IV:22:1 with numbers sufficient to analyse is the Pale Tussock whose 2 × 2 table appears thus:

	Males	Females	Totals
Upper Trap	(A)	(B)	(A + B)
	20	1	21
Lower Trap	(C)	(D)	(C + D)
	9	9	18
Totals	(A + C)	(B + D)	(N)
	29	10	39

Again the totals are too great to use Appendix G (p. 365), so compute

$$\chi^2 = \frac{N[|AD - BC| - \frac{1}{2}N]^2}{(A + B)(C + D)(A + C)(B + D)}$$

$$= \frac{39[|(20)(9) - (1)(9)| - \frac{39}{2}]^2}{(21)(18)(29)(10)}$$

$$= \frac{891,238}{54,810}$$

$$= 16 \cdot 3$$

As always with 2 × 2 tables, there is 1 d.f. and the χ^2 tables show the hypothesis that "males and females are evenly distributed with respect to height", to be less than 0·1% probable, or very highly improbable. From the table it can be seen that females are relatively more common near the ground than high in the air.

The next species that can be tested is the White Ermine:

	Males	Females	Totals
Upper Trap	(A)	(B)	(A + B)
	5	8	13
Lower Trap	(C)	(D)	(C + D)
	39	3	42
Totals	(A + C)	(B + D)	(N)
	44	11	55

with

$$\chi^2 = \frac{55[|(5)(3) - (8)(39)| - \frac{55}{2}]^2}{(13)(42)(44)(11)}$$

$$= \frac{55[(8)(39) - (5)(3) - 27·5]^2}{(13)(42)(44)(11)}$$

$$= \frac{55(269·5)^2}{(11)(240·24)}$$

$$= 15·1$$

From Appendix H (p. 369) it is less than 0·1%, or 1 in 1000, probable that males and females fly at the same height. In this instance the females fly higher.

The next species of interest is the Buff Ermine:

	Males	Females	Totals
Lower Trap	36	0	36
Upper Trap	0	0	0
Totals	36	0	36

Again the totals are too great to use Appendix G (p. 365) but Eq. 1 (p. 101) cannot be applied because the denominator is zero, as happens whenever zero occurs for a total down or across the table. Hence Eq. 1 is useless for this result which remains inconclusive.

This difficulty is avoided with the data for the Cinnabar:

	Males	Females	Totals
Lower Trap	24	0	24
Upper Trap	2	1	3
Totals	26	1	27

and the χ^2 value is 1·91, between 10 and 20% probable that the hypothesis "males and females are equally distributed with respect to height" is correct. However, this result should be regarded with caution because the simple test given by Eq. 1 is still not efficient when there are quantities as small as 1, 2 or 3 in the table. It does not become efficient until all the totals approach 10. The true equation for these lower values is more complex. Appendix G (p. 365) gives significance levels for tables with numbers up to 15 in the right-hand totals column, i.e. $A + B$ and $C + D \leqslant 15$. A more comprehensive table is listed in Appendix N (p. 380) Finney *et al.* (1964).

SIGNIFICANCE OF THE DIFFERENCE BETWEEN MEANS

Before proceeding further, there are two rules for variance to be accepted and used; their derivation will be found in textbooks of statistical theory.

Rule 1

The several individuals constituting a sample form a frequency distribution and the mean value depends upon this. If, for example, another supposedly similar batch of individuals were chosen, they would not have identical characteristics to the first batch. Though having a distribution that was very similar in general form, and so with the same variance, there would be slight differences in the individual measurements, and the means would be slightly different. Given enough such batches, there would be enough different means for these to form a sample with its own, very narrow, frequency distribution. Hence the means themselves have a variance. The variance of all these means can be predicted from the variance of any one of the original batches of individual measurements. One of the essential properties of variance that makes it such a useful statistic, is that *the variance of the mean is equal to the variance of the sample divided by the number in the sample*, i.e.

$$S^2_{\bar{y}} = \frac{s_y^2}{N}$$

where there are N "ys" in the sample and:

$$\bar{y} = \frac{\Sigma y}{N}$$

Rule 2

Imagine a set of measurements of length, or intervals of time, each of which is a y value. The mean of the lengths is $\bar{y}_1$ and the variance of the y_1s (not the $\bar{y}_1$s) is $s_{y_1}^2$. If we have a second set of ys, call these y_2s, that are longer than the first set, their mean is $\bar{y}_2$ and variance $s_{y_2}^2$.

The difference between the means, by subtraction, is

$$\bar{y}_3 = \bar{y}_2 - \bar{y}_1$$

The variance of the y_3s is evidently more than either the y_1s or the y_2s,

104   INTRODUCTION TO EXPERIMENTAL ECOLOGY

individually. In fact it is the same as if the y_1s and the y_2s were added together. *The variance of a difference (or sum) is equal to the sum of the variances,* i.e.

$$s^2_{(y_2-1)} = s^2_{(y_1)} + s^2_{(y_2)}$$

and this is equally applicable to the variance of the mean.

Note that this rule applies only to the variance, not to standard deviations or 95% limits; these cannot be added.

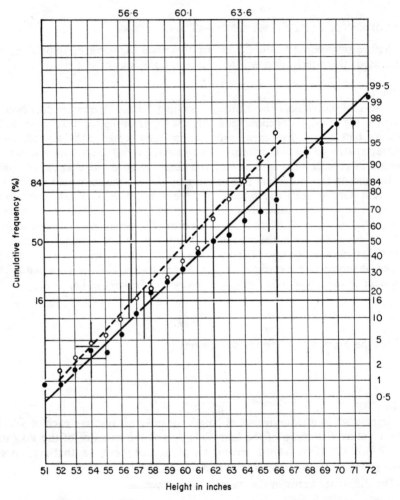

Fig. III:34. Cumulative frequency distributions of height for 238 schoolchildren. ● – – – male, ○ – – – female. (Data from Table IV:2:1). *N.B.* Symbols reversed in Fig. IV:2:1.

TEST FOR THE DIFFERENCE IN HEIGHT BETWEEN MALES AND FEMALES IN
EX. 2

The effect of age on height in boys and girls and of height on weight was
studied in Ex. 2 (p. 117) but there it was not established that there was any
statistical significance in the difference between the two sets of dimensions.
The simplest way to test this·possibility is to examine the difference between
the mean heights. Figure IV:2:1 shows that the ages of the two groups are
similar and so age may be ignored and all the individuals, 127 males and 111

TABLE III:16

Frequency distribution of height, in inches, of boys and girls
in Ex. 2

Height (in.)	N.	Females Cumul. N.	%	N.	Males Cumul. N.	%
50–50·9	0	0	0	1	1	0·8
51–51·9	2	2	1·8	0	1	0·8
52–52·9	1	3	2·7	1	2	1·6
53–53·9	2	5	4·5	2	4	3·2
54–54·9	2	7	6	0	4	3·2
55–55·9	4	11	10	4	8	6
56–56·9	8	19	17	6	14	11
57–57·9	4	23	21	12	26	20
58–58·9	7	30	27	6	32	25
59–59·9	11	41	37	9	41	32
60–60·9	9	50	45	13	54	43
61–61·9	22	72	65	10	64	50
62–62·9	12	84	76	6	70	55
63–63·9	9	93	84	11	81	64
64–64·9	9	102	92	7	88	69
65–65·9	5	107	96·4	8	96	76
66–66·9	4	111	100	15	111	87
67–67·9	0	111	100	7	118	93
68–68·9	0	111	100	3	121	95·3
69–69·9	0	111	100	3	124	97·6
70–70·9	0	111	100	0	124	97·6
71–71·9	0	111	100	2	126	99·2
72–72·9	0	111	100	1	127	100

females, treated as two frequency distributions of size (Table III:16). Probability plots of these distributions (Fig. III:34) can be fitted with regression
lines by the graphical method (p. 82) and the median and 16% and 84%
values read off to give the mean and S.D. The mean and variance may be
estimated as follows:

$$s^2 = \left[\frac{\textit{height at } 84\% \textit{ prob.} - \textit{height at } 16\% \textit{ prob.}}{2}\right]^2$$

Mean = median = height at 50% prob.

Hence:

	Females	*Males*

Variance $s_1^2 = \left[\dfrac{7}{2}\right]^2$ $s_2^2 = \left[\dfrac{8}{2}\right]^2$

$$= 12\cdot25 \qquad\qquad\qquad = 16\cdot0$$

Difference of mean heights $= \bar{h}_2 - \bar{h}_1$

$$= 62\cdot0 - 60\cdot1$$

$$= 1\cdot9$$

From Rule 1—Variance of means

$$s_{(\bar{h}_1)}^2 = \frac{12\cdot25}{111} \qquad\qquad s_{(\bar{h}_2)}^2 = \frac{16\cdot0}{127}$$

$$= 0\cdot120 \qquad\qquad\qquad = 0\cdot126$$

From Rule 2—Variance of the difference of means

$$s_{(\bar{h}_2 - \bar{h}_1)}^2 = 0\cdot126 + 0\cdot120 = 0\cdot246$$

The significance test relates the difference between means to the variability about them:

$$d = \frac{\text{Diff. of means}}{\text{S.D. of diff.}} = \frac{(\bar{h}_2 - \bar{h}_1)}{\sqrt[2]{s_{(\bar{h}_2 - \bar{h}_1)}^2}}$$

$$d = \frac{1\cdot9}{\sqrt[2]{(0\cdot246)}} \qquad = \frac{1\cdot9}{0\cdot50} = 3\cdot8$$

If $d > 2$, it is 95% probable that the two means differ, i.e. the difference is significant at the 5% level; if $d > 3$, the probability is 99%. In the present instance the difference is obviously highly significant.

A more rigorous test can be made by computing mean and variance, using the method in Appendix A. The results are as follows.

	Females		*Males*
Variance	$s_1^2 = 11\cdot30$		$s_2^2 = 18\cdot69$
Mean	$= 60\cdot53$	Mean	$= 62\cdot04$

$$d = \frac{1\cdot51}{0\cdot499} = 3\cdot03$$

In this instance, the graphical method overestimates the significance of the results and for this reason graphical methods cannot be relied upon when they give borderline results. Here both results show, beyond doubt, that the difference in height is highly significant and so the graphical result is quite adequate.

If there are less than 30 individuals in each part of the sample, a more complex test is needed. It is therefore best to enlarge the sample. If this is not possible and the test is critical, refer to Chapter 6 in Bailey (1959), "Statistical Methods in Biology".

COMMUNITY STRUCTURE

Exercises 32, 44 and 45 are concerned with the complex of species in a mixed community, in an attempt to define its structure and "diversity". Diversity is used to describe the relation between number of species and number of individuals. Some communities are richer than others in the number of species they contain, but these communities have not necessarily more individuals per unit area. The simple proportion "number of species per 1000 individuals", cannot be used to compare number of species in different communities because it changes with the size of the sample; for example, if all the species in a habitat are already represented in a collection, further collection reduces the proportion of species to individuals as more and more individuals, but no new species, are added. The measure of diversity must take account of this, and diversity, as a consistent, measurable characteristic of communities, can only exist if these communities have a definable structure which is a property of the communities as a whole and not of the separate species in it.

If the number of each species in a sample is recorded and then made into

TABLE III:17

Frequency distribution of abundance of Lepidoptera captured in a light trap at Rothamsted Experimental Station during 1935 (from Williams, 1964)

Individ. per sp.	Species obs.	log series	Individ. per sp.	Species	Individ. per sp.	Species	Individ. per sp.	Species
1	37	38·0	15	2	39	1	87	1
2	22	18·9	16	2	40	3	88	1
3	12	12·5	17	4	42	2	105	1
4	12	9·3	18	2	48	2	115	1
5	11	7·4	20	4	51	1	131	1
6	11	6·1	21	4	52	1	139	1
7	6	5·2	22	1	53	1	173	1
8	4	4·5	23	1	58	1	200	1
9	3	4·0	25	1	61	1	223	1
10	5	3·6	28	2	64	2	232	1
11	2	3·3	29	2	69	1	294	1
12	4	3·0	33	2	73	1	323	1
13	2	2·7	34	2	75	1	603	1
14	3	—	38	1	83	1	1799	1

Total individuals = 6814; total species = 197

a frequency distribution of the number of species with 1 individual, the number with 2, with 3, with 4 and so on, a hollow curve results with no entry in the zero column, the species with no individuals. The largest number of species have 1 individual; there are less with 2, less still with 3 and so on until a very few species contain most of the individuals and appear as single points at high values along the *x*-axis (Fig. III: 35). Notice that although it was necessary to identify to species to make this distribution, the species are not named in the analysis. In the same way that the identity of individuals

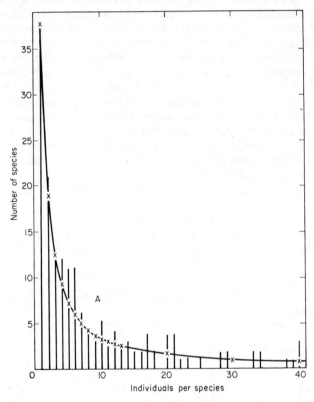

FIG. III:35.   Frequency distribution of species of Macro-Lepidoptera with different numbers of individuals, caught in a light trap at Rothamsted Experimental Station in 1935. (After Williams.)

was lost in abstracting the species characteristics in Exs. 1 and 2, so the identity of species is lost in defining the characteristics of the community.

The shape of the frequency distribution for species in communities is described well by the Logarithmic Series (Fig. III: 35), in the same way that other distributions are described by other mathematical series. A discussion of them will be found in Williams (1964) "Patterns in the Balance of Nature" and the series themselves are in most mathematical textbooks.

When the summation of the logarithmic series is written in the form:

$$S = \alpha \log_e \left[1 + \frac{N}{\alpha} \right]$$

where S = number of species,

N = number of individuals,

the quantity α defines the relationship between N and S and is independent of population density or sample size. Hence the quality of variability in different sized samples can be measured simply by α which is called the Index of Diversity. Changes in α between different sites and times then become a measure of the effect of environment on the community complex. Computation of α is avoided by the use of a graph from which α is obtained directly, given N and S (Appendix K, p. 372).

A second parameter, x, dependent upon the size of the sample, is given by

$$x = \frac{N}{N + \alpha}$$

The whole series, each term of which is the expected number of species with 1, 2, 3, 4, etc. individuals, can then be written as

$$\alpha x, \quad \frac{\alpha x^2}{2}, \quad \frac{\alpha x^3}{3}, \quad \frac{\alpha x^4}{4}, \quad \text{etc.}$$

and plotted alongside the measured frequency distribution (see Fig. III:35).

GROWTH RATES AND FORM

Many growth processes are more easily illustrated, and understood, when considered in ratios rather than absolute terms. It is therefore not surprising that the increments of growth of developing larvae depend upon their initial size. To expect the absolute rate of growth of a man to be similar to that of a mouse is unreasonable but the rate of growth relative to initial size may be comparable. For such relations of growth against time, single log plots are used. Exercise 14 (p. 149), which demonstrates this point, emphasizes that there is much in common between the growth of individuals and of populations.

When growth of organs is compared with growth of the whole body and the time element eliminated, growth may be expressed in its simplest form, Huxley's Power Law for allometric growth:

$$y = bx^\alpha$$

where y = weight or size of organ,

x = the body weight or size,

b = initial growth index,

α = growth rate.

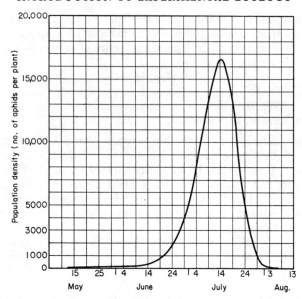

FIG. III:36.   The population growth curve for an aphid colony, e.g. of bean aphids on bean plants, shows a very slow increase in the first month and then an explosive increase in the second month, followed by an even more rapid decline as the migrants fly away.

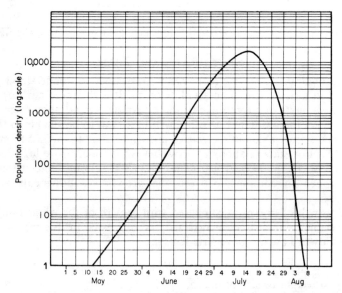

FIG. III:37.   On a logarithmic scale, the population growth *rate* of the aphid colony can be seen to be almost constant during the first two months. The arithmetic scale of Fig. III:36 misrepresents population growth which is a multiplicative process, not an additive one.

This gives a linear relationship on double log coordinates, as given by the Power Law for aggregation (p. 89). When both organ and body retain the same ratio, $\alpha = 1$ and the adult looks like the infant. Usually heads grow more slowly than bodies and for them $\alpha < 1$.

The rate of increase of an aphid population is so great in the third generation that the changes in earlier generations are dwarfed on the arithmetic scale (Fig. III:36). The growth of such populations is usually plotted on single log coordinates (Fig. III:37).

In theory a new population developing in a limited environment should increase slowly at first, then more rapidly, then slowly again until it reaches a maximum tolerable level for the environment. This theoretical growth curve, the logistic, is sigmoid and symmetrical (Fig. III:38). Its derivation is quite different from other sigmoid curves met so far, and is more complex.

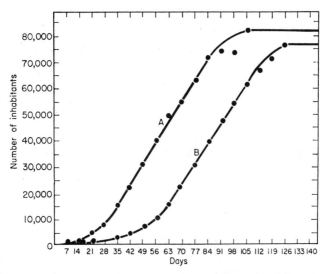

FIG. III:38. The growth curve (logistic curve) of two different bee colonies in the same apiary (redrawn after Bodenheimer).

The usual form of this equation is

$$P(t) = \frac{P_\infty}{1 + e^{a-bt}}$$

where $P(t)$ is the population at time t,

P_∞ is the maximum supportable population,

a/b is the time at the inflexion ($P_\infty/2$),

b is the degree of acceleration of the growth rate,

e is the base of the Naperian, or Natural logarithms.

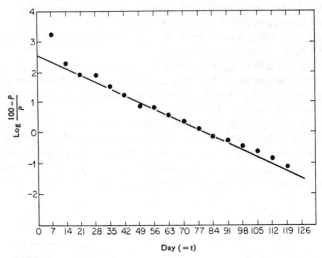

FIG. III:39.   Data for line A (Fig. III:38) plotted on a semi-logarithmic scale.

Rearranging:

$$e^{a-bt} = \frac{P_\infty - P(t)}{P(t)}$$

If the population growth curve is now converted, like accumulative probability curves, to a percentage, i.e. $N_\infty = 100$, the equation becomes:

$$\frac{100 - P}{P} = e^{a-bt}$$

and taking logs to base e (Natural Logs)

$$\mathrm{Log}_e\left(\frac{100 - P}{P}\right) = a - bt$$

which gives a linear distribution of points when plotted on semi-log paper, log $[(100 - P)/P]$ against time. a and b can be read off as intercept and. tan β respectively, as for other regressions (Fig. III:27). They must, however, be corrected to natural logs from the base 10 logs used in the log paper. This is done by multiplying both a and b by 2·303.

Note that in Ex. 14, where rate of growth is considered, rate is the reciprocal of the time taken for completion of the process. The fastest rate, when the temperature or other environmental factors are ideal, is often taken to be 100% and other, slower rates, expressed as percentages of this.

CHAPTER IV

Exercises in Ecology

CLASSIFICATION OF EXERCISES BY SUBJECT

Subjects frequently occurring in school and university Zoology, Biology and General Science syllabuses are listed below, followed by exercises relating to them.

Variability of living things, Exs. 1, 7.

Development of animals, Exs. 2, 13, 14.

Responses and adaptations of animals to the physical environment, Exs. 8, 9, 10, 11, 12, 13, 15, 16, 17, 18, 21, 22, 23, 35, 38.

Animal behaviour, Exs. 7, 21, 24, 26, 28, 29, 30, 31.

Habitats, Exs. 19, 25, 36, 37, 39.

Interdependence of plants and animals in habitats, Exs. 30, 33, 36, 37, 39, 43.

Animal abundance and distribution, Exs. 3, 4, 5, 6, 7, 24, 25, 27, 35, 36, 37, 38, 40.

Distribution maps and gradients, Ex. 39.

Daily and seasonal cycles of activity and abundance, Exs. 25, 27, 31, 32.

Fluctuation in population density, Exs. 32, 33.

Biological equilibrium, Ex. 34.

Dispersal, Exs. 22, 23, 24.

Interspecific competition, Exs. 20, 41, 42.

Food relationships, predation, parasitism, Exs. 33, 34.

Community structure and diversity, Exs. 44, 45.

Evolution, Exs. 18, 19, 30.

CLASSIFICATION OF EXERCISES BY REQUIREMENTS

The following list is to help readers choose exercises that are suitable to the facilities they have, and to the time and the number of students available. Details of some exercises may need modifying in different circumstances or localities.

APPARATUS

Mostly simple, e.g. jars, metal trays, boxes, grids, nets, canes, gardening tools, rulers, scales, paint, stop watch. Exs. 2, 3, 4, 5, 6, 7, 8, 9, 15, 16, 17, 20, 21, 23, 24, 25, 26, 27, 28, 29, 30, 31, 33, 34, 35, 36, 37, 38, 39, 40, 41, 42, 43.

Thermometers, thermocouple, temperature-controlled cabinets or ovens, refrigerators. Exs. 1, 10, 11, 12, 13, 14.،

pH indicator. Exs. 35, 39, 43.

Light trap. Exs. 11, 22, 32, 44, 45.

SITES

Indoors. Exs. 1, 2, 8, 9, 10, 12, 13, 14, 16, 17, 21.

Outdoors with little vegetation nearby. Exs. 11, 12, 22, 25, 32, 44, 45.

Town gardens, parks, waste land, commons. Exs. 3, 4, 5, 6, 7, 11, 15, 18, 22, 23, 24, 25, 27, 28, 29, 30, 31, 32, 33, 34, 35, 37, 38, 39, 41, 42, 44, 45.

Farmland. Exs. 3, 4, 5, 6, 7, 11, 15, 18, 20, 22, 23, 25, 27, 32, 34, 35, 37, 38, 39, 40, 41, 42.

Woodland. Exs. 7, 11, 20, 33, 35, 38, 39, 40, 41, 42.

Downland, heath. Exs. 19, 35, 39, 43.

Soils. Exs. 9, 34, 35, 36, 37, 38.

TIME OF YEAR

* Indicates that, in U.K., an exercise is likely to be most successful at this time of year, though possible at other times.

Throughout the year. Exs. 1, 2, 9, 16, 17, 20, 21, 32, 34, 39, 44, 45.

Spring. Exs. 5, 6, 7, 15, 18, 19, 21*, 26, 27, 30, 33, 35*, 36, 37, 38*, 39*, 40, 43.

Summer. Exs. 3*, 4*, 5*, 6*, 7*, 8*, 9*, 10, 12, 13, 15*, 17*, 18, 19, 21*, 22, 23, 24, 26, 27*, 28, 29, 30, 31, 33*, 35, 38, 39*, 43*.

Autumn. Exs. 3, 4, 5, 6, 7, 8, 9* 10, 12, 13, 15*, 18, 19, 21*, 22, 23, 26, 27, 35*, 36, 37, 38*, 39*, 41, 42, 43.

Winter. Exs. 8, 11, 33, 39, 40.

LABOUR

For individual students. Exs. 8, 9, 10, 13, 14, 16, 17, 29.

For pairs. Exs. 6, 12, 21, 28, 38.

For groups. Exs. 1, 2, 3, 4, 5, 7, 11, 15, 18, 19, 20, 22, 23, 24, 25, 26, 27, 30, 31, 32, 33, 34, 35, 36, 37, 39, 40, 41, 42, 43, 44, 45.

Many of the group exercises are suitable for individuals as projects if enough time is available.

Requiring a few hours. Exs. 1, 2, 3, 4, 5, 6, 8, 12, 15, 16, 18, 19, 20, 21, 23, 25, 26, 27, 28, 33, 35, 36, 37, 38, 40, 41, 42, 43.

Requiring a few days. Exs. 7, 9, 10, 17, 29, 30, 31, 39.

Long-term. Exs. 11, 13, 14, 22, 24, 32, 34, 44, 45.

EXERCISE 1. Variability Between Individuals

Introduction:

Certain characters are commonly regarded as fixed for particular species. Man is homoiothermic and we regard his normal body temperature as 98·4°F. Quite small deviations from this normal often indicate sickness. However, every individual differs, even in such seemingly constant physiological attributes, and 98·4°F is only an approximation to the normal temperature for most individuals. The way in which individuals of a population differ from each other in such common characters can be defined statistically. This definition links the description of the size or behaviour of individuals to that of the whole population. All specific characters can be defined in this way.

TABLE IV:1:1

Frequency distribution and accumulated percentage distribution of the body temperature of 276 students between 0800 and 0900 hr. (See Table III:7 for details of lay-out.)

Temperature °F	Number of students	Cumul. ΣN	Cumul. %
96·5–96·6	1	1	0·4
96·7–96·8	1	2	0·7
96·9–97·0	4	6	2·2
97·1–97·2	5	11	3·8
97·3.97·4	10	21	7·6
97·5–97·6	28	49	18
97·7–97·8	38	87	31
97·9–98·0	41	128	46
98·1–98·2	44	172	62
98·3–98·4	57	229	83
98·5–98·6	19	248	90
98·7–98·8	17	265	96
98·9–99·0	7	272	98·6
99·1–99·2	3	275	99·6
99·3–99·4	1	276	100

Hypothesis:

The characteristics that distinguish individuals approximate to a mean value typical of the species.

Apparatus:

Several clinical thermometers.

Site and Time of Year:
Any time, any place.

Procedure:

Measure the oral temperature of as many students as possible to 0·1°F accuracy. Make a frequency distribution of the results (Table IV:1:1). List the temperatures in 0·2° intervals (column 1) and the number of people in each interval (column 2). List the cumulative sum and convert these to percentages on arithmetic probability paper (Fig. IV:1). Draw a straight line through the points; read off the mean, standard deviation and 95% limits (see Tables III:7, 8, Fig. III:12, p. 56).

Results:

The results given here are the temperatures of 276 students. It is unlikely that such large numbers will be available, so the results will be more erratic. This illustrates that the larger the sample the more nearly is the ideal,

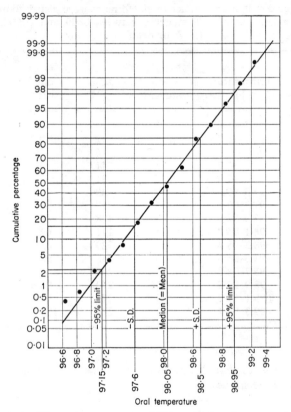

Fig. IV:1:1.   The oral temperature of 276 medical students seated in class, 0800–0900 hr.

theoretically regular, distribution approached. Thus it may be assumed that if a whole population could be used, instead of a small sample, a very regular result would be obtained. The line drawn represents this ideal.

Individual temperatures ranged from 96·5 to 99·3°F. The mean was 98·05°F, measured at 0800–0900 hr. Later in the day it would be higher, for there is a slight diurnal cycle of temperature. The temperature of each individual is a single value, at any one time, but for the whole sample, and probably for the population it represents, the temperature is 98·05 $\pm$ 0·45°F where these are the mean and standard deviation. Alternatively, we could say 95% of the population had temperatures between 97·15 and 98·95°F.

Conclusions:

The temperatures of individuals differ at the same moment in time. The range of individual temperatures in a population can be described by a probability distribution, defined by its mean and standard deviation.

EXERCISE 2. RELATED CHARACTERS

Introduction:

In addition to individual variability in characters usually regarded as constant for a species (Ex. 1), there are many characters that are apparently associated in a general way with other characters, but for which the association does not hold good for each individual. For instance, young people generally grow taller as they become older and grow heavier as they grow taller. However, it is possible to find a person of 16 years who weighs less than one of 14 years. It is therefore necessary to be able to extract the general relation between weight, height and age, without being misled by individuals who do not conform to this rule.

Hypothesis:

Weight increases with height and height increases with age in most growing animals. These basic relationships may be obscured by the variability between individuals in a small sample.

Apparatus:

Scales and rulers.

Site and Time:

Any time; any place.

Procedure:

Record the age and sex, and measure the height and weight of as many students, aged 11–18, as possible. If necessary, supplement the data with measurements of younger students from medical records to increase the ranges of ages. Tabulate the results (Table IV:2:1). Plot height against age, marking the sexes by different symbols (Fig. IV:2:1). Plot weight against height (Fig. IV:2:2). Note that height is dependent upon age, not age upon height, so age is plotted as the horizontal x-axis. This is very important because the statistical relationship called the regression, between these two quantities, or variables, is not interchangeable. It is easily obtained by graphical methods, and its calculation is not necessary to appreciate the points made here. The regression is discussed in more detail in Chapter III, p. 78.

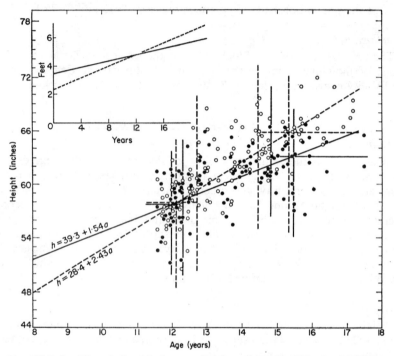

FIG. IV:2:1.   The relationship between height and age for 238 schoolchildren. ○ – – – – male, ● ——— female. (Data from Table IV:2:1.) Inset shows the regression lines extrapolated to the y axis.

Results:

Between the ages of 11 and 18 years size increases in general but there is a wide range of individual variability. At the age of about 11–12 years, males and females do not differ greatly in height. However, the height–age relationship increases more rapidly in males than females so that by the age of 15 to 18, segregation in the scatter of the two kinds of symbols is evident (Fig. IV:2:1). This segregation is capable of statistical definition; the regression coefficient of height on age is greater for males than females.

The relationship between weight and height is even more clearly defined

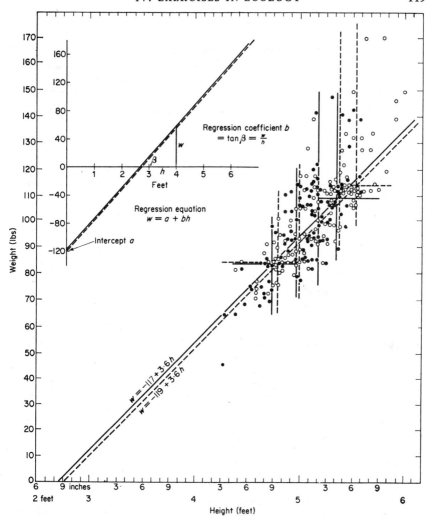

FIG. IV:2:2. The relationship between weight and height for 238 schoolchildren.
○ – – – – male, ● ——— female. (Data from Table IV:2:1). Inset shows the
regression line extrapolated to the y axis. w = weight; h = height.

because the individual deviations vary less; weight is more closely dependent
upon height than height is upon age. The segregation between sexes, however,
is less.

Conclusions:

Generally men eventually grow taller than women, but for a given height
they tend to weigh much the same or sometimes slightly less. Basic relations
between characters, often obscured by the variability of individuals, are
revealed in a population when a large enough sample is analysed.

TABLE IV:2:1

Tables of age/height/weight of a group of children

Male			Female		
Age (months)	Height (in.)	Weight (lb)	Age (months)	Height (in.)	Weight (lb)
203	66·5	117·0	210	62·0	116·0
200	71·0	147·0	210	65·5	140·0
206	69·5	171·5	193	59·8	115·0
206	68·3	134·0	197	64·8	112·0
205	67·5	171·5	189	64·3	113·5
188	67·3	112·0	190	56·8	98·5
188	63·3	115·5	191	65·3	107·0
189	67·0	128·0	186	57·0	83·5
196	64·5	98·0	197	61·5	121·0
189	65·0	114·0	191	63·3	113·5
193	67·8	127·5	190	66·8	140·0
189	66·3	112·0	191	62·5	112·5
188	65·8	150·5	177	61·3	81·0
193	66·3	133·0	179	63·0	98·5
193	72·0	150·0	178	66·5	117·5
194	65·3	134·5	176	61·3	112·0
188	71·0	140·0	185	63·3	101·0
178	63·5	102·5	180	63·3	114·0
182	67·0	133·0	175	60·8	93·5
177	63·0	111·0	178	61·5	103·5
177	60·5	112·0	183	64·3	109·5
175	65·5	114·0	175	60·3	86·0
176	63·8	98·5	180	61·3	110·5
175	64·0	92·0	185	59·0	104·0
175	68·0	112·0	185	65·3	118·0
175	63·5	98·5	182	58·3	104·5
186	66·5	112·0	180	59·0	112·0
178	63·8	112·0	177	61·8	142·5
183	64·8	111·0	177	61·3	112·0
184	66·5	112·0	183	66·5	112·0
176	61·5	81·0	178	63·5	148·5
176	65·0	118·5	183	64·8	102·5
180	61·8	104·0	185	60·0	106·0
185	66·0	105·0	182	62·0	91·5
186	66·0	112·0	184	62·3	108·0
183	66·0	105·5	182	65·5	133·0
178	67·3	119·5	182	64·0	111·5
162	64·5	119·0	186	63·5	108·0
164	60·5	95·0	186	57·8	95·0
173	69·0	112·5	164	65·3	98·0

	Male			Female	
Age (months)	Height (in.)	Weight (lb)	Age (months)	Height (in.)	Weight (lb)
170	63·8	112·5	169	62·3	99·5
174	66·0	108·0	173	62·8	102·5
164	63·5	108·0	166	59·3	89·5
173	61·3	93·0	168	61·5	95·0
165	64·8	98·0	165	61·3	106·5
163	65·3	117·5	165	55·5	67·0
164	57·8	95·0	171	61·5	91·0
168	60·0	93·5	172	64·8	142·0
164	66·5	112·0	171	63·0	84·0
164	61·5	140·0	167	61·0	93·5
167	62·0	107·5	164	63·3	108·0
169	62·0	100·0	169	61·5	85·0
172	65·0	112·0	167	62·3	92·5
163	66·0	112·0	170	64·3	90·0
174	63·0	112·0	164	58·0	83·5
166	67·3	121·0	166	61·5	103·5
168	66·5	111·5	169	62·0	98·5
166	62·5	84·0	171	62·5	112·0
164	58·0	84·0	155	66·0	144·5
171	61·8	112·0	155	62·3	105·0
162	63·0	91·0	153	63·3	108·0
174	69·8	119·5	150	61·3	94·0
173	66·0	112·0	162	58·0	84·0
166	62·0	91·0	163	56·5	84·0
174	53·8	196·0	155	61·3	107·0
153	57·8	79·5	154	61·0	122·5
159	63·3	112·0	156	54·5	75·0
151	61·0	81·0	154	60·0	114·0
156	66·3	106·0	157	60·5	112·0
156	58·3	92·5	161	59·0	92·0
160	64·0	116·0	154	62·8	93·5
155	57·3	80·5	152	60·5	105·0
155	61·8	91·5	160	62·0	94·5
162	60·0	105·0	157	64·5	123·5
161	56·8	75·0	150	59·5	78·5
153	64·8	128·0	145	59·0	91·5
153	60·0	84·0	140	60·0	77·0
151	58·3	86·0	148	60·5	84·5
160	59·3	78·5	141	61·8	85·0
157	58·0	80·5	149	64·3	110·5
156	61·5	108·5	147	58·3	111·5
158	65·0	121·0	143	51·3	50·5
157	60·5	105·0	145	58·8	89·0
152	60·8	97·0	149	58·3	93·0

Male			Female		
Age (months)	Height (in.)	Weight (lb)	Age (months)	Height (in.)	Weight (lb)
159	62·8	99·0	142	56·0	72·5
156	68·5	114·0	141	61·3	85·0
152	59·5	105·0	143	55·5	84·0
160	60·5	84·0	143	56·3	85·0
156	61·8	112·0	147	61·3	115·0
139	55·0	73·5	148	59·0	95·0
140	59·5	94·5	150	54·5	74·0
142	55·0	76·0	147	51·5	64·0
150	59·5	84·0	141	56·0	72·5
146	55·0	71·5	148	56·3	77·0
150	59·0	99·5	144	59·5	93·5
144	59·5	88·0	144	61·0	92·0
149	56·3	72·0	146	60·0	109·0
149	57·0	92·0	142	56·5	69·0
144	60·0	117·5	147	55·8	75·0
144	57·0	84·0	145	57·8	84·0
142	55·0	70·0	140	53·5	81·0
145	56·5	91·0	139	52·8	63·5
143	57·5	101·0	143	58·3	77·5
150	59·0	98·0	139	57·5	96·0
150	61·8	118·0	147	59·8	84·5
·142	56·0	87·5	139	61·5	104·0
146	57·3	83·0	143	61·5	116·5
151	59·3	87·0	140	53·8	68·5
144	57·3	76·5	147	59·5	101·0
139	60·5	87·0	144	55·8	73·5
150	59·5	84·0	146	56·3	83·5
150	60·8	128·0			
147	50·5	79·0			
151	66·3	117·0			
141	53·3	84·0			
140	56·5	84·0			
147	57·0	84·0			
141	57·5	85·0			
148	60·5	118·0			
140	56·8	83·5			
143	57·5	75·0			
146	57·5	90·0			
144	62·8	94·0			
140	58·5	86·5			
144	60·0	89·0			
149	52·5	81·0			
142	58·8	84·0			

EXERCISES 3 AND 4. ESTIMATING THE SIZE OF POPULATIONS

Introduction:

It is rarely possible to count all the individuals in an area, but the size of populations can be estimated using techniques suitable to the numbers, size and behaviour of the animals and their habitat. For species living in fairly uniform and accessible habitats the populations can be estimated by the marking and recapture methods or by using square grids of known area. Each method may be used separately but it is preferable to use both as a check on each other. They are particularly suitable for fairly long-lived species that are not very active flyers, such as frog hoppers, grasshoppers, ladybirds, etc. *Philaenus spumarius*, a frog hopper (Fig, C2a, p. 283), was used in the exercises described.

Site:

The exercise is suitable for meadowland, heath, rough grass or wasteland but not for trees and bushes. The area should not exceed 1 acre and is best defined by hedges or buildings. It is sound practice to record the locality accurately, e.g.

Flatford Mill: Grid reference TM077333.
Area: $\frac{1}{4}$ acre.
Vegetation: meadowland.

Time of Year:

June to September. It is important to choose a warm dry day and wait until the dew has dried, otherwise marking-paint will not stick and insects will be damaged.

EXERCISE 3. MARKING AND RECAPTURE METHOD

Hypothesis:

The total number of animals in a population can be estimated by repeated sampling for marked individuals.

Apparatus:

Sweep net, pooter, oil paint (Windsor and Newton) or quick-drying cellulose paint.

Procedure:

Sweep one-fifth of the area and try to collect 200 froghoppers. Mark these by placing a *small* spot of paint on the forewings or pronotum with the point of a pin. Allow it to dry but do not cage insects in small tubes where they can walk over each other, because the paint sticks to their legs. If the insects prove too active to mark, anaesthetize them with a *whiff* of ethyl acetate, mark them and let them recover in fresh air. Reject any damaged specimens, then count and scatter the marked individuals $(= F_1)$ throughout the sampling area.

On a warm day 3–4 hours is sufficient for marked individuals to redistribute themselves throughout the population. In cooler weather it is advisable to wait until the next day before collecting again. On the second occasion sweep the area and try to collect at least the same number of individuals as on the first occasion. The more insects collected, the more accurate the result. Count the total number caught on the second occasion $(= F_2)$ and the number of these that are marked $(= F_3)$.

Since
$$\frac{F_1}{N} = \frac{F_3}{F_2},$$

$$\therefore N = \frac{F_1 F_2}{F_3}$$

where N = total population.

Results:

No. marked and released on 1st occasion $(F_1) = 247$

No. recaptured on 2nd occasion $(F_2) = 259$

No. of marked individuals recaptured $(F_3) = 16$

$$\therefore \text{Population of froghoppers } (N) = \frac{247 \times 259}{16} = 3998$$

For this method to succeed, the marking technique must not harm the insect nor affect its behaviour and the sampling properties of the population must remain constant. If individuals spend part of the day resting among the bases of plants and at other times are on the tips of the vegetation, results will be inconsistent. It is assumed that marked insects mix at random through the whole population and that there is neither immigration nor emigration, birth or death between the time of release and recapture. For the insects mentioned above, these are not misleading assumptions providing recaptures are made within 24 hours of release.

Conclusions:

The population can be estimated by this method but another independent estimate is needed to verify the results.

EXERCISE 4. QUADRAT SAMPLING METHOD

Hypothesis:

Populations can be estimated by sampling from a known proportion of the area they inhabit.

Apparatus:

At least ten, 33 cm × 33 cm (Area = 0·1 sq. m), wire quadrats and canes to use as markers.

Procedure:

Measure the plot and calculate its area. Distribute 20 quadrats in the plot early in the morning, covering the whole area as evenly as possible (see p. 93), and mark each with a cane. Allow at least one hour to elapse before counting the insects. Approach each quadrat cautiously without allowing a shadow to fall on it. Lightly brush the vegetation with a hand and count the froghoppers as they jump out of the quadrat. Search for any remaining.

When there is more than one species or instar involved, it is often impossible to identify these without catching some and if identification is important, supplement counts with sweeps to estimate the proportions of each species or instar present.

Results:

Area of plot $\quad\quad\quad\quad\quad$ = 733 sq. m.

No. of insects seen in 20 quadrats = 9

Population (N) $\quad\quad\quad\quad$ = $\dfrac{733 \times 9}{0{\cdot}1 \times 20}$ = 3299

Conclusions:

Estimates obtained by using quadrats are usually lower than those obtained from marking and recapture methods because a single insect may be overlooked however carefully the quadrats are observed. With only nine insects seen, one insect overlooked is equivalent to a 10% underestimate.

EXERCISES 5–7. COMPARISON OF SAMPLING METHODS

Introduction:

To study animal populations in different habitats, different sampling techniques are required. No sampling method is ideal and each has characteristic advantages and limitations. The information collected may be either qualitative, listing the species present in the habitat; quantitative, counting the numbers present, or semi-quantitative, providing rough estimates of relative commonness.

One single method is rarely "best" and when possible in any ecological study, two complementary methods should be employed.

Three pairs of quantitative techniques are described in Exs. 5–7 to illustrate the variability between techniques, observers and animal's responses. Comparison of results from the different methods may reveal information on the habits of animals (Ex. 19).

Sites:

Exercise 5. The *air*, preferably over a meadow or garden.

Exercise 6. The *vegetation*, in meadows, on wayside verges, on waste land.

Exercise 7. The *ground* surface, in gardens, meadows, ploughed fields, or on wayside verges.

Time of Year:

April to October; in dry weather.

EXERCISE 5. VARIABILITY BETWEEN TECHNIQUES: A COMPARISON OF WATER TRAPS AND STICKY TRAPS

Hypothesis:

Some sampling methods give more variable results than others.

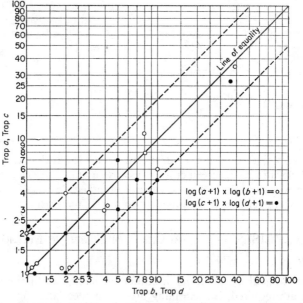

FIG. IV:5:1. Comparison of samples from pairs of sticky traps, plotted as log (N + 1), for trap a × trap b and trap c × trap d. The solid line of equality represents points where the two samples are exactly alike. The dotted parallels represent the $\times \atop \div$ 2 limits.

Apparatus:

Water trap (see p. 250). Shallow tins of any shape, 5–8 cm deep and about 0·1 sq. m in area; forceps, specimen tubes, water and liquid detergent.

Sticky trap (see p. 250). Glass boiling tubes or jam jars and 1 m canes. Use the same number of pairs of each type of trap.

Procedure:

Arrange the tins in pairs on stands or boxes, 20 cm above the surface of the vegetation. Traps in each pair should be no more than 1 m apart. Half fill each tin with water and add a few drops of detergent, so that trapped insects sink and cannot escape. Prepare the sticky traps, leaving part of each trap clean to hold. Hang these traps vertically on canes in pairs at a similar height to the water traps and at least 5 m from them.

Expose the traps until there is a catch of 50 to 100 insects in each water trap and 30–50 on each sticky trap. Remove the insects using forceps and transfer them to specimen tubes. Preserve in alcohol if necessary. Insects may be removed from sticky traps indoors, using a brush dipped in a solvent, e.g. xylol, benzene or petroleum ether, to avoid damaging them excessively. Keep catches from individual traps separate.

Determine the insects to families and tabulate the numbers under traps and families in orders. Sum each family for each kind of trap and compute the ratio of the water trap to sticky trap catches (Table IV:5:1).

Plot the log $(N + 1)$ for pairs of like traps keeping the two kinds of traps on separate graphs (Figs. IV:5:1 and IV:5:2) (see Chapter III, p. 73).

Results:

From Table IV:5:1: On average, the water traps caught 2·8 times as many insects as the sticky traps. This is the ratio of sample size for all insects and

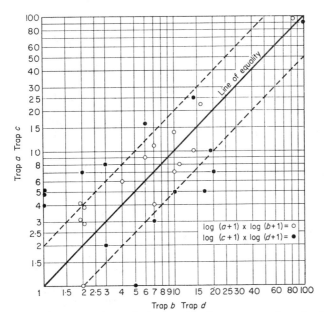

FIG. IV:5:2. Comparison of samples from pairs of water traps, plotted as log $(N + 1)$ for trap a × trap b and trap c × trap d. The solid line of equality represents points where the two samples are exactly alike. The dotted parallels represent the ×2 limits.

the expected ratio for separate taxa. By comparison, however, there were more cercopids, delphacids, Thysanoptera, muscids, calliphorids, Coleoptera and Apoidea in the water traps than expected, whilst aphids, syrphids and the acalypterates were caught in about the expected proportions. Jassids and

TABLE IV:5:1

Determination (column 1) of samples from four replicate water traps (columns 2, 3, 4, 5) and four sticky traps (columns 7, 8, 9, 10); the sums for each kind of trap (columns 6, 11) and the ratio of water/sticky trap samples (column 12)

Column 1	2	3	4	5	6	7	8	9	10	11	12
	Water Traps					Sticky Traps					
	a	b	c	d	Σ	a	b	c	d	Σ	$\dfrac{\Sigma \text{ Water*}}{\Sigma \text{ Sticky}}$
HEMIPTERA											
Cercopidae	8	5	4	9	26	3	1	0	2	6	*4·3*
Jassidae	6	9	4	0	19	2	3	4	1	10	1·9
Delphacidae	3	1	2	6	12	0	1	0	0	1	*12·0*
Aphididae	10	6	6	19	41	2	3	4	6	15	2·7
THYSANOPTERA											
Thripidae	9	13	15	5	42	2	3	2	4	11	*3·8*
Aeolothripidae	0	1	1	2	4	0	0	1	0	1	*4·0*
DIPTERA											
Syrphidae	3	1	0	4	8	0	1	1	1	3	2·7
Acalypterates	21	15	9	23	68	10	7	4	9	30	2·3
Muscidae	5	3	7	2	17	1	2	1	0	4	*4·3*
Calliphoridae	3	6	6	1	16	0	0	0	1	1	*16·0*
COLEOPTERA											
All Families	2	1	4	0	7	1	0	1	0	2	*3·5*
HYMENOPTERA											
Apoidea	2	1	3	0	6	0	0	0	0	0	*>6·0*
Parasitica	7	11	4	16	38	5	9	6	4	24	1·6
OTHER INSECTS	13	9	24	13	59	7	7	3	8	25	2·4
TOTAL	92	82	89	100	363	33	37	27	36	133	2·8
	174		189			70		63			

* Ratios in italics are above average (2·8).

parasitic Hymenoptera were possibly slightly less common in the water traps than expected.

From Fig. IV:5:1: (Sticky traps). Out of 15 pairs of samples, all 15 from trap a lie within $\overset{\times}{\div}2$ (see p. 75) of trap b; 12 out of 15 from trap c lie within $\overset{\times}{\div}2$ of trap d; in other words, both pairs of traps gave similar results.

From Fig. IV:5:2: (Water traps). Out of 15 pairs of samples all 15 from trap a fell within $\overset{\times}{\div}2$ of trap b; only 5 from trap c fell within $\overset{\times}{\div}2$ of trap d; traps a and b were more consistently alike than traps c and d.

Comparing Figs. IV:5:1 and IV:5:2: (Sticky traps vs water traps).

With 27 out of 30 pairs within the limits $\overset{\times}{\div}2$, sticky traps gave more consistent results than water traps, which had only 20 out of 30 within the limits.

Conclusions:

The water traps caught 2·8 times as many insects as the sticky traps, but in spite of this, the numbers on the sticky traps were less variable. This may be because attraction to colour varies as light quality and intensity changes and also because air currents are more turbulent over angular tins than over upright cylinders. Catches in one pair of water traps were much more variable than in the other, perhaps because of differences between local air currents. Generally sticky traps gave more consistent samples than water traps.

EXERCISE 6. VARIABILITY BETWEEN OBSERVERS: A COMPARISON OF SEARCHING AND SWEEPING

Hypothesis:

Samples that depend upon personal ability or choice may be biased by the observer.

Apparatus:

Sweep net, pooter (see p. 253), specimen tubes, forceps.

Procedure: Searching

Select an area about 1 m square containing vegetation representative of the complete area. Watch the vegetation in each square carefully for 5 or 10 min without casting a shadow on it and, with as little disturbance as possible, catch all insects. Then systematically search the flowers, leaves and stems for 15 min, catching everything. This may be repeated by a different person on another plot.

Procedure: Sweeping

Traverse the area taking ten sweeps at first and collect the insects from the net with a pooter. Make more sweeps as necessary to collect a total of 100–200 insects. This may also be repeated by a different person. Determine insects to families and other invertebrates to classes or orders. Tabulate numbers under taxon, sampling method and person. Sum each taxon for each sampling method and compute the ratio of sweeping/searching (Table IV:6:1).

Plot log (N + 1) for one person against another, putting sweeping and searching on the same graph represented by different symbols (Fig. IV:6:1) (see Chapter III, p. 67).

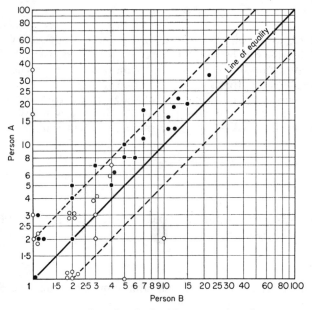

Fig. IV:6:1. Comparison of results obtained by person A and person B. Log (N + 1) for each taxon with different symbols for each sampling method. The solid line signifies equality and the broken lines the $\times\atop\div$ 2 limits. Searching ○; sweeping ●.

Results:

From Table IV:6:1: On average, sweeping produced 2·8 times as many insects as searching.

From Fig. IV:6:1: Results obtained by searching were erratic with 5 out of 22 points outside the limits, $\times\atop\div$2. Sweeping also had 4 out of 22 points outside the $\times\atop\div$2 limits. This is because the whole scatter is shifted toward the y axis, not because the points are widely scattered. New limits, parallel to the present ones, but $\times\atop\div$2 from the mean of the scatter [$\overline{\log(x+1)}$, $\overline{\log(y+1)}$] will show this. Find this mean and draw these limits (see p. 75).

TABLE IV:6:1

INVERTEBRATES ON VEGETATION

Determination (column 1) of samples obtained by searching, persons A and B (columns 2, 3) and by sweeping, persons A and B (columns 5, 6). Sums for searching (column 4) and sweeping (column 7) and ratio of sweeping/searching (column 8)

Column 1	2	3	4	5	6	7	8
	Searching			Sweeping			$\dfrac{\sum \text{Sweeping}}{\sum \text{Searching}}$
	A	B	Σ	A	B	Σ	
ARANEIDA	1	0	1	6	2	8	8·0
ORTHOPTERA	Nymphs and adults seen but not caught						
HEMIPTERA							
Nabidae	0	1	1	5	3	8	8·0
Miridae	3	2	5	4	3	7	1·4
Cercopidae	2	1	3	19	14	33	11·0
Jassidae	1	3	4	21	12	33	8·3
Delphacidae	0	1	1	10	6	16	16·0
Aphididae	35	0	35	12	11	23	0·7
THYSANOPTERA							
Thripidae	3	2	5	15	10	25	5·0
Aeolothripidae	0	1	1	4	1	5	5·0
DIPTERA							
Chironomidae	1	0	1	9	4	13	13·0
Syrphidae	5	3	8	1	0	1	0·1
Acalypterates	6	3	9	32	21	53	5·9
Muscidae	2	1	3	7	4	11	3·7
Calliphoridae	1	0	1	1	1	2	2·0
COLEOPTERA							
Staphylinidae	2	1	3	3	1	4	1·3
Coccinellidae	2	0	2	18	11	29	14·5
Nitidulidae	16	0	16	7	5	12	0·8
HYMENOPTERA							
Tenthredinoidea	0	1	1	1	0	1	1·0
Apoidea	1	9	10	2	0	2	0·2
Parasitica	2	1	3	17	6	23	7·7
LEPIDOPTERA							
All Families	0	4	4	0	0	0	0·0
OTHER INSECTS	0	1	1	12	10	22	22·0
TOTAL	83	35	118	206	125	331	2·8

* Ratios in italics are above average.

132 INTRODUCTION TO EXPERIMENTAL ECOLOGY

Conclusions:

The erratic results obtained by searching may have resulted from the different ability of searchers and from the chance occurrence of a few dandelion flowers on one plot and a knapweed plant on the other, each of which attracted different insects. Sweeping covers a much wider area, so samples a better cross-section of the herbage. Whereas sweeping gave more consistent results, the difference between sweepers was more pronounced, perhaps because one used longer sweeps than the other. As expected, searching was generally most effective for the larger, brighter insects, which are easily seen but which often evade the sweep net.

EXERCISE 7. VARIABILITY IN ANIMAL'S RESPONSES

Hypothesis:

Sampling methods that depend on sensory responses to attract the catch are biased in favour of certain species.

Apparatus:

Eight to sixteen 1 lb honey jars, stale meat or fish for bait, forceps, specimen tubes.

Procedure:

Arrange four to eight unbaited traps and the same number of baited traps in pairs throughout the available area. Sink the jars to ground level, so that no rim protrudes and protect from rain with a cover of board, or glass raised above the level of the ground with stones (p. 252). Individual jars in each pair should be within 1 m of each other; baited traps should be separated by at least 4 m from unbaited, but both baited and unbaited traps should be scattered throughout the whole area. Label the traps in each pair (i) and (ii).

Empty the traps each morning and continue the experiment until a total of 50–100 insects have been caught in all the unbaited traps. Collect the animals in four lots, unbaited (i) and (ii) and baited (i) and (ii). Determine the catch as far as possible. Sum catches from unbaited and baited traps and compute the ratio of baited/unbaited catches for each taxon. Tabulate (Table IV:7:1) under taxon (column 1), unbaited traps (i), (ii) and $\sum$ (columns 2, 3, 4), baited traps (i), (ii) and $\sum$ (columns 5, 6, 7) and ratio (column 8). Plot log (baited + 1) × log (unbaited + 1) for traps (i), (ii) and the $\sum$ (i + ii). (Fig. IV:7:1) (see Chapter III, p. 67).

Results:

From Table IV:7:1: Dermaptera, Calliphoridae, Carabidae, Silphidae, Staphylinidae and Formicidae were all more common in baited than unbaited traps. All other taxa were nearly equally common. Note that the ratio for Acari is 0·0 because only one was caught. Note also that the ratio for the whole catch (2·5) is greater than 1·0.

TABLE IV:7:1

INVERTEBRATES ON THE GROUND

Catches from pairs of unbaited and baited traps, the sums
for each type, and the ratio of baited/unbaited catches

Column 1	2	3	4	5	6	7	8
	Unbaited traps			Baited traps			$\frac{\sum \text{Baited*}}{\sum \text{Unbaited}}$
Trap no.	i	ii	$\sum$	i	ii	$\sum$	
ISOPODA	0	2	2	1	0	1	0·5
OPILIONES	1	0	1	0	1	1	1·0
ACARI	0	1	1	0	0	0	0·0
ARANEIDA	2	6	8	3	7	10	1·3
COLLEMBOLA	17	30	47	21	29	50	1·1
DERMAPTERA	0	0	0	3	0	3	>3·0
HEMIPTERA							
Cercopidae	2	3	5	1	5	6	1·2
DIPTERA							
Phoridae	1	0	1	0	1	1	1·0
Calliphoridae	0	0	0	10	16	26	>26·0
COLEOPTERA							
Carabidae	2	3	5	19	13	32	6·4
Silphidae	0	0	0	1	2	3	>3·0
Staphylinidae	0	1	1	4	7	11	11·0
HYMENOPTERA							
Formicoidea	3	1	4	17	24	41	10·3
TOTAL	28	47	75	80	105	185	2·5

* Ratios in italics are above average.

From Fig. IV:7:1: Series (i), Series (ii) and Series (i + ii) are all equally scattered. Taken altogether the points lie either within the limits or towards the y-axis, i.e. catches from the baited traps are equal to or greater than catches from the unbaited traps. Only one point falls towards the x-axis and this is for a very small sample.

Conclusions:

Individual traps within a pair behaved similarly whether baited or unbaited but the bait increased the catch of Dermaptera, Calliphoridae, Carabidae, Silphidae, Staphylinidae and Formicidae, all flesh-eating insects which were attracted by the smell. Numbers of other insects caught were similar in both kinds of traps; their common ratio is therefore nearly 1·0 (65/69). The ratio of catches for baited/unbaited traps including all animals was 2·5, indicating that bait attracted additional animals but did not repel any.

134   INTRODUCTION TO EXPERIMENTAL ECOLOGY

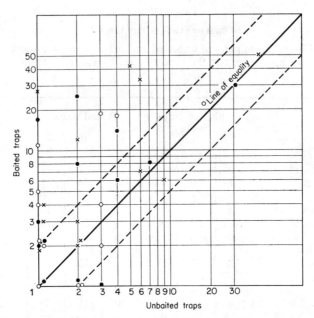

FIG. IV:7:1.   Comparison of samples, log ($N + 1$), from baited × unbaited pitfall traps. The solid line of equality represents points where the two samples are alike. The $\frac{x}{4}2$ limits are dotted. Series i traps ○; series ii traps ●; series (i + ii) traps ×.

EXERCISE 8. RESPONSE TO HUMIDITY

Introduction:

Small differences in microclimate frequently determine whether or not a habitat is suitable for a species. Closely related species may often be found in very different situations because of their differing requirements for temperature, humidity, wind speed or light (see p. 5).

The two common spiders, *Zygiella x-notata* (Fig. IV:8:1a) and *Z. atrica* are closely allied morphologically (Fig. IV:8:1b). Both are orb-spinners and make a characteristic web in which there is a radius devoid of spiral threads, with a signal line in the middle of this region leading from the centre of the web to the spider's retreat (Fig. IV:8:1c). *Z. atrica* may occasionally spin a web without an empty sector and then the signal thread forms an angle of 90° with the plane of the web. Their webs are found in different situations. This may be caused by different responses to microclimate.

Hypothesis:

The humidity requirements of different, even closely related, species may differ.

Apparatus:

A long, narrow box or tin about 50 × 5 × 5 cm, with a tight-fitting lid is required, with two small dishes to fit inside each end, some granular calcium chloride and water.

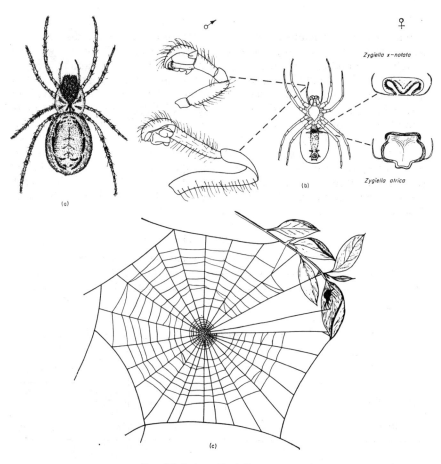

FIG. IV:8:1a. *Zygiella x-notata.*
FIG. IV:8:1b. Morphological features for determining the species and sex of *Z. x-notata* and *Z. atrica.* Left-hand drawings show the shape and relative proportions of male palps, upper = *Z. x-notata;* lower = *Z. atrica.* Right-hand drawings show shape of the female epigyne; upper = *Z. x-notata*; lower = *Z. atrica.* The positions of these structures on the whole animal are indicated by the dotted lines.
FIG. IV:8:1c. Orb web of *Zygiella* spp.

Sites:

Z. atrica spins on shrubs and bushes and *Z. x-notata* on walls, sheds and in houses, often on staircases and in the corners of window frames.

Time of Year:

Both species are common during the summer and autumn, and small numbers of *Z. x-notata* may be found in houses and sheds throughout the winter.

Procedure:

Place a dish of calcium chloride at one end of the box and a dish of water at the other. When the lid is replaced this arrangement will create a gradient of relative humidity inside the box ranging from maximum vapour pressure at one end to a small value at the other. Capture a spider, put it into a box and replace the lid. Keep the box in a room at 20–25°C and at intervals of 2 to 3 hr, open the box and note whether the spider is resting in the more humid or the drier half of the box. Remove the silk threads, renew the calcium chloride if necessary, replace the lid and repeat the experiment four to six times for each spider. Alternatively, if the spider is on the lid, instead of removing the threads and disturbing the spider turn the lid around before replacing it. Repeat the experiment using at least three individuals of each species. Tabulate the results by species (column 1) and position (columns 2, 3, 4, Table IV:8:1).

TABLE IV:8:1

Typical results for four replicates with each of four spiders of each species, *Zygiella x-notata* and *Z. atrica*, in a humidity choice chamber

Column 1	2	3	4
Species	Observed positions of resting spiders in the box		
	Dry end	Moist end	Indeterminate
Z. x-notata	14	0	2
Z. atrica	1	13	2

Results:

From Table IV:8:1: *Z. x-notata* was found predominantly at the dry end; *Z. atrica*, at the wet end.

Conclusions:

The two species react differently to water vapour in the air; *Z. x-notata* preferred the dry air and *Z. atrica* the moist. This suggests an explanation for the distribution of the two species in the field. The continuous transpiration of water vapour among the leaves of shrubs and bushes where *Z. atrica* lives must produce a greater relative humidity in the air there than in the air adjacent to dry window-panes and walls inhabited by *Z. x-notata*. Thus their distribution is possibly a direct response to different relative humidities.

Rider:

Humidity preferences may be affected by temperature.

Procedure:

Repeat the experiment with *Z. x-notata* at less than 5°C.

Result:

The spiders go to the moist end.

Conclusion:

Temperature reverses the mechanism of humidity preference in *Z. x-notata*. This may induce the spiders to move from their normally dry sites before the temperature falls low enough to inactivate them. Otherwise they may desiccate in a long, cold, dry period.

EXERCISE 9. RESPONSE TO SOIL MOISTURE

Introduction:

Like many invertebrates, slugs and their eggs are sensitive to changes of humidity in their surroundings and adults live in moist, shady places and retire into seclusion to avoid excessive loss of water in dry weather (see p. 11).

The Grey Field or Common Field Slug (*Agriolimax reticulatus*) is a convenient species to study both in U.K. and U.S. It is pale white to brown, mottled or speckled and extremely slimy to the touch. The breathing pore is surrounded by a clear cream or whitish circle. When handled, specimens exude a milky white slime. Mature specimens are 4–5 cm long when moving, and weigh 0·7–0·8 g.

Hypothesis:

Animals usually select oviposition sites suitable for the development of their eggs. Available soil moisture may limit egg development.

Apparatus:

Ten glass jars with lids, a burette or graduated cylinder, a narrow tray, loam soil and peat are required. Sieve the peat and mix one part of peat to two parts of soil to produce a water-retaining compost.

Site:

Agriolimax reticulatus and other species of slugs may be collected in gardens and fields, in damp places especially where there is an abundance of rotting vegetation. They can often be found on the soil surface beneath matted grass.

Time of Year:

All ages of *A. reticulatus* may be found throughout the year and eggs are laid in each month. Maximum numbers of mature slugs occur from July to September.

Procedure:

Place known weights of dried compost, to about 10 cm deep, into each of the jars and adjust their percentage saturation to 20, 25, 50, 75 and 100% by adding the appropriate weight of water calculated from the formula given below.

$$\text{Saturation percentage} = 100 \times \frac{\text{(weight of water added)}}{\text{(weight of oven-dry compost)}}$$

Collect 25 mature slugs each weighing 0·7–0·8 g. Put five into each jar and replace the lids. Examine the compost daily for 5 days for egg masses; note whether the eggs are buried or on the surface and the total number in each jar. The individual eggs are spherical, gelatinous, clear and whitish with a dark centre and about 2 mm diameter. Tabulate and sum total egg production at each humidity (Table IV:9:1).

Slugs laying in the jars have no choice between compost of different percentage saturation. Present mature adults with a choice by slanting a narrow tray containing compost with the lower end immersed in water. In this way the compost at the lower end will remain flooded and there will be a complete graduation to almost air-dry compost at the upper end. Place ten slugs in the tray, cover with muslin, and place in darkness. After 24 hr note where the slugs have congregated and search the soil in transects parallel to the water level, for egg masses. It may be necessary to repeat this experiment for a few days. Take samples of compost from an area containing eggs and determine the mean saturation percentages of compost by weighing, drying to constant weight at 105°C and re-weighing.

$$\text{Saturation percentage} = \frac{\text{(loss of weight on drying)}}{\text{(weight of oven-dry compost)}} \times 100$$

Place the eggs laid each day on moist cotton wool in another jar. Keep these at 20°C for 3 weeks. After 10 days examine daily and count the number of eggs hatched. When all eggs have hatched, calculate the mean length of the incubation period at 20°C.

Results:

From Table IV:9:1: Most eggs were laid on the compost at 25–75% saturation. Eggs are unlikely to be laid at 10% and few at 100%. Note also that the depth at which the egg masses are laid varies according to the amount of moisture present; they are laid deeper in the drier compost.

When slugs were offered a choice of oviposition site in a tray, adult and egg masses were found concentrated in a narrow band in compost that was between 40 and 80% saturated.

The mean incubation period at 20°C was about 15 days (min. 11 days, max. 21 days).

Four days prior to håtching, the eyespots with lesser pigmented areas around them may be seen and the head and tail distinguished. About 24 hr before hatching, the embryos become very active and eventually head and mouthparts are thrust through the shell and the whole body is slowly drawn out. The average length at emergence is about 2·7 mm .

Conclusions:

The eggs develop only in contact with a moist surface and die if too dry. Oviposition sites are selected where the soil moisture content is high, but not excessive. Breeding in the field is thus only likely to occur in soils which are 25–75% saturated with water; flooding or desiccation would destroy eggs in nature. Control of slugs in fields or gardens is facilitated by dry conditions and the removal of unwanted rotting vegetation, which retains moisture and also raises the temperature slightly as it decomposes.

TABLE IV:9:1

The number of eggs laid daily by five mature slugs in composts with different water contents

Percentage saturation	10	25	50	75	100	Total
Days			No. of eggs laid			
1	0	0	32	41	0	73
2	0	0	48	72	0	120
3	0	45	62	36	0	143
4	0	0	0	51	41	92
5	0	56	23	0	0	79
Total	0	101	165	200	41	507

EXERCISE 10. MORTALITY AND TEMPERATURE

Introduction:

Life depends on heat and the tolerable range of temperatures is narrow. Although different insect species may be found in places with a wide range of temperatures, from snow fields to hot springs, no individual species can tolerate the whole range. Some species develop special resistant stages to survive severe conditions, but normally, tolerable temperatures range from near freezing to about 40°C (see p. 8) (see Ex. 13 before doing this exercise).

Hypothesis:

Life persists only within a narrow range of temperatures.

Apparatus:

Thermostatically controlled oven and refrigerator, insect cage such as a wooden box covered with gauze, small glass tubes with a plaster of paris layer in the bottom.

Site:

Laboratory.

Time of Year:

July, August, September.

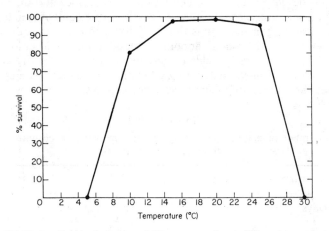

FIG. IV:10:1.   Survival of eggs of *Noctua pronuba* at different temperatures.

Procedure:

Collect living Large Yellow Underwing (*Noctua pronuba*) U.K., Corn Earworm (*Heliothis zea*) U.S., or other Noctuid moths at lights or by searching for them. During the day they often rest in long grass. Keep in a gauze-covered cage in a sheltered place outside; feed with sugar and water solution absorbed on a cotton wool pad. Collect eggs laid on the walls of the cage twice a day. Place them in the glass tubes and moisten the plaster of paris with distilled water. Place the tubes in such constant temperatures from 0° to 30°C as are available. Inspect twice daily, remove egg shells and larvae as they hatch and record numbers. Tabulate temperature, number of eggs used, number hatched and percentage survival, i.e. hatched (columns 1, 2, 3, 4, Table IV:10:1). Plot percentage survival against temperature (Fig. IV:10:1).

Results for Large Yellow Underwing:

Below 5°C and above 30°C no eggs hatched. Between these temperatures most of the eggs survived, and larvae emerged.

Conclusions:

Eggs of this moth survived to hatch only within the range of 5° to 30°C. If other species are used, the range may differ slightly, but many species that live in temperate climates and lay eggs in summer time, tolerate a similar range of temperatures. The eggs of species that lay in the autumn are usually resistant to freezing, but in the Large Yellow Underwing, the last instar larva is the frost resistant stage which diapauses during the winter. For the eggs of this species, the lower and upper thermal death points, below and above which few survive, lie between 5° and 10° and between 25° and 30°, respectively.

TABLE IV:10:1

Survival of eggs of *Noctua pronuba* at different temperatures

Column 1 Temperature °C	2 No. of eggs incubated	3 No. hatched	4 % survival
5	226	0	0
10	65	52	80
15	106	103	97
20	59	58	98
25	153	145	95
30	92	0	0

EXERCISE 11. TEMPERATURE THRESHOLDS

Introduction:

Poikilotherms require heat from outside their bodies to live, but different bodily functions respond differently to change in temperature. Some functions, such as flight, are only possible when body temperature is above a certain so-called threshold value. Small insects cannot raise the temperature of their bodies above that of the surrounding air, so they can only fly when the air temperature exceeds their temperature threshold for flight. However, once their threshold is exceeded and they are flying, further increase in temperature does not appreciably improve their ability to fly; indeed, excessive heat may inhibit flight. Temperatures high enough to do this are rare in equable climates, so, in general, two ranges of temperature are distinguishable; those below the threshold when few, if any individuals fly, and those above, when the number flying is independent of temperature (see p. 10).

Hypothesis:

Air temperatures above a certain threshold value are necessary for flight in small insects.

Apparatus:

One or more light traps (p. 250) [or suction traps (p. 253) if available].

Site:

Almost anywhere near vegetation.

Time of Year:

Mid November to mid February.

Procedure:

Empty the trap each morning at 0900 hours and count the Winter Gnats which in S. England were mainly *Trichocera annulatus* (Diptera: Trichoceridae)

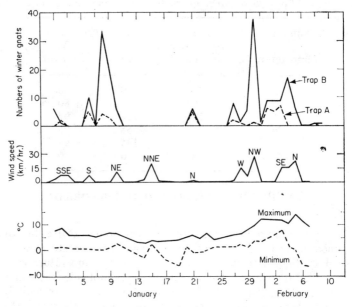

FIG. IV:11:1.   Wind speed and direction, daily maximum and minimum temperature and nightly catch of *Trichocera annulatus* in two light traps near Burleigh School, Hatfield, England, in January and February, 1964.

(Fig. D14a, p. 293). This species is recognizable by the bands of grey-brown around the abdomen. When Winter Gnats are not available, almost any common dipterous fly will serve. Make sure the light comes on each after-noon before sunset. Record daily maximum and minimum temperature. Plot the daily catch and the temperatures as well as any other meteorological records taken (Fig. IV:11:1). When about 40 nights' catches are available, with temperatures ranging from −1 to 16°C, plot the catch for each night against the maximum temperature for the afternoon of the day *before* (Fig. IV:11:2).

Results:

   The day by day diagram illustrating catch will probably show similar rises and falls to several meteorological factors, especially wind speed, but careful examination shows that the rises and falls do not coincide exactly. This type of diagram is very misleading (Fig. IV:11:1). From Fig. IV:11:2: when

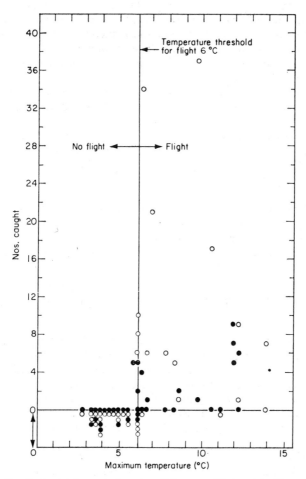

FIG. IV:11:2.   Each night's catch from two light traps (distinguished by solid and open circles) plotted against the maximum temperature of the preceding day.

catch is plotted as a scatter diagram against maximum temperature, two groups of points are clearly distinguishable. These have been separated by a vertical line at 6°C. Below this value zero catches predominate in both traps. Above 6°C, zero catches form only a small proportion of the total. However, the size of the catch bears no relation to the temperature. Evidently temperature either permits or prevents flight, but does not govern the number of insects flying.

As the insects are trapped overnight, it may at first seem logical to use the minimum rather than the maximum temperature for this analysis. In this instance, however, the insects are not flying at night, when the minimum temperature occurs, but just at sunset. Owing to the asymmetrical diurnal cycle of temperature in northern latitudes, the maximum daily temperature in mid-winter occurs in mid-afternoon and is therefore only an hour or two before the trap is catching the midges. The maximum thermometer reading, therefore, is an approximation to the temperature at the time of flight. Some error in temperature recording is introduced which may result in a less clear-cut threshold than in this example; a band of temperatures, 1 or 2 degrees wide, may then be necessary to represent the threshold, instead of a single line.

In other latitudes and seasons, when the times of maximum temperature and flight do not approximately coincide, temperature must be recorded at the time of day when most of the chosen insects fly.

Conclusion:

In poikilotherms, some activities, such as speed of walking, increase proportionately as temperature increases. Other activities, such as flight, are possible only above a threshold temperature. A further increase in temperature, above this threshold, does not produce a corresponding increase in numbers flying.

EXERCISE 12. THE BODY TEMPERATURE OF POIKILOTHERMS

Introduction:

The most important physiological attribute influencing the ecology of insects is their inability to control their body temperature. Thus, their rate of metabolism and activity is dependent upon external temperatures. A few insects increase body temperature by orienting themselves to receive maximum solar radiation, e.g. locusts, and some cool themselves by vibrating their wings and evaporating moisture, e.g. bees. But for most insects, body temperature is dependent upon that of the air in their microhabitats (see p. 9).

Hypothesis:

The body temperature of invertebrates is usually the same as that of the air around them.

Apparatus:

A thermocouple (see p. 260) and insect cage.

Site:

Field and laboratory.

Time of Year:
    May to October, whenever full grown larvae of Noctuid moths, Cabbage White (U.K.) or Alfalfa (U.S.) butterflies are available.

Procedure:
    Collect or rear larvae of Noctuid moths, Cabbage White or Alfalfa butterflies. Put larvae, on their food plant in a cage, in a variety of air temperatures obtained by placing the cage in sunshine, heating it with electric lamps* in the laboratory, cooling it in a refrigerator, etc.

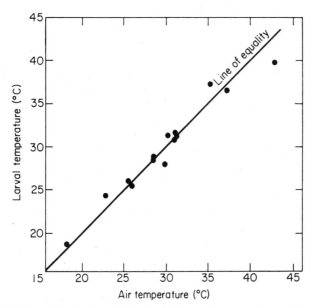

Fɪɢ. IV:12:1. The relationship between the body temperature of larvae and air temperature, showing how larval temperatures depend on air temperature (see Fig. III:1, p. 41).

    In each place allow time for the temperature to settle down. Then measure the air temperature close to the larvae and close to the reference junction at the thermocouple; using a mercury thermometer. Finally, measure the insect's temperature by gently pressing the other junction of the thermocouple into a lateral fold of the cuticle. (See p. 261 describing thermocouple calibration.)
    Record a wide range of air temperatures (Table IV:12:1). Plot insect temperature against air temperature (Fig. IV:12:1).

* The radiant heat received by an insect at a point 7·5 cm from a 100 W tungsten bulb is approximately equivalent to the heat received from the mid-day June sun, 50–55°N.

There will be individual differences in body temperature and in the efficiency of measuring it. These may be reduced by taking a mean of several readings.

Results:

The general trend of body temperatures conforms almost exactly with the air temperatures, despite the small scatter of points due to individual variation and inefficient measuring techniques.

Conclusions:

The body temperature of these insects was similar to that of the air around them, except when they were exposed to strong, direct radiation.

TABLE IV:12:1

Larval temperatures taken by pressing the probe of a thermocouple in lateral folds of the cuticle. Each reading is the average from six larvae

Air temperature °C	Larval temperature °C
25·6	26·2
43·2	40·0
28·7	28·9
31·2	31·0
31·5	31·5
26·2	25·6
30·1	28·2
31·4	31·8
30·4	31·5
28·8	28·9
37·5	37·0
35·6	37·5
23·0	24·5
18·2	18·7

EXERCISE 13. TEMPERATURE AND RATE OF DEVELOPMENT

Introduction:

Besides defining limits within which life is possible (Ex. 10), temperature also determines the rate at which cold-blooded animals grow. This is because the many metabolic processes which constitute life are all temperature dependent. Collectively they function most rapidly at some optimum temperature. Below and above the optimum, one or many of them function less effectively, thus slowing down the whole metabolism. Usually the time taken for development decreases gradually as temperature increases above the lower thermal death point until the optimum value, just below the upper thermal death point, is reached. Above this optimum value, time required for development again increases, usually rapidly, until excessive heat kills the animal (see p. 8).

Hypothesis:

Temperature controls the rate of development in poikilotherms.

Apparatus, Site, Time of Year:

As for Ex. 10. These two exercises are best done together.

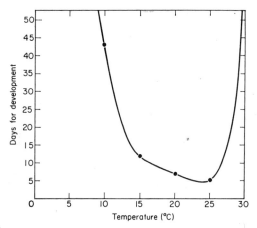

FIG. IV:13:1.   Time required for the development of eggs of *Noctua pronuba* at different temperatures.

Procedure:

Record the time taken for development of the eggs from laying to hatching during Ex. 10. Tabulate temperature, number of eggs and development time (columns 1, 2, 3, Table IV:13:1). Compute the reciprocal of the development time in days (1/time). This is the rate of development (column 4, Table IV:13:1). Plot time for development (Fig. IV:13:1) and rate of development (Fig. IV: 13:2) against temperature (see Chapter III, p. 112).

Results:

The time taken for development of the eggs of the Large Yellow Underwing is 43 days at 10°C and only 5 days at 25°C. At 5° and 30° the eggs never hatch. The developmental rate is therefore highest at 25°C, decreases slowly down to zero at 5°C, and decreases rapidly down to zero at 30°C.

Conclusions:

The shape of this curve for rate of development of moth eggs is typical of curves describing the developmental processes in poikilotherms. It shows clearly the optimum temperature for rate of development, which is important in the life of the animal. However, it does not follow that the fastest developers are the largest and most viable individuals. Eventual size is often greatest in individuals developing more slowly at a lower temperature.

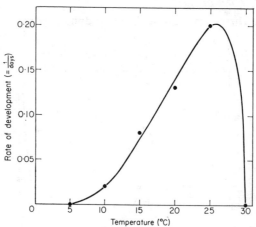

FIG. IV:13:2.   Rate of development of eggs of *Noctua pronuba* at different temperatures.

TABLE IV:13:1

Time and rate of development of eggs of *Noctua pronuba* at different temperatures

Column 1	2	3	4
Temperature °C	No. of eggs	Time for development (days)	Rate of development $\left(\dfrac{1}{days}\right)$
5	226	∞	0·00
10	65	43·1	0·02
15	106	11·7	0·08
20	59	7·5	0·13
25	153	5·0	0·20
30	92	∞	0·00

EXERCISE 14. SIZE AND RATE OF DEVELOPMENT

Introduction:

The size of insects, which ranges from that of the largest Protozoan to that of the smallest Mammals, is limited by their exoskeleton and respiratory system.

As a larva develops and increases in size the exoskeleton must be shed periodically because it does not grow with the larva. Each new skin must accommodate the growing larva during the interval, or instar, between moults and hence it is loose at the beginning of each instar and tight at the end. In Macrolepidoptera the several instars last about the same length of time and in the Large Yellow Underwing moth (*Noctua pronuba*) there are seven instars. The eventual form of the seventh instar larva is different from that of the newly hatched first instar. The first instar has a large head and small body whilst the seventh instar has a large body and small head. This is because different organs have different rates of growth during development. For the same reason, the heads of adult men are smaller, relative to the body, than the heads of infants (see p. 109).

Hypothesis:

Rate of growth is proportional to the size of the organism. The rate of growth of different organs may differ.

Apparatus:

An insect rearing cage, occasional use of a refrigerator, binocular microscope with scale.

Time of Year:

Whenever eggs of a large butterfly or moth are available. *Noctua pronuba* was used in this example.

Site:

Laboratory.

Procedure:

Incubate about 20 eggs of a large butterfly or moth. Immediately after they hatch, and *after* each successive moult, measure the width of the head capsule and the overall body length. To facilitate measurement chill the larvae for ten minutes to immobilize them. Remember that at the time of moulting the larva is vulnerable and disturbance during moult may be fatal. If larvae of a hawk moth can be used, measure also the length of the abdominal horn. Measure several larvae, about ten if possible, and compute the means of each measurement. Tabulate the means (Table IV:14:1). Plot head width and body length against instar number, allotting equal intervals for instar number on the horizontal *x*-axis (Fig. IV:14:1). Join the two sets of points with a

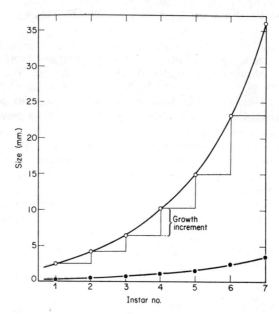

FIG. IV:14:1. The relationship between instar and body length (○), and head width
(●), in *Noctua pronuba*.

TABLE IV:14:1

Growth of larva of *Noctua pronuba*

Instar	Mean head width, mm	Increment of growth mm	Mean body length, mm	Increment of growth mm
1	0·33		2·52	
		0·17		1·78
2	0·50		4·30	
		0·28		2·32
3	0·78		6·62	
		0·37		3·63
4	1·15		10·35	
		0·49		4·79
5	1·64		15·14	
		0·81		8·22
6	2·45		23·36	
		1·20		12·54
7	3·65		35·90	

smooth curve. Repeat these plots using single log paper, putting the measurements, i.e. the y-axis, on the logarithmic scale (see Chapter III, p. 109). Draw straight lines through the points. Using parallel rulers, draw a line parallel to the head width line to intersect the body length line (Fig. IV:14:2).

Results:

The increment of growth of both body length and head width increases with each instar (Fig. IV:14:1). Each increment of growth is proportional to initial size (Fig. IV: 14:2). The rate of growth is slightly greater in body length than in head width.

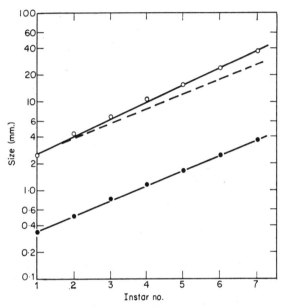

Fɪɢ. IV:14:2. The relationship between instar and body length (○), and head width (●) of *Noctua pronuba* plotted on a log scale. This demonstrates (a) that increments of growth are proportional to initial size and (b) that the rate of growth of body length is slightly greater than that of head width. – – – –, line parallel to line for head width.

Conclusions:

The experiment confirmed the hypothesis that rate of growth is proportional to the size of the insects. The rate of growth is slightly greater for body length than for head width, although the difference is less than in some other species, e.g. hawk or sphinx moths.

EXERCISE 15. Colour and Concealment

Introduction:

There is sometimes a clear resemblance between the colour of animals and their background; the foxes and owls in the Arctic are white, foxes and owls in deserts are brown. There is no simple rule, however, relating colour to habitat; for example, treetop monkeys are never truly green, whilst parakeets, tree snakes and tree frogs often are. The Common Cinnabar moth (U.K.) exposes red and black wings when settled, colours that never combine in its habitat, whilst the Green Hairstreak butterfly (U.K.) and the Olive Hairstreak (U.S.) have wings as green as the vegetation on which they rest. Pattern further complicates the issue. In their normal habitat, zebra are camouflaged by their sombre off-white and near-black tones, but their striped pattern makes them appear vivid away from this habitat (see pp. 24, 27).

Hypothesis:

The colours and patterns of animals have ecological functions of conceal-ment, warning or sexual attraction.

Apparatus:

Sweep net, kite net, sieve, white tray, forceps and specimen tubes.

Sites:

Grassy heath or meadow; hedge bottom litter; pond, river or estuarine mud; wooden fences and tree bark; clumps of flowering Umbelliferae or Compositae.

Time of Year:

Late spring to autumn, during the daytime.

Procedure:

Collect about 50 insects from each habitat, except from fences and trees where 10 to 20 will suffice, by sweeping the grass, sieving litter on to the tray and catching insects on flowers, mud, fences and bark with either forceps, tubes or kite nets.

Few insects are uniformly coloured but decide which of the following one or two colours predominate in each insect, black, grey, brown, buff, green, blue, red, orange, yellow, white, and enter it in a table (Table IV:15:1–5). For example, five green grasshoppers with buff stripes would be scored as 5 green and 5 buff, one seven-spot ladybird as 1 red and 1 black, or two orange soldier beetles as 2 orange.

In the examples below, the insects have been identified to facilitate refer-ence to other adaptive features, but this is not necessary to understand the principles of adaptive colouration. Summarize the detailed tables to emphasize gross features of colour distribution, collecting together the indeterminate colours grey and brown, the vegetation colours buff, green and blue and the bright colours red, orange, yellow and white (Table IV:15:6).

Results:

From Table IV:15:6: In the first four sites, grass, litter, mud and bark, the predominant colour group corresponds with the background (italics in Table IV:15:6). On flowers, black was the commonest colour (bold type) and the bright colours second (italics).

From Tables IV:15:1–5: Many of the insects on flowers were conspicuous and probably objectionable to predators; bees and wasps sting, ladybirds and stink bugs exude offensive fluids when molested and Cinnabar cater-pillars are hairy. The Syrphids may be without these objectionable features, but their bright colours probably confer some immunity from predators which have learnt to associate yellow and black stripes with distasteful food. The Red Underwing and Yellow Underwing moths both have bright colours that show only when the forewings are raised.

Conclusions:

At rest on grass, litter, mud and bark, the insects matched their back-grounds. Active, on flowers, they tended to contrast with them but the black insects, when not feeding on the flowers, would probably be hiding in some dark place where their colours contrasted less. Evidently it is better to be protected when at rest than when active.

Considered collectively, the insects' colours blended with those of their habitats; colour is predominantly for concealment. However, there were many exceptions, in which brilliant colours contrasted with the background and were flaunted, or else were hidden until the insect was disturbed.

Discussion:

Such general colour resemblances of animals and their habitats enhance camouflage but there also exist many specialized patterns, independent of colour perceptions, which have the same concealing effect. The irregular light and dark patches on the wings of many Geometrid moths, e.g. Carpet Moths (*Xanthorhoe* spp.) distract attention from the true outlines of the insects as they rest on a background of similar colour. This disruptive colouration and concealment of contour occurs in many groups of animals; in the eggs and chicks of ground-dwelling birds, fishes, snakes, gazelle and zebra, serving to prevent, or delay as long as possible, the first recognition of the animal by sight.

In some animals, e.g. grasshoppers and frogs, a pattern appears to join separate parts of the body, and bands of colour often pass along the side of the head and abdomen distracting attention from the eyes. The Comma (U.K.) or Anglewing (U.S.) butterflies rest with wings closed, and the pattern, which continues line for line across the gap between the wings, camouflages their outline. Moths are often difficult to see on bark because they rest with their own markings aligned with the striations on the background.

The function of these concealing devices is to provide partial protection from predators like birds and reptiles which hunt by sight (see Ex. 17, p. 158).

Many insects resemble parts of plants. When not feeding, larvae of many Geometridae grip the vegetation with their prolegs and erect the rest of the body rigidly to resemble twigs. The buff head and wing-tips, and silver-grey wings of the Buff Tip Moth (*Phalera bucephala*) (U.K.), resemble freshly broken silver-birch twigs, some Geometrid and Drepanid moths resemble bird droppings, the Lunate Moth (U.S.) and Oak Beauty (U.K.) are easily confused with lichens, and the tree hoppers *Centrotus cornutus* (U.K.) and *Stictocephala bubalis* (U.S.) are like small thorns (Figs. C1, a, b, p. 283).

Frequently behaviour is allied to colour to aid concealment or mimicry. In many moths, like the Red and Yellow Underwings, grasshoppers and lantern flies, the forewings are coloured to conceal, but the hind wings are brilliant when exposed. This combination of colours is called flash coloration. When at rest the colour conceals the insect, but when it moves, there is a momentary flash of conspicuousness, which again merges into the habitat as movement ceases. The sudden disappearance of colour and movement confuses and misdirects predators.

The hind wings of the Eyed Hawk Moth (U.K.) and Twin-spotted Sphinx (U.S.) are magenta, each with a single large blue and grey intimidatory eye-spot. At rest, the brown fore-wings cover these eyespots, but if the insect is disturbed they may be suddenly exposed as a warning.

TABLE IV:15:1

Insects on grass

	Black	Grey	Brown	Buff	Green	Blue	Red	Orange	Yellow	White
5 Grasshoppers	0	0	0	5	5	0	0	0	0	0
1 Nabidae	0	0	0	0	1	0	0	0	0	0
10 Aphididae	3	0	0	0	7	0	0	0	0	0
6 Cercopidae	0	0	3	3	0	0	0	0	0	0
3 Jassidae	0	0	0	1	2	0	0	0	0	0
1 Yellow Underwing Moth	0	0	1	0	0	0	0	0	1	0
1 Grass Moth	0	0	0	1	0	0	0	0	0	0
1 Sawfly larva	0	0	0	0	1	0	0	0	0	0
2 Ants	2	0	0	0	0	0	0	0	0	0
1 Ladybird	1	0	0	0	0	0	0	0	1	0
9 Acalypterae	6	0	0	5	0	0	0	0	0	0
1 Muscidae	0	1	0	0	0	0	0	0	0	0
2 Spiders	0	0	2	0	0	0	0	0	0	0
10 Others	4	2	4	6	0	0	0	0	0	0
Totals	16	3	10	21	16	0	0	0	2	0

TABLE IV:15:2

Insects on litter

	Black	Grey	Brown	Buff	Green	Blue	Red	Orange	Yellow	White
2 Carabidae	2	0	0	0	0	0	0	0	0	0
8 Staphylinidae	8	0	3	0	0	0	0	0	0	0
1 Chrysomelidae	1	0	0	0	0	0	0	0	0	0
1 Curculionidae	0	1	1	0	0	0	0	0	0	0
3 Thysanoptera	3	0	0	0	0	0	0	0	0	0
5 Mites	0	0	5	0	0	0	0	0	0	0
1 Harvestman	0	0	1	0	0	0	0	0	0	0
2 Others	2	0	0	0	0	0	0	0	0	0
Totals	16	1	10	0	0	0	0	0	0	0

TABLE IV:15:3

Insects on mud

	Black	Grey	Brown	Buff	Green	Blue	Red	Orange	Yellow	White
9 Collembola	0	9	0	0	0	0	0	0	0	0
1 Gerridae	0	0	1	0	0	0	0	0	0	0
7 Saldidae	0	7	0	0	0	0	0	0	0	0
6 Psychodidae	0	6	0	0	0	0	0	0	0	0
7 Scatopsidae	7	0	0	0	0	0	0	0	0	0
3 Phoridae	3	0	0	0	0	0	0	0	0	0
5 Muscidae	0	5	0	0	0	0	0	0	0	0
Totals	10	27	1	0	0	0	0	0	0	0

TABLE IV:15:4

Insects on fences and bark

	Black	Grey	Brown	Buff	Green	Blue	Red	Orange	Yellow	White
2 Collembola	0	2	0	0	0	0	0	0	0	0
4 Dermaptera	0	0	4	4	0	0	0	0	0	0
1 Red Underwing	0	0	1	0	0	0	1	0	0	0
1 Carpet Moth	0	1	1	0	0	0	0	0	0	0
1 Wasp beetle	1	0	0	0	0	0	0	0	1	0
2 Wasps	2	0	0	0	0	0	0	0	2	0
5 Woodlice	0	5	0	0	0	0	0	0	0	0
3 Spiders	0	0	3	0	0	0	0	0	0	0
Totals	3	8	9	4	0	0	1	0	3	0

TABLE IV:15:5

Insects on flowers

	Black	Grey	Brown	Buff	Green	Blue	Red	Orange	Yellow	White
1 Miridae	0	0	0	0	1	0	0	0	0	0
5 Cantharidae	0	0	0	0	0	0	0	5	0	0
3 Coccinellidae	3	0	0	0	0	0	3	0	0	0
1 Tenthredinidae	1	0	0	0	0	0	0	0	1	0
4 Bumble bees	4	0	0	0	0	0	0	3	1	0
4 Wasps	4	0	0	0	0	0	0	0	4	0
8 Parasitic Hymenoptera	8	0	0	0	0	0	0	0	0	0
2 Ants	2	0	0	0	0	0	0	0	0	0
6 Cinnabar larvae	6	0	0	0	0	0	0	6	0	0
2 Pieris sp.	2	0	0	0	0	0	0	0	0	2
3 Syrphidae	3	0	0	0	0	0	0	0	3	0
6 Acalypterate flies	6	0	0	0	0	0	0	0	0	0
1 Bluebottle	0	0	0	0	0	1	0	0	0	0
8 Others	3	0	5	0	0	0	0	0	0	0
Totals	42	0	5	0	1	1	3	14	9	2

TABLE IV:15:6

Summary of insect colour and habitat

	Black	Grey Brown	Buff Green Blue	Red Orange Yellow White	Σ
Grass	16	13	37	2	68
Litter	16	11	0	0	27
Mud	10	28	0	0	38
Bark	3	17	4	4	28
Flowers	42	5	2	28	77
	87	74	43	34	238

There is experimental evidence that birds find red, orange and yellow insects least acceptable when presented with a choice. The insect's bold, bright patterns and colours facilitate their recognition by predators. Such insects are usually distasteful and it is to their advantage to be easily recognized; birds that try them once do not make the same mistake again. However, not all bright insects are obnoxious. Some, such as the yellow striped Hover flies (U.K.) or Flower Flies (U.S.), merely imitate wasps, and thereby obtain some gratuitous immunity from predation. This immunity depends upon the wasps being unpleasant and common enough for young, learning birds to be impressed by their appearance, and on the birds having colour vision. The bright colour would be ineffective against mammals, which, except for the Primates, are colour blind. Man uses colours and patterns both to warn and advertise and to camouflage, especially in warfare.

EXERCISES 16 AND 17. ADAPTIVE COLOUR CHANGES

Introduction:

Many animals have evolved colouration to match their background (see Ex. 15) and thus derive partial protection from predators that hunt by sight. The background of insects from many orders changes during their lives, or in successive generations, and many species have evolved colour changes to match those that occur in their habitat. The time scale of these changes differs greatly. The rapid, reversible, colour changes of the chameleon and octopus are widely known; the slower, but equally reversible seasonal change of the lacewing (*Chrysopa vulgaris*) (U.K.) (Fig. B9, p. 273) is less so. The change of this species from green in summer to red brown in winter is accomplished by the accumulation of carotenoids, which disappear after hibernation, so that the insect becomes green again. A rapid, but irreversible, change occurs in many arboreal hawk moth larvae which change from green to brown when they leave the leaves to make the journey down the brown bark to pupate in the soil. The slowest scale of colour change is genetical and it appears only in different generations (see p. 24).

EXERCISE 16. REVERSIBLE COLOUR CHANGES

Different colour forms of the stick insect may be various shades of brown or green depending on the quantity of brown, orange, yellow and green pigment in the epidermal cells. Each of these stable "morphologically" coloured forms, except the green, also exhibits variable "physiological" colour changes produced by the rapid clumping or dispersion of brown and orange pigments within the cells.

Hypothesis:

Colouring of some animals becomes adapted to their background by reversible physiological processes.

Apparatus:

Adult stick insects (U.K.) or walking sticks (U.S.) and two boxes are required, one with a glass lid and one with an almost completely darkened interior.

Site:

Indoors or outdoors.

Time of Year:

Any time.

Procedure:

Select a few similar, fairly dark, adult stick insects and place half in the darkened box and half in the glass-topped box on dark leaves, like ivy or holly. Expose this group to sunlight for a few hours, then examine the insects in each box.

Results and Conclusions:

The insects in the sunlight on the green leaves develop darker colouration than those kept in subdued light.

The effect of light is exerted through the eyes and is determined by contrast between the background and the surroundings. If the ventral half of the eyes only is blackened with paint the insect still darkens, but if the eyes are completely covered all colour change ceases. The movement of pigment is controlled by a hormone produced in a ductless gland near the brain. High humidity also produces dark colouring, and dry air, pallor.

EXERCISE 17. IRREVERSIBLE COLOUR CHANGES

Colour changes are usually influenced most by light. The colour of Large White Butterfly (*Pieris brassicae*) pupae is produced by the combined effect of black melanin in the superficial layer of the cuticle, white pigment in the epidermal cells and the green colour of the deeper tissues. The amount of black or white pigment depends on the quality of light falling on the larva as it rests before pupation. The colour of Peacock Butterfly (*Vanessa io*) pupae is influenced in a similar manner.

Hypothesis:

Permanent deposition of cuticular pigment may depend upon background colour.

Apparatus:

About ten last instar larvae of Large White or Peacock butterflies are required and two boxes about 25 × 25 × 25 cm with a glass top, one painted dark green inside and the other cream or buff.

Eggs of these two species are obtainable from entomological suppliers and the larvae may be reared on cabbage and nettles, respectively.

Site:

Indoors.

Time of Year:

Any time when larvae and food plants are available.

Procedure:

When the last instar larvae cease feeding, allocate half to each box with a few bare twigs and a pad of moist cotton wool. If Peacock larvae are used, put fresh nettles in water pots in the green box and a few twigs in the cream one.

Results and Conclusion:

In *Pieris*, the reflected green light suppressed the formation of black and white pigment and the pupae appear green. Pupae in the cream box were a normal ivory white, with their surface speckled and reticulated with black.

Peacock pupae among the nettle leaves were light greenish yellow to pinkish grey, but those suspended on twigs were yellowish with head, thorax and wings washed with gold and sparsely speckled with black and reddish markings.

Discussion:

Permanent colour changes which aid concealment are often inherited. In particular the dark melanic forms of many Geometrid moths, which have increased rapidly in some industrial areas, provide a spectacular example of the evolution of adaptive, morphological colouration. Large proportions of Peppered Moth (*Biston betularia*) populations inhabiting industrial districts in England now consist of a black variety (var. *carbonaria*) or a more rare, dark, slightly speckled variety (var. *insularia*) instead of the normal form, white, speckled thinly with black. The spread of these melanic forms is partly due to their selective advantage over the normal form when seen by insectivorous birds like English Robins, Hedge Sparrows and Great Titmice against soot-blackened surroundings. A gene conferring some physiological advantage is also probably involved.

On uniformly sooty oak bark, *carbonaria* is generally invisible at 2–5 m but the normal form remains visible at 30 m or more; *insularia* is visible at intermediate distances. Mottled birch trunks offer concealing backgrounds to all three forms, but least for *carbonaria*.

The percentage survival per day in a soot-polluted wood for male *carbonaria* may be 17% higher, and for *insularia* 11% higher, than in the normal form.

Similar protection is probably derived in industrially polluted districts by melanic forms of the Scalloped Hazel (*Gonodontis bidentata*), Waved Umber (*Hemerophila abruptaria*), Willow Beauty (*Cleora rhomboidaria*) and Mottled Beauty (*Alcis repandata*).

EXERCISE 18. ADAPTATION TO THE AERIAL ENVIRONMENT

Introduction:

Exercises 5, 6 and 7 provide material to investigate the relative abundance of various invertebrate orders within three important environments; the air, the vegetation and the ground (see p. 125).

Hypothesis:

Aerial life is more common in the most highly evolved terrestrial invertebrates.

Procedure:

Use the samples obtained in Exs. 5, 6 and 7. Sum the catches obtained by all sampling methods in each environment, air, vegetation and ground surface. Tabulate the summed catches under order of insects, or the nearest suitable taxon for the other invertebrates (Table IV:18:1) (column 1) and environment (columns 2, 3, 4). Sum columns 2, 3 and 4 (column 5). The total catch for each environment should be of the same order of magnitude (e.g. 496, 449, 260). If they are not, divide the largest by 10 or 100 to bring them together.

Plot the distribution of each order, within the three habitats, as percentage histograms, arranging the orders in sequence from least to most aerial (Fig. IV:18:1).

Results:

From Table IV:18:1. The numbers of Arthropods other than insects are very small because the methods were not designed specifically to sample them.

If the underground environment had been included, by taking soil samples, the Acari especially would have been much more common. The sample for

TABLE IV:18:1

Summed samples of invertebrates from water and sticky traps in the air, from searching and sweeping in vegetation and from baited and unbaited pit-fall traps on the ground

Column 1	2	3	4	5
	In the air	On the vegetation	On the ground	Total
Isopoda	0	0	3	3
Opiliones	0	0	2	2
Acari	0	0	1	1
Araneida	0	9	18	27
Collembola	0	0	97	97
Dermaptera	0	0	3	3
Hemiptera	130	169	11	310
Thysanoptera	58	36	0	94
Diptera	147	102	28	277
Coleoptera	9	66	52	127
Hymenoptera	68	40	45	153
Lepidoptera	0	4	0	4
Others	84	23	0	107
Total	496	449	260	1205

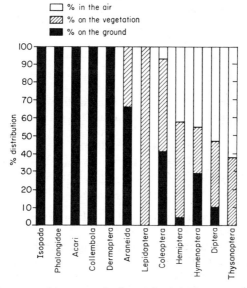

□ % in the air
▨ % on the vegetation
■ % on the ground

FIG. IV:18:1. Percentage histograms for invertebrate taxa arranged according to their distribution in the air, in vegetation and on the ground.

Lepidoptera is misleading because they happened to be seen sitting on vegetation; they might just as easily have been seen in the air, if they were day-flying species.

From Fig. IV:18:1: Most of the non-insect Arthropods live on the ground. Some spiders were found on vegetation; some might have been caught in the air, floating on silken threads. The more primitive insects, Collembola and Dermaptera were also on the ground; the more advanced orders, Thysanoptera, Diptera, Hymenoptera, Hemiptera and Coleoptera were found mainly in the air or on vegetation.

Conclusion:

The collective sample, using six different sampling methods, shows that evolutionary progress in arthropods has tended to produce an increasingly aerial life.

EXERCISE 19. EVOLUTION IN ECOLOGY

Introduction:

Exercise 18 compares the ground fauna with that on vegetation and Ex. 43 the fauna on chalk down with that on heath land. In each instance a more moderate and permanent habitat is compared with a more extreme and unstable habitat. Protected by the layer of vegetation, the ground surface varies less in temperature and humidity than the vegetation itself. With a more diverse and luxuriant growth of vegetation, the ground surface of a chalk downland has a more protected microclimate than that of a sparsely covered heath. Evolutionary trends have been to develop more resistance to extreme habitats, e.g., Man's invasions of deserts, Polar regions and space. We may therefore compare the success with which the various invertebrate taxa have colonized each of these exacting habitats (see p. 24).

Hypothesis:

Evolutionary trends are towards the colonization of wider and therefore more exacting habitats.

Sites:

Ground surface with the covering vegetation (Ex. 18). Chalk down and greensand heath (Ex. 43).

Procedure:

Extract columns 1, 3 and 4 from Table IV:18:1. Extract columns 1, 2 and 3 from Table IV:43:2. Use only those taxa that occur in both tables. Convert to log $(N + 1)$ (Chapter III, p. 66) and tabulate (Table IV:19:1) taxa (column 1), animals caught (log $(N + 1)$) for vegetation, ground, greensand heath, chalk down (columns 2, 3, 4, 5). Take the differences of logs of (vegetation—ground) and (greensand heath—chalk down) (columns 6, 7). Plot diff. of logs (vegetation—ground) × (heath—chalk); fit regression (see p. 78).

Results:

In Table IV:19:1: The differences of logs range from about $-1\cdot5$ to $+1\cdot5$ in each instance.

In Fig. IV:19:1: The scatter of the points shows a positive relationship meaning that as the animals become more sub-aerial, they also become associated more with heath.

The difference of logs is a measure of the relative numbers found in each of the sites of a pair. If subtracting the logs of numbers on the ground from those on vegetation gives a positive result (i.e. more on vegetation), this indicates greater adaptability to the more extreme environment. A positive result from the difference of logs (heath—chalk) means the same thing.

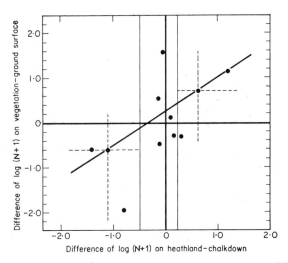

FIG. IV:19:1. The ability of different invertebrate groups to colonize difficult habitats, expressed as a regression of the difference of log density (vegetation—ground surface) against the difference of log density (heathland—chalkdown). Regression fitted by medial cross method.

The wide range of results ($-1\cdot5$ to $+1\cdot5$; remembering that 2 in difference of logs means 100 times) shows that the degree of adaptability differs widely between taxa.

The positive relationship shows that those taxa that are more highly adapted to the climatically variable sub-aerial environment are also more highly adapted to the rigorous heathland.

Conclusion:

Amongst the invertebrates, insects have most successfully evolved species versatile in their ability to colonize the more extreme environments.

TABLE IV:19:1

Log (N + 1) of samples from the ground surface and covering vegetation, and from chalk down and greensand heath

Column 1	2	3	4	5	6	7
	Logarithm of (N + 1)				Diffs. of logs	
Taxon	vegetation	ground	heath	chalk	(vegetation— ground) y	(heath— chalk) x
Isopoda	0·00	0·60	0·95	2·34	− 0 60	− 1·39
Opiliones	0·00	0·48	1·08	1·20	− 0·48	− 0·12
Acari	0·00	0·30	1·26	1·00	− 0·30	0·26
Araneida	1·00	1·28	2·02	1·87	− 0·28	0·15
Collembola	0·00	1·99	0·60	1·41	− 1·99	− 0·81
Dermaptera	0·00	0·60	0·00	1·11	− 0·60	− 1·11
Hemiptera	2·23	1·08	1·88	0·70	1·15	1·18
Thysanoptera	1·57	0·00	1·04	1·11	1·57	− 0·07
Diptera	2·01	1·46	1·20	1·34	0·55	− 0·14
Coleoptera	1·83	1·72	1·80	1·70	0·11	0·10
Lepidoptera	0·70	0·00	1·08	0·48	0·70	0·60

EXERCISE 20. SIZE AND COMPETITION

Introduction:

Two species of beetle *Scolytus destructor* and *S. multistriatus*, feed in the sapwood, between bark and wood, of living elm trees in the U.K. (Fig. E 30, p. 315). They transmit a fungal disease, Dutch Elm Disease, which kills infected trees within a few years, death usually starting in the crown and working downwards. When trees are dead their bark can be peeled easily to expose old vacated galleries produced by female beetles and their larvae.

Each female cuts a central gallery, the mother gallery, about 2 mm diam. for *S. destructor* and 1–1·2 mm for *S. multistriatus*, along whose sides a number of eggs are laid in August. After hatching, young larvae bore outwards from the mother gallery as they feed, producing a pattern of radiating galleries (Fig. IV:20:1) at the end of which they pupate and eventually emerge as adults in May and June. The exit holes, indicating that the attack has passed, appear as pin holes in the bark, and are easily seen in dead trees. In the U.S. the place of *S. destructor* is taken by *Hylurgopinus rufipes* (see p. 14).

Hypothesis:

The size of animals, evolved by ecological pressures, limits the number of suitable habitats available to them.

Apparatus:

Penknife, axe or chisel, chalk, tape measure.

Site:

Dead elm trees killed by Dutch elm disease.

Time of Year:

Any time.

Procedure:

Mark five separate areas, 20 × 20 cm, on the trunk and five similar squares, or their equivalent area, on smooth side branches.

Cut and remove the dead bark from the marked sites to reveal the galleries either on the surface of the wood or the bark. Those on the trunk are usually more easily distinguished as the frass and dust, produced during tunnelling,

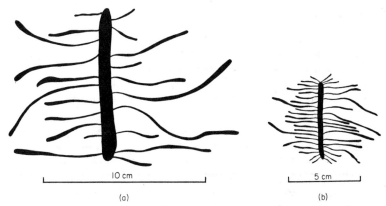

(a) (b)

FIG. IV:20:1. Mother gallery and radiating larval galleries of (a) *Scolytus destructor*; (b) *S. multistriatus*.

sticks to the bark. However, once collected, the bark can be taken indoors, cleaned of debris and used repeatedly. For each insect species, count and record the number of mother galleries on each sample of bark from rough trunk and smooth branches, and the number of larval galleries for each of ten mother galleries (Table IV:20:1). Compute the mean number of mother galleries per 400 sq. cm and the mean number of larval galleries per mother gallery, for each species. Measure the length of the larval tunnels in cm and make distribution histograms of tunnel length for each species. Convert to accumulated percentages and plot on probability paper. Read off the mean, standard deviation and 95% limits (see Chapter III, p. 54). Compare the relative numbers of eggs laid by the two species in the two habitats assuming each side tunnel represents one egg laid.

Results:

From Table IV:20:1: In this sample from medium sized branches, 15 times as many *S. multistriatus* females made egg-laying galleries as females of *S. destructor* (30·4 as compared with 2·0), and each laid twice as many eggs (46·6 compared with 22·9). The larval galleries were only one third as long (1·5 cm compared with 5·0 cm).

Discussion:

Notice how many larval galleries begin perpendicular to the main gallery and become more longitudinal as they are extended. This tends to shape the whole gallery structure to the cylindrical plane it occupies. Even so, the large galleries of *S. destructor* are unsuited to small branches of 5–8 cm diam. which will contain only *S. multistriatus*.

Conclusion:

Size influences the spatial distribution of these beetles and thereby reduces competition.

Riders:

1. Measure the cross-sectional area of the larval burrows and compare the volume of food eaten per beetle by the different species.
2. Assess the surface area of the trunk and limbs of a big elm tree and estimate the total population of each species it would support.

TABLE IV:20:1

Comparison of the density of mother galleries and length of larval tunnels for two species of bark beetles in branches 10–15 cm diameter.

Scolytus destructor			Scolytus multistriatus		
Mother galleries	Larval tunnels		Mother galleries	Larval tunnels	
No./400 sq. cm	No./gallery	Length cm	No./400 sq. cm	No./gallery	Length cm
2	15; 35		37	46; 61	
1	27; 12		27	43; 29	
0	29; 31		45	94; 55	
3	19; 17		43	30; 38	
4	16; 28		0	56; 14	
Mean 2·0	22·9	5·0	30·4	46·6	1·5

EXERCISES 21–23. ENVIRONMENTAL CONTROL

Introduction:

The most variable and unstable medium for life is the air. Although it is not a self-contained environment, since even the most persistently airborne birds such as the albatross and vulture must settle to sleep and lay eggs, the air is, nevertheless, a vital medium for the movement of many animals and especially many adult insects. Because the medium is so unstable, the ability of insects to control their movement is dependent upon their flight capacity. Tiny insects like Fairy Flies (Mymaridae (Fig. F 7b. p. 323) must go where the slightest breath of air blows them, but large insects such as Dragonflies can control their aerial movements with great precision, except in high winds. Thus insects, large enough to fly faster than 9 k/hr, a common wind speed above the level of sheltering vegetation, can distribute themselves at will in the air. Small slow-flying insects tend to be distributed more randomly by air currents. Size is thus an important factor affecting the control an animal has over its environment.

EXERCISE 21. THE SPEED OF FLIGHT OF INSECTS

Many insects can fly, but small ones, which constitute the majority, fly slowly. The speed of flight of insects determines how much they control their own movement within the constantly moving aerial environment.

Hypothesis:
Speed of movement is related to size.

Apparatus:
A narrow room with a southern aspect, a stop-watch, and a millimetre scale.

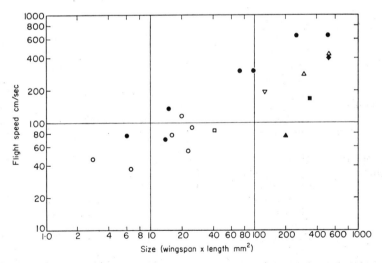

FIG. IV:21:1. The relationship between flight speed and size for specimens from eight insect taxa. ○ Homoptera, □ Psocoptera, ▲ Neuroptera, △ Lepidoptera, ■ Ichneumonoidea, ▽ Coleoptera, ● Diptera, ◆ Apoidea.

Site and Time of Year:
Indoors, any time when insects are flying.

Procedure:
Catch, very carefully so as not to damage them, insects flying in bright light. Bring them into a room in separate glass containers. Try to catch as wide a range of sizes and families as possible. Release them at a measured distance of about 5 m from a window. Time their flight to the window. Repeat the flight several times with each insect and choose the shortest time.

Some nocturnal moths may fly direct to a bright light in a dark room if kept in a dark box beforehand for half an hour. Reject any that fly in large spirals and try to get at least two direct flights before accepting the results. The room must be kept warm (above about 20°C) for most species to fly strongly.

TABLE IV:21:1

The relationship between flight speed and size for various insects

Column 1	2	3	4	5	6	7
Order	Species (or family)	Length mm	Wingspan mm	Wingspan × length	Time, sec	Flight speed cm/sec
Homoptera	Aphididae	2·5	10·0	25·0	5·0	91
	Aphididae	2·5	9·3	23·0	8·3	55
	Aphididae	2·5	7·8	16·0	6·0	76
	Aphididae	1·2	5·5	6·6	12·5	37
	Aleyrodidae	1·0	2·8	2·8	10·0	46
	Jassidae	2·8	7·0	20·0	4·0	116
Psocoptera	*Stenopsocus*	3·5	11·6	41·0	5·3	85
Neuroptera	*Chrysopa*	8·0	25·0	200·0	6·0	76
Lepidoptera	*Agrochola*	14·0	37·0	520·0	1·1	435
	Lycaena	10·0	30·0	300·0	1·6	286
Coleoptera	*Hister*	7·0	18·0	126·0	2·4	192
Diptera	Cecidomyiidae	2·0	7·0	14·0	6·5	70
	Empididae	5·5	13·0	72·0	1·5	305
	Drosophilidae	2·7	5·5	15·0	3·3	137
	Chloropidae	1·7	3·5	6·0	6·0	76
	Muscidae	7·0	14·0	98·0	1·5	305
	Tabanidae	16·0	32·0	512·0	0·7	650
	Tabanidae	11·0	23·0	253·0	0·7	650
Hymenoptera	Ichneumonoidea	11·0	31·0	340·0	2·7	167
	Apoidea	16·0	32·0	512·0	1·3	414

Identify the insects to family and measure the overall length and wingspan in mm. Tabulate the results (Table IV:21:1, columns 1, 2, 3, 4). List the product of length × wingspan (column 5) and the minimum time to fly 5 m (column 6). Compute the flight velocity in cm/sec (column 7). Plot flight velocity against size as measured by length × wingspan on log × log paper (Fig. IV:21:1), marking the different families with different symbols. Fit a regression (Fig. III:29 and p. 78).

Results:

The heavy bodied Diptera (Tabanidae), and Hymenoptera (Apoidea) flew much faster (400–650 cm/sec) than the slender-bodied Neuroptera, Ichneumonoidea and Cecidomyiidae (70–167 cm/sec). Most other insects flew at speeds between these extremes.

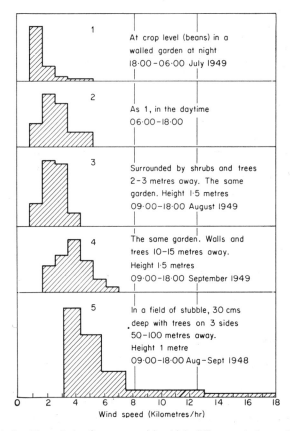

Fig. IV:21:2. The relative frequency with which different wind speeds occur at different sites, at different times of the day and heights above the ground.

Conclusions:

In spite of the family differences, speed of flight is proportional to size, such that the large insects have better flight control in moving air (p. 83).

Riders:

Measure wind speed with a small cup anemometer at different heights, places, and times, for several weeks, and make frequency distributions of wind speed for each site (Fig. IV:21:2). Then compare the measured speed of flight of different insect species with the wind speed to show that large insects can control their flight direction at most times and in most places, but that small ones lose control when they leave the shelter of vegetation.

Compare height and running speed of pupils from different classes using records from school sports.

EXERCISE 22. The Aerial Distribution of Large Insects

The Order Lepidoptera includes insects with a wing span ranging from minute to about 20 cm. Moths belonging to the rather loosely defined Macrolepidoptera and included in South's "Moths of the British Isles", mostly · have a wingspan greater than 2 cm and form a convenient group of "large" insects. They fly at speeds exceeding 7 km/hr, so can control their dispersal in moving air better than can very small insects (see p. 29).

Hypothesis:

Large insects can usually control their aerial distribution.

Apparatus:

Two light traps and means of identifying Macrolepidoptera (see Appendix N).

Site:

One trap on top of a building and one within 1 m of the ground. Try to arrange the traps so that they each have a similar field of view in the same direction, but are not visible each from the other.

Time of Year:

Any time from May to September. About 1 month's trapping should give adequate numbers.

Procedure:

Collect the catch daily. Identify the moths to species and sex if possible. List the catch (Table IV:22:1) under species (column 1) male, female and total for the lower trap (columns 2, 3, 4) and upper trap (columns 5, 6, 7). In the table, those insects whose sex was not known are listed only in the "total" column. Sum all columns.

Inspection of the table may be adequate to decide that some species are more commonly caught at one height than the other. Where the number is large enough, a statistical test, χ^2 (chi-squared) may be applied (see Chapter III, p. 99) to test the validity of the results. To facilitate inspection, extract the species with obviously different distributions and make a summary table (Table IV:22:2). Make notes on the weather whilst trapping continues (Table IV:22:3).

Results (from a school in S.W. England):

Table IV:22:2 shows the difference in height distribution of the different species.

The total number of species (64) was the same at both heights and the ratio of males to females was about the same at 1 m as at 27 m.

Two species (Pale Tussock and White Ermine) were common enough to test statistically the ratio of males to females at the two heights. Using the χ^2 (chi-squared) test, the Pale Tussock males were found to be flying significantly higher than the females and the female White Ermines significantly higher than the males (see p. 101).

From Table IV:22:3 there were catches of 0 to 80 individuals on warm nights. Numbers caught decreased from 77 to 3 with no change in weather and increased from 2 to 21 with no change. On clear nights, 5 out of 6 catches were small whilst on cloudy nights 3 out of 4 catches were large.

Conclusions:

The view from the upper trap was unlimited for well over 4 km whilst that from the lower trap was partially obscured by trees about 80 m away. This may have increased the sampling area; investigate its size by marking and recapture (see Ex. 3).

Cold, rain and moonlight tended to reduce the catch, but this correlation between catch and weather is not good and there is considerable unexplained variation cf. Fig. IV:11:1.

The ratio of males to females in the total catch was similar in both traps, but this comparison is unsatisfactory because of the small number of females in both traps. Where numbers were adequate for a valid comparison between height of flight of different sexes, the males flew higher than the females in one species, the Pale Tussock, and lower in the other, the White Ermine.

The number of species caught in each trap (64), was surprisingly the same. All three species of Arctiidae, White Ermine, Buff Ermine and Cinnabar, flew low. Of the three Plusiids, the Golden Y and Burnished Brass flew low and the Silver Y flew high. The Silver Y is a known migrant and the high fliers may have been migrating. The one Geometrid, the Peppered Moth, and the three other Noctuids were also high-fliers. The Peppered Moth lays its eggs on trees, which may account for its high flight and the Dark Sword Grass is another migrant. No certain explanation is known for their choice of height of flight; this offers scope for enquiry. However, it is clear that these moths have much control over their flight path and hence their aerial environment.

TABLE IV:22:1

Macrolepidoptera in light traps. June 10–July 15

Column 1	1 m trap			27 m trap		
	♂	♀	Σ	♂	♀	Σ
Poplar Hawk-moth (*Laothoe populi*)	6	—	6	—	—	—
Eyed Hawk-moth (*Smerinthus ocellata*)	1	—	1	1	1	2
Privet Hawk-moth (*Sphinx ligustri*)	7	—	7	—	—	—
Elephant Hawk-moth (*Deilephila elpenor*)	1	—	1	1	—	1
Puss (*Cerura vinula*)	—	—	—	2	—	2
Swallow Prominent (*Pheosia tremula*)	—	—	—	1	1	2
Lesser Swallow Prominent (*Pheosia gnoma*)	—	—	—	—	1	1
Pebble Prominent (*Notodonta ziczac*)	1	—	1	—	—	—
Iron Prominent (*Notodonta dromedarius*)	1	—	1	—	—	—
Buff-tip (*Phalera bucephala*)	1	—	1	1	—	1
Figure of Eighty (*Tethea ocularis*)	1	—	1	2	—	2
Common Lutestring (*Tethea duplaris*)	—	—	—	1	—	1
Pale Tussock (*Dasychira pudibunda*)	9	9	18	20	1	21
Lackey (*Malacosoma neustria*)	—	—	—	1	—	1
Lappet (*Gastropacha quercifolia*)	—	—	1	—	—	—
Green Silver Lines (*Bena fagana*)	—	—	—	1	1	2
White Ermine (*Spilosoma lubricipeda*)	39	3	42	5	8	13
Buff Ermine (*Spilosoma lutea*)	36	—	36	—	—	—
Muslin (*Cycnia mendica*)	2	—	2	—	—	—
Garden Tiger (*Arctia caja*)	—	—	2	—	—	—
Cinnabar (*Callimorpha jacobaeae*)	24	—	24	2	1	3
Common Footman (*Lithosia lurideola*)	3	—	3	—	—	—
Dark Dagger (*Apatele tridens*)	—	—	—	1	—	1
Grey Dagger (*Apatele psi*)	2	—	2	4	—	4
Knotgrass (*Apatele rumicis*)	6	—	6	9	—	9
Coronet (*Craniophora ligustri*)	1	—	1	1	—	1
Marbled Beauty (*Cryphia perla*)	3	—	3	7	—	7
Heart and Dart (*Agrotis exclamationis*)	15	—	15	84	—	84
Dark Sword Grass (*Agrotis ipsilon*)	3	—	3	19	—	19
Setaceous Hebrew Character (*Amathes c-nigrum*)	1	—	1	1	—	1
Flame Shoulder (*Ochropleura plecta*)	—	—	2	11	—	11
Flame (*Axylia putris*)	6	—	6	24	3	27
Large Yellow Underwing (*Nocta pronuba*)	4	—	4	6	—	6
Broad-Bordered Yellow Underwing (*Lampra fimbriata*)	—	—	—	1	—	1

Column	1	2	3	4	5	6	7
		1 m trap			27 m trap		
		♂	♀	Σ	♂	♀	Σ
Green Arches (*Anaplectoides prasina*)		—	—	—	2	—	2
Cabbage (*Mamestra brassicae*)		—	—	—	9	—	9
Dot (*Melanchra persicariae*)		1	—	1	17	—	17
Bright-line Brown-eye (*Diataraxia oleracea*)		2	—	2	1	1	2
Light Brocade (*Hadena w-latinum*)		1	—	1	—	—	—
Pale-shouldered Brocade (*Hadena thalassina*)		—	—	—	5	—	5
Broom Moth (*Ceramica pisi*)		—	—	—	2	—	2
Glaucous Shears (*Hadena bombycina*)		—	—	—	1	—	1
Lychnis (*Hadena bicruris*)		1	—	1	—	—	—
Silver Cloud (*Xylomyges conspicillaris*)		1	—	1	13	—	13
Procus spp.		—	—	4	24	5	29
Clouded-bordered Brindle (*Apamea crenata*)		3	—	3	68	—	68
Light Arches (*Apamea lithoxylaea*)		—	—	—	4	5	9
Dark Arches (*Apamea monoglypha*)		1	—	1	2	—	2
Union Rustic (*Apamea pabulatricula*)		—	—	—	1	1	2
Angle Shades (*Phlogophora meticulosa*)		—	2	2	14	8	22
Common Wainscot (*Leucania pallens*)		—	—	—	1	—	1
Shoulder-striped Wainscot (*Leucania comma*)		3	—	3	—	—	—
Treble Lines (*Meristis trigrammica*)		2	—	2	1	—	1
Satellite (*Eupsilia transversa*)		—	—	—	1	—	1
Shark (*Cucullia umbratica*)		—	—	—	1	—	1
Burnished Brass (*Plusia chrysitis*)		—	—	23	1	—	1
Golden Y (*Plusia jota*)		7	—	7	—	—	—
Beautiful Golden Y (*Plusia pulchrina*)		25	—	25	—	—	2
Silver Y (*Plusia gamma*)		9	—	9	31	—	31
Gold Spot (*Plusia festucae*)		1	—	1	—	—	—
Spectacle (*Abrostola triplasia*)		4	—	4	1	—	1
Snout (*Hypena proboscidalis*)		—	—	—	1	—	1
Beautiful Hook-tip (*Laspeyria flexula*)		1	—	1	—	—	—
Blood-vein (*Calothysanis amata*)		3	—	3	—	—	—
False Mocha (*Cosymbia porata*)		—	1	1	—	—	—
Broken-barred Carpet (*Electrophaes corylata*)		1	1	2	—	—	5
Common Marbled Carpet (*Dysstroma truncata*)		4	—	4	4	—	4
Red Twin-spot Carpet (*Xanthorhoe spadicearia*)		1	—	1	—	—	—
Pine Carpet (*Thera firmata*)		—	—	—	4	—	4
Flame Carpet (*Xanthorhoe designata*)		—	—	—	1	—	1
Green Carpet (*Colostygia pectinataria*)		—	—	—	1	—	1
Silver-ground Carpet (*Xanthorhoe montanata*)		3	—	4	2	—	2
Common Carpet (*Epirrhoe alternata*)		—	—	1	—	—	—
Galium Carpet (*Epirrhoe galiata*)		—	—	1	—	—	—
Garden Carpet (*Xanthorhoe fluctuata*)		—	—	—	5	—	5
May High-flyer (*Hydriomena coerulata*)		2	—	2	—	—	5
Clouded Border (*Lomaspilis marginata*)		6	—	6	—	—	—
Green Pug (*Chloroclystis rectangulata*)		—	—	—	1	—	1

Column	1	2	3	4	5	6	7
		1 m trap			27 m trap		
		♂	♀	Σ	♂	♀	Σ
Barred Red (*Ellopia fasciaria*)		2	—	2	1	—	1
Light Emerald (*Campaea margaritata*)		1	—	1	1	—	1
Lunar Thorn (*Selenia lunaria*)		1	—	1	—	—	—
Scalloped Hazel (*Gonodontis bidentata*)		1	—	1	—	—	—
Swallow-tailed moth (*Ourapteryx sambucaria*)		1	—	1	1	—	1
Scorched Wing (*Plagodis dolabraria*)		2	—	2	12	—	12
Brimstone (*Opisthograptis luteolata*)		1	—	1	—	—	—
Peacock moth (*Semiothisa notata*)		—	—	—	1	—	1
Peppered Moth (*Biston betularia*)		11	—	11	49	—	49
V-moth (*Itame wauaria*)		1	—	1	—	—	—
Leopard moth (*Zeuzera pyrina*)		—	—	—	2	—	2
Ghost Moth (*Hepialus humuli*)		—	2	2	—	—	—
Total		276	18	329	492	37	541

TABLE IV:22:2

The distribution between two heights of selected species

Species	No. at 1 m	No. at 27 m
White Ermine	42	13
Buff Ermine	36	0
Cinnabar	24	3
Heart and Dart	15	84
Clouded-bordered Brindle	3	68
Dark Swordgrass	3	19
Silver Y	9	31
Beautiful Golden Y	25	2
Burnished Brass	23	1
Peppered Moth	11	49

TABLE IV:22:3

Weather and total catch of moths on selected nights

Column 1	2	3	4
Date	Weather	Sky	Total catch (two traps)
June 16	Warm	Cloudy	80
June 17	Warm	Cloudy	77
June 18	Warm	Cloudy	3
June 30	Warm	Clear	3
July 1	Warm	Clear	5
July 2	Warm	Clear	2
July 3	Warm	Clear	21
July 4	Warm	Some cloud	18
July 5	Warm	Clear	0
July 6	Warm	Clear	6

EXERCISE 23. THE AERIAL DISTRIBUTION OF SMALL INSECTS

Introduction:

Insects less than about 5 mm. long usually fly at speeds less than the speed of the wind in which they are flying. They are thus blown about in the air and quickly become dispersed by its turbulent motion. Turbulent air movement is complex and mixes together all particles in it, including small insects. They, therefore, cannot determine their individual flight paths, unlike the moths considered in Ex. 22, and instead, they become randomly distributed in the air, like inanimate particles such as seeds, spores and dust (see p. 23).

Hypothesis:

Small flying insects tend to become randomly distributed by air currents.

Apparatus:

Pairs of sticky traps with a large surface area (see p. 250).

Site:

Place the pairs of traps in different situations, some sheltered by buildings, some near vegetation and some in open places. Traps should be 1 m apart with the exposed surfaces 1 m above ground.

Time of Year:

May to October.

Procedure:

Pick off the insects from the trap with a needle and count them, disregarding any insect as big as or bigger than 5 mm long. Record the total catch.

TABLE IV:23:1

Samples of the aerial population of small insects; simultaneous sets of paired catches on sticky traps

Col 1	2	3	4	5	6	7	8	9	10	11	12	13	14	15	16	17	18	19	20	21	22
Sample No.		1		2		3		4		5		6		7		8		9		10	
Time (hr)		½		½		1		1		2		2		4		4		8		8	
Site	Trap	c	d*	c	d	c	d	c	d	c	d	c	d	c	d	c	d	c	d	c	d
A	1	0	0	1	0	12	8	3	1	6	3	3	3	12	4	85	37	45	10	127	21
A	2	0		1		4		2		3		6		16		48		35		106	
B	1	1	0	0	0	0	3	2	1	2	1	10	1	5	9	119	10	153	58	120	2
B	2	1		0		3		3		1		9		14		109		95		118	
C	1	1	1	2	2	1	1	5	5	6	2	1	0	22	11	35	10	136	4	134	7
C	2	0		0		0		10		4		1		33		25		132		141	
D	1	0	1	0	0	0	0	8	4	4	1	4	9	4	1	68	4	14	5	63	16
D	2	1		0		0		4		3		13		3		72		19		79	

Sample		Sp 1 c	Sp 1 d	Sp 2 c	Sp 2 d	Sp 3 c	Sp 3 d	Sp 4 c	Sp 4 d	Sp 5 c	Sp 5 d	Sp 6 c	Sp 6 d	Sp 7 c	Sp 7 d	Sp 8 c	Sp 8 d	Sp 9 c	Sp 9 d	Sp 10 c	Sp 10 d
E	1	237	4	67	2	29	4	33	4	11	12	8	1	2	2	2	5	0	1	1	1
	2	241		69		33		29		23		7		4		7		1		0	
F	1	14	19	147	121	4	1	1	0	10	1	10	4	1	1	0	2	0	0	1	3
	2	33		268		3		1		11		6		2		2		0		0	
G	1	55	5	99	27	33	11	2	0	2	1	3	7	1	1	0	0	2	1	0	2
	2	50		126		22		2		1		10		2		0		1		3	
H	1	108	26	72	37	14	9	12	5	3	2	15	4	2	2	1	1	1	0	4	2
	2	82		109		5		7		1		11		0		0		0		2	
I	1	188	18	60	52	16	4	8	2	12	5	1	0	0	3	1	4	0	0	0	0
	2	206		112		12		10		17		1		3		5		0		2	
J	1	429	12	67	1	20	7	34	10	16	1	1	2	2	2	0	0	1	1	0	1
	2	441		66		13		24		17		2		2		0		1		1	
Σc		2972		1891		765		272		171		104		56		38		11		18	
Σd			130		317		97		46		35		24		22		24		5		12

*c = catch; d = difference between catches.

For small samples leave the traps for 30 min; for large samples leave them for 4 hr or a day.

Collect ten samples, each consisting of ten pairs of observations. Tabulate the observations by site (Table IV:23:1, column 1), trap (column 2) and catch (columns 3, 5, 7, etc.). Sum the catch for each sample (20 observations) and take the mean ($\bar{c}$: Table IV:23:2). This gives an estimate of population density for each sample.

Subtract the smaller catch of each pair from the larger (columns 2, 4, 6, etc., Table IV:23:1) to obtain a measure of the spatial differences in distribution. Use the differences (ten observations per sample) to compute spatial variance, s_d^2 (Table IV:23:2) (see Chapter III, p. 54) and divide this by 2, because 2 traps were used for each observation. Plot spatial variance against mean population density on log × log paper (Fig. IV:23:1 *cf.* Fig. II:19, p. 23).

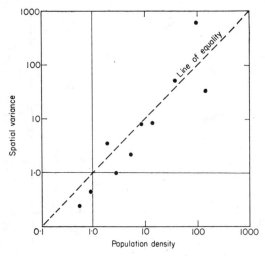

FIG. IV:23:1.   Spatial variance (s^2) is related to mean population density (m) by a power law which gives a linear regression in logs: $\log s^2 = \log a + b \log m$. For small flying insects s^2 and m are similar, approaching the line of equality which represents randomness.

Draw in the 45° line of equality, through the origin, along which variance and mean are equal. This line represents randomness in the spatial disposition of the insects.

Results:

Although there is great variability between samples, in general the variance remains similar to the mean.

Conclusions:

Small flying insects are distributed almost at random in the air, away from the immediate vicinity of vegetation. This means that they have little control

TABLE IV:23:2

Calculation of mean population density and spatial variance from the data in Table IV:23:1

$$n_d = 10, \quad n_c = 20, \quad S_d^2 = \frac{\sum (d^2) - \dfrac{(\sum d)^2}{n_d}}{(n_d - 1)}$$

Sample No.	1	2	3	4	5	6	7	8	9	10
$\sum d$	12	5	24	22	24	35	46	97	317	130
$\sum (d^2)$	22	7	120	66	98	267	364	1869	22,953	2296
$(\sum d)^2$	144	25	576	484	576	1225	2116	9409	100,489	16,900
$\dfrac{(\sum d)^2}{n_d}$	14	2·5	57·6	48·4	57·6	123	212	941	10,049	1690
$\sum (d^2) - \dfrac{(\sum d)^2}{n_d}$	8	4·5	62·4	17·6	40·4	144	152	928	11,904	606
Sd^2	0·89	0·50	6·9	1·96	4·49	16	17	103	1321	67·3
$Sd^2/2$	0·45	0·25	3·5	1·0	2·2	8·0	8·5	51	660	33·7
$\sum c$	18	11	38	56	104	171	272	765	1891	2972
$\bar{c} = \dfrac{\sum c}{n_c}$	0·90	0·55	1·9	2·8	5·2	8·6	13·6	38·3	94·6	148·6

over their aerial environment and it is generally true that the smaller the organism the less control it has over its environment; Man has the most.

Whilst aggregations of animals or plants are easy to recognize and are quite familiar, random distribution is not only difficult to recognize but is also unfamiliar and it requires considerable statistical effort to demonstrate it.

EXERCISE 24. AGGREGATION AND MIGRATION

Introduction:

Although small flying insects are often randomly distributed by air currents (Ex. 23), random distributions are otherwise rare. Most animals have a tendency to aggregate, either actively, by collecting together, or passively, the young accumulating around their parents. The degree of aggregation is characteristic of the species, like many other behaviour patterns. It may also be characteristically different in certain generations or age groups (see p. 23).

Another behaviour pattern characteristic for different species is migration. Some species are commonly found flying strongly far from their known habitats, like the Monarch Butterfly which regularly migrates North and South in North America and occasionally appears in Britain, where its food plant does not grow. Other species are rarely seen outside the shelter of the vegetation upon which they feed.

Together, *aggregation and migration determine the spatial distribution of populations within the area where the environment permits life.* For example, the migrations of men have distributed human populations over much of the world, but the tendency to aggregate results in very irregular distribution in the area occupied, with concentrations in cities. Aggregation and regular migration can both be clearly seen in the life cycle of the Black Bean Aphid, *Aphis fabae* (Fig. II:27, p. 33) in U.K. and the Sunflower Aphid, *Aphis helianthi* in U.S. Pea Aphid, *Macrosiphum pisum*, occurs in both regions on alfalfa, peas and beans, but must be handled more carefully because it is easily disturbed (see p. 29).

Hypothesis:

Aggregation and migration are species characteristics determining spatial distribution.

Apparatus:

Broad bean seeds.

Site:

Any convenient plot of land about 20 m. square.

Time of Year:

May, June and July.

Procedure:
Sow the beans, two at a time, 0·5 m apart in rows 1 m apart, in late March. If both seeds germinate, remove the weaker plant in April. Bean aphids will migrate into the plot and winged migrants appear in the crowns of the plants in May.

As soon as winged aphids are seen on the plants, cut off 10–20 alternate plants from one row. Take indoors in boxes or bags without disturbing the

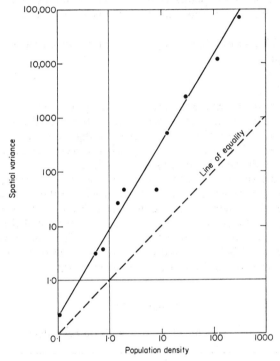

FIG. IV:24:1. Bean aphids are highly aggregated on bean plants and the spatial variance is much greater than mean population density, but the power law, $s^2 = am^b$, still represents the relationship (see Fig. IV:23:1).

aphids. Count the aphids on each stem separately. Every 2 or 3 days, collect another sample row and repeat until ten rows have been sampled and recorded (Table IV:24:1).

Calculate the mean and variance of each sample (see Ex. 23, p. 175 and Chapter III, p. 54).

Plot $s^2 \times m$ on double log paper. Draw a straight line through the points and also draw in the 45° line of equality through the origin (Fig. IV:24:1). Compare the resulting scatter with Fig. IV:23:1 and Fig. II:19, p. 23.

When counting the aphid colonies early in the season, note that most of the aphids are wingless except for the original winged migrants. Later when the samples have been taken, observe the colonies on the remaining plants. When the colonies grow larger in July, new winged adults will appear.

Results:

The winged adults do not accumulate on the plant, but fly away as soon as their wings are dry. About 09.00 h these migrants can be seen flying up into the air. Some of the plants may be killed by aphids in years when the colonies grow large, but note that the aphids migrate whether or not the plant is dying.

Figure IV:24:1 shows that s^2 is greater than m, which means these aphids are aggregated.

Conclusions:

The dispersal of the winged migrants, even from living plants, shows that migration is a part of the normal behaviour of the winged generation of the Bean Aphid; as in Man it is commonest in young adults.

As mentioned in Ex. 23, aggregation is easier to recognize than randomness and it can be recognized by looking at the colonies.

The statistical analysis described makes it possible to compare the degree of aggregation in different years or between species. If the exercise is repeated on one species the same slope of the line through the scatter will occur each time. The degree of aggregation is therefore characteristic of the species, and so is the tendency of the winged aphids to migrate. This ensures that new host plants will be found before the old ones are destroyed by too many aphids. Consider this problem in Man using census records.

TABLE IV:24:1

Sample counts of Bean Aphids on bean stems

Sample no.	Aphids per stem	No. of stems	M	S^2
1	2, 0, 0, 0, 0, 0, 0, 0, 0, 0, 0, 0, 0, 0, 0, 0, 0	17	0·11	0·23
2	7, 2, 0, 0, 0, 0, 0, 0, 0, 0, 0, 0, 0, 0, 0, 0, 0	17	0·52	3·01
3	6, 3, 0, 0, 0, 0, 0, 0, 0, 0, 0, 0	12	0·75	3·47
4	22, 2, 2, 0, 0, 0, 0, 0, 0, 0, 0, 0, 0, 0, 0, 0, 0, 0	18	1·44	26·7
5	27, 4, 0, 0, 0, 0, 0, 0, 0, 0, 0, 0, 0, 0, 0, 0	16	1·93	45·7
6	20, 15, 13, 13, 12, 12, 7, 6, 5, 1, 0, 0, 0	13	8·0	44·2
7	76, 64, 23, 21, 17, 12, 0, 0, 0, 0, 0, 0, 0, 0, 0, 0, 0	17	12·4	522·3
8	152, 89, 66, 7, 5, 4, 2, 0, 0, 0, 0	11	29·5	2587
9	318, 242, 196, 145, 119, 106, 39, 11, 6, 0	10	118·2	11,832
10	891, 653, 364, 280, 157, 238, 173, 124, 97, 38	10	301·5	72,402

EXERCISE 25. BREEDING POPULATIONS, AGGREGATION AND ACTIVITY

Introduction:

Earwigs are nocturnal insects and during the daytime collect to hide in dark crevices, usually near suitable food. Females, which overwinter as adults, lay 20–30 eggs in March and April, and these hatch in May. Adults of the new generation appear from July until early October. The eggs are laid in a shallow pit or sometimes down to 2·5 cm. below the ground and are protected by the female who partly covers them with her body. The nymphs remain in the nest for a few days after hatching. The mother may fight intruding insects. From June onward, daytime aggregations contain nymphs and adults, with increasing proportions of adults until October. Only *Forficula auricularia* (Common European Earwig) (Fig. B 14a, p. 275) is likely to be encountered in U.K. In U.S., *Euborellia annulipes* (Ring-legged Earwig) may be more common. The different instars can be distinguished by counting the number of antennal segments. Instars I, II, III, IV and adults have 8, 10, 11, 12 and 14 antennal segments (see p. 275), respectively, in *F. auricularia* (Fig. B 14c).

Hypothesis:

Aggregation by breeding and feeding animals is accentuated by hiding during periods of inactivity.

Apparatus:

Stoppered tubes or pill boxes, forceps, a white sheet. A spray gun, or aerosol, containing pyrethrum, will kill the insects rapidly. Hand lens or dissecting microscope.

Sites:

Cracks and crevices in fences, logs, refuse tips, beneath loose stones and tiles, door lintels, loose bark, ivy-covered tree stumps; flower petals, potato peelings, etc.

Time of Year:

July to October.

Procedure:

Record the site of aggregation and in the laboratory count the numbers of males and females (Fig. B 14b, p. 275) and each instar nymphs (Table IV: 25:1). For aggregations above ground level place the sheet beneath the site before disturbing the earwigs.

Compute percentage age structure in the total population and mean size of aggregation in each habitat.

Results:

From Table IV:25:1, the size of aggregations differs in different sites but differs more between habitats. The significance of differences between means for different habitats may be tested statistically, but more material should be collected than is given under "bricks" before doing so (see Chapter III, p. 103). There is no significant difference between numbers of males and females.

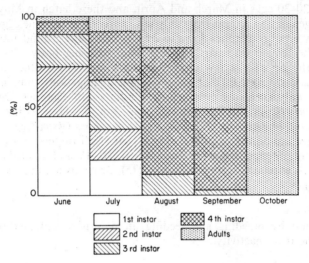

FIG. IV:25:1. Age structure in typical aggregations of earwigs (*Forficula auricularia*) from June to October in England.

TABLE IV:25:1

Numbers, instars and sex of Earwigs, *Forficula auricularia*, in different habitats

Site	Under loose bark on posts						Under dead tree trunk			Under bricks			Age structure	Sex
Sample no.	1	2	3	4	5	6	1	2	3	1	2	Σ	%	%
Stage														
Instar 1	0	0	0	0	0	0	0	0	0	0	0	0	0	
Instar 2	0	0	0	0	0	0	1	0	0	0	0	1	0·2	
Instar 3	13	2	0	0	3	31	25	0	0	3	0	77	12·2	
Instar 4	26	5	30	22	65	143	121	3	24	3	0	442	69·9	
Total nymphs	39	7	30	22	68	174	147	3	24	6	0	520		
Adult males	5	0	6	3	8	7	12	0	12	0	1	58		51·8
Adult females	12	1	7	1	6	3	13	5	8	1	1	54		48·2
Total adults	17	1	13	4	14	10	25	5	20	1	2	112	17·7	100·0
Total numbers in aggregations	56	8	43	26	82	184	172	8	44	7	2	632	100·0	
Mean size of aggregations				66				72			4·5	57·5		

Conclusions:

Earwigs aggregated to breed and feed. The size of aggregations depended upon the site and the local population. Population density was greater near food, hence habitats distant from a food supply contained smaller aggregations. Earwigs do not feed during the daytime, so the aggregations were created partly by the need to hide and were encouraged by common choice of damp, dark holes and crevices. However, if there were no tendency to aggregate, individuals would not accept a site already occupied and the daytime population would be more dispersed.

Riders:

Examination of known sites of aggregation at night will show that activity produces dispersal.

Repeat the exercise in different months and make a percentage histogram of the seasonal cycle of population structure (Fig. IV:25:1).

EXERCISE 26. AGGRESSION

Introduction:

Aggressive behaviour between individuals of the same or different species is common. Social insects often fight vigorously to defend their nests from other species. Honeybees deter intruders attempting to enter the hive by stinging them. They can be induced to sting cotton wool spheres wrapped in muslin displayed near the hive, especially if these are jerked or impregnated with odour to simulate an aggressor. Bee's stings have backward-sloping barbs, so when a soft object is stung the sting remains embedded in it (p. 21).

Hypothesis:

Behaviour patterns result from combinations of basic behavioural elements, such as aggressiveness, that are released by specific situations.

Time of Year:

April–September when bees are flying.

Site:

Anywhere with a beehive.

Apparatus:

Spheres of cotton wool about 2 cm in diameter, wrapped in black muslin; protective clothing for the observers.

Procedure:

(a) *Aggression stimulated by threatening behaviour*

Hold one sphere motionless on a wire 5 cm in front of a hive entrance. Suspend another sphere from a cotton thread and jerk it up and down 15 cm further along the hive entrance. Leave the spheres in position until 4–8 bees have stung them. Repeat the trials about 20–30 times. (One trial could be done by each member of a group.)

Count the number of stings in the motionless and the jerked spheres for each trial and tabulate results (Table IV:26:1).

(b) *Aggression stimulated by apparent presence of natural enemies*

Place a number of experimental spheres in glass dishes with small mammals for 1–3 hr before presenting them to a colony of bees, together with clean, control spheres, or, rub experimental spheres with human sweat from the forehead and control spheres with a few drops of distilled water. Attach the spheres, alternating experimental and control spheres, to the four corners of a piece of board 15 cm. square. Jerk the spheres on front of the hive entrance until a total of 4–8 bees have stung in each trial. Repeat 20–30 times and tabulate results (Tables IV:26:2 and IV:26:3).

Conclusions:

The movement of foreign bodies or the suspected presence of natural enemies (voles, shrews or men) stimulates the release of the stinging response in honeybees.

Riders:

Test the effect on bees of different coloured spheres, previously stung spheres, and spheres moving at different speeds.

TABLE IV:26:1

The effect of movement on the releasing of the stinging response of honeybees

Jerked spheres stung more	29 trials
Motionless spheres stung more	1 trial
Both spheres stung equally	0 trials
Mean no. of stings in jerked spheres	5·0
Mean no. of stings in motionless spheres	0·3

TABLE IV:26:2

The effect of shrew and vole scent on the releasing of the stinging response of honeybees

	Experimental spheres kept with	
	Voles	Shrews
Experimental spheres stung more	19 trials	24 trials
Control spheres stung more	5 trials	1 trial
Both spheres stung equally	1 trial	0 trials
Mean. no. of stings in experimental spheres	2·24	4·00
Mean no. of stings in control spheres	1·00	0·40

TABLE IV:26:3

The effect of human sweat on the releasing of the stinging
response of honeybees

Experimental spheres stung more	21 trials
Control spheres stung more	2 trials
Both spheres stung equally	2 trials
Mean no. of stings in experimental spheres	3·64
Mean no. of stings in control spheres	0·64

EXERCISE 27. DISTRIBUTION IN TIME

Introduction:

The total numbers of animals on meadow or waste-land vegetation is
generally similar during the day and night, but the composition of the fauna
varies immensely. This exercise illustrates the changes that occur within a
24 hr cycle and shows how different groups of animals behave. It also emphasizes the necessity to sample at a fixed time during the 24 hr if the faunal
composition is to be compared on different days.

Hypothesis:

Animal activity is distributed throughout the available time.

Apparatus:

Sweep net, pooter, specimen tubes, and forceps.

Site:

Meadow, wayside or waste-land vegetation.

Time of Year:

April to October, in fine weather.

Procedure:

The samples should be taken once between 1100 h and 1600 h and
repeated 1–3 hr after sunset. Traverse the area taking ten sweeps at first
and collect and count the invertebrates from the net. Estimate, then make,
the number of sweeps necessary to collect 500–1000 animals. Sort and identify
these indoors using keys in Chapter VI. Include families represented by no
more than five specimens, together with the unidentified insects, in the
"others" category. Such small samples are too liable to chance error to be
reliable. Tabulate the results under order and family (columns 1, 2), total
catch by day and night (columns 3, 4) and the ratio day catch/night catch
calculated from logarithms as described in Chapter III (p. 68), so that the
taxa are listed in sequence with the greatest ratio first (Table IV:27:1).

Results:

From Table IV:27:1 the total catch was almost equal by day and night, but many taxa showed large differences in distribution. Dermaptera, Crambidae, Nabidae, Staphylinidae, Culicidae and Araneida were most commonly caught at night; Miridae, Cercopidae, Jassidae, Thysanoptera, Coccinellidae, Parasitica, Apoidea, and Chironomidae were found by day.

Animals do not just disappear by either day or night, but redistribute themselves. The Dermaptera, Crambidae, Nabidae, Culicidae and Araneida were all active at night, but hidden at the base of vegetation, on trees or in buildings by day. Syrphids were active by day, but too fast to be caught in a sweep net. They were caught at night when they rest on the tops of vegetation. Bumble bees (Apoidea) were also active by day, but were more easily caught than Syrphids. At night bees rest too, but protected in their nests. Chirono-

TABLE IV:27:1

Anthropods collected in 1000 sweeps in a flowering meadow in August

Column 1	2	3	4	5
Order	Taxon	Day	Night	Ratio day/night
Hymenoptera	Apoidea	8	0	8·9
Hymenoptera	Parasitica	87	9	8·7
Collembola		95	18	5·0
Thysanoptera		30	6	4·4
Hemiptera	Miridae	88	21	4·1
Diptera	Chironomidae	17	5	3·0
Coleoptera	Coccinellidae	41	16	2·5
Hemiptera	Jassidae	41	19	2·1
Hemiptera	Cercopidae	35	17	2·0
Hemiptera	Aphididae	8	4	1·8
Diptera	Acalypterates	50	33	1·5
Hemiptera	Delphacidae	21	15	1·4
Coleoptera	Cantharidae	101	83	1·2
Others		91	80	1·1
Diptera	Muscidae	48	47	1·0
Coleoptera	Curculionidae	5	6	0·85
Diptera	Dolichopodidae	12	21	0·59
Lepidoptera	Crambidae	2	12	0·23
Dermaptera		1	8	0·22
Hemiptera	Nabidae	64	324	0·20
Diptera	Culicidae	1	9	0·20
Diptera	Syrphidae	17	118	0·15
Araneida		0	7	0·13
Coleoptera	Staphylinidae	5	53	0·11
Total		868	931	

midae were probably caught resting by day and at night were flying too high to be swept. Parasitica, Coccinellidae, Cercopidae, and Jassidae were caught when active by day but not at night when they were lower down the stems of the vegetation. Cantharidae, by contrast, were also active by day, but stayed on the flowers at night when they were caught equally easily. This is probably true also for Delphacidae, Aphididae, Curculionidae, Dolichopodidae, the Acalypterates and Muscidae. The Thysanoptera were active by day and probably stay in flower heads at night also, but then they hide deep in crevices and are difficult to dislodge.

Conclusion:

By spreading out its activity over the available time, as well as the available space, the whole community of animals makes more effective use of the environment. Motorists and industries increasingly do the same.

EXERCISES 28–30. BEHAVIOUR IN A CHANGING ENVIRONMENT

Introduction:

Animals are so closely adapted to their environment that they can only with difficulty adjust to permanent changes in it. Temporary fluctuations in the environment are, however, quite normal and the behaviour of animals and plants is adjusted to them. The physical environment is largely inflexible and few animals, except man and on a smaller scale social insects, can do much to affect it. In contrast, the biological environment is flexible on an evolutionary scale and whilst one species is making adjustments to its environment, the other animals and plants which contribute to that environment, are also adjusting to it. Thus an insect species and the flower on which it feeds evolve as an interdependent partnership. This involves synchronization of annual and daily activity, and specialization of behaviour of the individual insect, the population and the species (see p. 25).

EXERCISE 28. THE BEHAVIOUR OF INDIVIDUALS

There are many necessary functions within a social colony such as a beehive and these are most efficiently performed when individuals specialize in certain jobs. This is commonly referred to as "division of labour". Perhaps the most specialized job is egg laying, performed by the queen. Less specialized are the separate jobs of pollen and nectar collection, within the general function of foraging. A single individual may do several such jobs during its adult life, but it usually does only one at a time, thereby synchronizing most effectively the running of the colony.

Hypothesis:

Invertebrate behaviour patterns are usually stereotyped but the behaviour of individuals may be modified to suit the requirements of social groups.

Apparatus:

A stop-watch, or watch with a second hand.

Site:

As large a plot of Field or Garden Beans or Hairy Vetch as possible.

Time of Year:

On summer days when bees are foraging.

Procedure:

It is best to work in pairs to make these observations, one person to watch and one to record.

Follow individual honeybees foraging on the crop for as long as possible and record whether they visit the extra-floral nectaries (*a*), the flower bases (*b*), enter the flower (*c*) or whether they combine these different types of visits. Time the visits to each flower with a stopwatch. Construct a table for bees seen visiting ten or more flowers to show constancy of behaviour (Table IV:28:1) and calculate the mean time spent on each type of visit.

Results (from Field Beans):

From Table IV:28:1, 86% of the bees were constant to one type of behaviour and most of those with varied behaviour combined visits to extra-floral nectaries (*a* type) with feeding through the holes at the flower bases (*b* type). Few visited extrafloral nectaries and entered flowers. Honeybees that entered flowers, mostly scrabbled on the anthers to collect pollen and pushed their tongues towards the nectaries.

Most time was spent on *c* visits (mean 12 sec), less on *b* visits (mean 8 secs) and least on *a* visits (mean 5 secs).

Conclusions:

Individual behaviour tends to become stereotyped over short periods and to occupy time appropriate to the task. This is probably less true of vertebrates, especially Man, in which greater mental development produces added distractions and complications.

TABLE IV:28:1

Number of bees making various types of visits to bean flowers

| | | | Types of visit* | | | |
	a	*b*	*c*	*a* and *b*	*a* and *c*	*b* and *c*	*a, b* and *c*
22 June	7	20	12	0	0	4	2
23 June	3	25	7	3	0	0	0
29 June	18	14	8	10	0	0	0
30 June	23	11	2	3	0	2	0
Totals	51	70	29	16	0	6	2

a, extrafloral nectaries; *b*, holes in flower bases; *c*, entering flowers.

EXERCISE 29. The Behaviour of Populations

The extrafloral nectaries continue to secrete nectar after secretions from the floral nectaries have ceased and little pollen remains in the old flowers. Hence the foraging behaviour of the whole population must be adjusted to a changing food supply.

Hypothesis:

Seasonal changes in the environment produce corresponding changes in population behaviour, even though the behaviour of individuals remains constant.

Apparatus:

None.

Site:

As large a plot of Field or Garden Beans or Hairy Vetch as possible.

Time of Year:

Observations should be taken for several weeks over most of the beans' flowering period.

Procedure:

Compare the number of bees making *a*, *b*, or *c* type visits (see Ex. 28) on days at the beginning of the flowering period, with the number towards the end of the period. Record the number of bees entering flowers which have pollen in the corbiculae, on their hind legs (Table IV:29:1).

TABLE IV:29:1

Changes in the behaviour of the honeybee population during flowering

	Number of bees counted	Percentages of bees making visits of types*		
		a	*b*	*c*
				Pollen in corbiculae
				Present	Absent
22 June	49	12·2	53·1	34·7	0
23 June	349	16·0	58·2	22·1	3·7
29 June	425	58·3	33·3	7·8	0·7
30 June	267	79·8	14·6	4·9	0·7

*a, extrafloral nectaries; *b*, holes in flower bases; *c*, entering flowers.

Results (from Field Beans):

From Table IV:29:1, the percentage of honeybees visiting the extrafloral ' nectaries increased and the percentage of other types of visits decreased, towards the end of flowering.

Conclusions:

The tendency of individual bees to have stereotyped foraging behaviour helps to keep the proportion of each type of foraging visit constant, but the colony succeeds in adjusting the number of visits to the seasonal availability of food. These changes in type of visit with season, may result partly from changes in the behaviour of individuals but mainly from new bees taking to the currently most needed occupation.

EXERCISE 30. The Behaviour of Species

Honey bees and bumble bees rely upon flowers for food and are responsible for much of their cross-pollination, without which fruits and seeds of many plants would not set. The flowers are pollinated as the bees probe the nectaries at the base of the corolla tube and there is a relationship between the length of the corolla tube and the length of the tongue of the bumble bees that visit them. The different species of bumble bee behave characteristically when visiting flowers; the shorter-tongued species often bite holes in the corolla tube to steal nectar and are thus less likely to pollinate the flowers than the long-tongued species which enter the flower normally.

Hypothesis:

The behaviour of species evolves with their environment and is inseparable from it.

Apparatus:

Tubes, a net, a ruler scaled in mm and a razor blade.

Site:

Observations may be made on bumble bees visiting garden or wayside flowers.

Time of Year:

There is a sequence of species and castes appearing through the spring and summer. In U.K., queens of *Bombus pratorum* appear first in late March or early April; in April and May, *B. pratorum* workers, and queens of *B. lucorum*, *hortorum* and *agrorum* are common, followed by the workers from June onwards. *B. agrorum* workers are usually latest.

Bees may be studied on any given day but more comprehensive data will be collected over a period as different flowers, bee species and castes appear.

Procedure:

Collect ten specimens of each flower being visited by the bees and measure the length of the corolla tubes. In flowers in which the tube gradually converges to the base, assess how far the bee is able to push its head into the tube and measure the distance from there to the base. On some flowers the measurements can be made externally; with others it is easier to split them longitudinally with a razor blade. Calculate the average length of the corolla tube for each flower species.

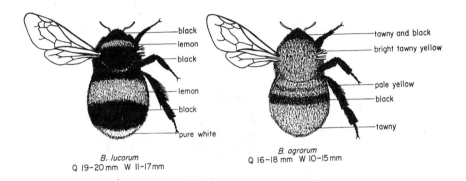

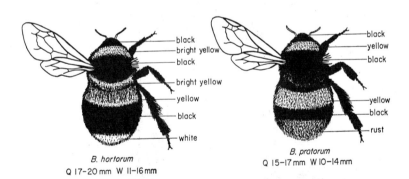

FIG. IV:30:1. Queens of four British species of Bumble Bee; workers of each species have similar colours and patterns but are smaller. The length of queens (Q) and workers (W) is given beneath each figure.

Count and identify the queens and workers (Fig. IV:30:1) visiting different flowers. With practice this should be possible without catching the bees, but providing they are held by the wings, when they cannot sting (Fig. IV:30:2) they may be caught for examination, then released.

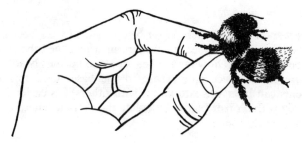

FIG. IV:30:2.   Method for holding bumble bee to avoid being stung.
(Redrawn after Sladen.)

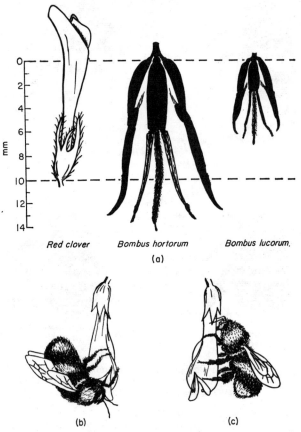

FIG. IV:30:3.   (a) Tongues of a long-tongued species of bumble bee (*Bombus hortorum*)
and a short-tongued species (*B. lucorum*) compared with a Red-clover flower. *B. hortorum* could reach the nectaries via the corolla tube [as in (b)] but *B. lucorum* must rob
the nectaries by biting through the corolla tube near the base (c).
(Redrawn after Meidell.)

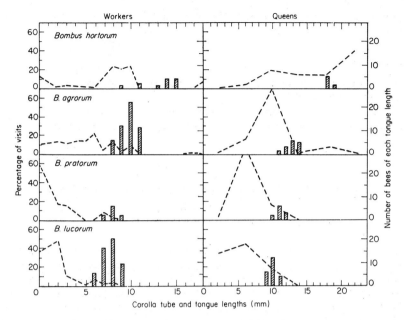

FIG. IV:30:4. Dotted lines show the percentage of visits paid by workers and queens to flowers with corolla tubes of different lengths. Tongue lengths are shown as histograms.

Collect about four workers of each species to measure their tongue lengths (Fig. IV:30:3a). *Bombus hortorum* and *agrorum* must be killed before they can be measured, but if female *lucorum* and *pratorum* are kept in semi-darkness, they may be induced to drink syrup from a capillary tube graduated in mm fixed in the apex of a paper funnel. As they extend their tongue into the tube, the tongue-length can be measured. Avoid collecting too many queens; only one sample should be taken by each class. The tongue of one queen could be measured to illustrate a typical example. The figures for queens quoted on Fig. IV:30:4 may then be used to relate to the lengths of corolla tubes.

Note also how the bees in the flowers collect nectar, especially *B. lucorum* and *B. hortorum*. Record the number of "robber" nectar visits made to flowers with corolla tubes of different lengths.

Results:

The short-tongued bees, *B. lucorum* and to a lesser extent *B. pratorum* are mostly the species which bite the bases of the corollae for "illegal" collection of nectar (Fig. IV:30:3b, c). *B. hortorum* always visits flowers by the obvious and correct entrance, but *B. lucorum* is an opportunist and obtains nectar by various methods. The difference in behaviour between a normal visitor and a thief may be seen most easily on the long-spurred *Aquilegia* where *B. lucorum* lands on the base of the flower, then runs away from the normal opening, down the spur, to bite the tip.

Conclusion:

The foraging behaviour is different in each species of bee and is adjusted to different species of flowers.

Since many flowers depend for their pollination on the visits of certain species of insects, the flower and the insect must have evolved together.

EXERCISE 31. SYNCHRONIZATION OF ACTIVITY AND ENVIRONMENT

Introduction:

Hive bees collect honey and pollen from flowers for food and, when visiting bean flowers, may pollinate them. Only bees that enter the flowers touch the anthers and stigma, although bees collecting nectar through holes previously bitten by bumble bees at the base of the corollae (see Ex. 30) may pollinate self-fertile flowers indirectly, by shaking pollen from the anthers on to the stigma. Honeybees visiting extrafloral nectaries will not pollinate. The abundance and concentration of floral and extrafloral nectar and of pollen varies with time of day (see p. 25).

Hypothesis:

Behaviour patterns become adjusted to synchronize activity most efficiently with availability of resources.

Apparatus:

Three different colours of quick-drying paint and a brush.

Site:

As large a plot of Field or Garden Beans or Hairy Vetch as possible.

Time of Year:

Fine days during the flowering period, which may extend from late May until early July for different varieties and seasons.

Procedure:

The time of maximum pollen availability may be determined by observations on bean or vetch flowers made between 8:00 a.m. and 6:00 p.m. Examine ten chosen plants at hourly intervals for seven days and paint the standard petal of each flower that has appeared in the previous hour, using a different colour paint each day.

Bean flowers usually open on three successive days and close on the first two nights, vetch flowers on 2 to 6 days. Using data collected on the third day, when the ages of all the flowers are known, plot the time at which flowers of different ages open throughout 1 day (Fig. IV:31:1). For 15 min of each hour during the day record the number of bees (a) visiting the extrafloral nectaries, (b) the flower bases and (c) entering the flowers; plot as Fig. IV:31:1.

Results (from Field Beans):

From Fig. IV:31:1, 2-day-old flowers opened in the morning; 1-day-old about noon or early afternoon and fresh flowers in mid-afternoon. Thus most pollen was available in mid-afternoon when the fresh flowers are opening. Similarly more bees entered the flowers at this time of day.

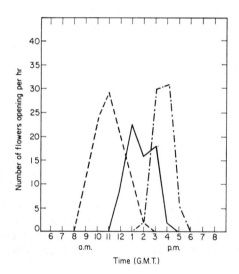

FIG. IV:31:1. Time of opening of *Vicia faba* flowers of different ages. Fresh flowers —·—·—; one-day-old flowers ————; two-day-old flowers – – – –. (After Synge.)

Conclusion:

The bees enter the flowers most frequently when pollen is most freely available.

EXERCISE 32. ANNUAL POPULATION CYCLES

Introduction:

Unlike homoiotherms, which can control body temperature and keep active throughout the year, poikilotherms depend upon heat from the sun to keep them warm enough to be active. Insects usually pass the winter in an inactive state in cool climates and have a life history adapted to this. There may be more than one generation whilst the temperature is high enough for development and activity, but a single generation each year is more common. Individuals of this generation may be active any time from early spring to late autumn. There is, however, a preponderance of species active in summer and this produces an annual cycle of total numbers of insects flying and also of numbers of species active (see p. 28).

Hypothesis:

In temperate climates most species of insect have one generation per year, timed so that the adults are active in summer. The diversity of species in the flying population is greatest in summer.

Apparatus:

A light trap (Chapter V, p. 250) and means of identifying Macrolepidoptera, e.g. South (1961) or Holland (1913) (see p. 377) or a reference collection.

Site:

Anywhere outdoors.

Time of Year:

Throughout the year.

Procedure:

Make a chart for each of the 25 most common moths, marking each night on which a catch occurred (Fig. IV:32:1). Arrange the species in seasonal order of appearance.

Identify all Macrolepidoptera and work out the Index of Diversity for each month (see Chapter III, p. 107), tabulate the results (Table IV:32:1) and plot α against month (Fig. IV:32:2). If not all the Macrolepidoptera can be identified (U.S.) select one large family, e.g. Noctuidae, and use this for the diversity study.

Results (from S. England):

From Fig. IV:32:1, the first species, *Selenia bilunaria*, has two generations. There are fewer individuals in the first generation in April and May than in the second in July and in some years the first generation may be too small to detect. The second species, *Plutella maculipennis*, the Diamond Backed Moth, a Microlepidopteran and not included in South (1961), is included here because it was caught in every month from April to October though no separate generations were distinguishable. *Xanthorhoe fluctuata* and *Opisthograptis luteolata* are similar. In contrast, *Spilosoma lubricipeda* has one distinct generation. In all there are 17 single-generation species, four with two generations and four indeterminate.

Figure IV:31:2 shows that population diversity is negligible through the winter, when the numbers of insects flying is also small. It reaches a maximum of 22·5–28·0 in July.

Conclusions:

About three-quarters of the common species listed had one generation per year; the remainder had either two generations or an overlapping series of

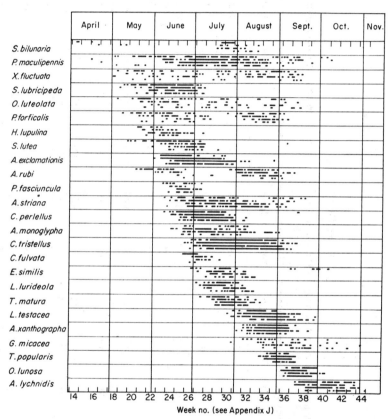

FIG. IV:32:1. Dates of occurrence of 25 species of Lepidoptera in a light trap at Rothamsted Experimental Station, England, for 1933, 1934, 1935 and 1936. (After Williams.)

generations that makes accurate determination difficult. There was a sequence of species in flight from May to October, with most in July and there was a tendency for those species that have one late generation, to fly for a shorter period of time. Why is this?

Diversity was also much greater in July with an average Index of 26, and

TABLE IV:32:1

Seasonal changes in numbers of individuals, species and Index of Diversity in captures of Lepidoptera in 8 yr at Rothamsted

		Jan.	Feb.	Mar.	Apr.	May	June	July	Aug.	Sept.	Oct.	Nov.	Dec.
1933	no.	—	—	26	71	328	701	1003	719	407	214	89	27
	spp.	—	—	6	12	39	77	94	61	34	14	7	2
	α	—	—	3·0	3·5	11	22	25	16	8·8	7·0	2·0	<1
1934	no.	0	13	13	55	136	545	1082	650	528	152	181	28
	spp.	0	1	4	8	29	56	100	54	32	14	8	2
	α	0	<1	<1	2·9	12	18	28	14·5	7·5	4·0	2·0	<1
1935	no.	5	1	25	24	127	1083	3164	1688	647	171	29	2
	spp.	4	1	8	9	27	69	122	86	36	18	2	2
	α	<1	0	5·0	5·1	11	16	25	10·5	8·0	5·0	<1	0
1936	no.	0	0	0	16	38	437	454	506	447	52	30	19
	spp.	0	0	0	5	16	47	69	47	34	11	3	2
	α	0	0	0	3·0	12	12·5	22·5	12·5	8·5	3·7	<1	<1
4-yr mean	no.	1·6	4·6	16	41·5	157·2	691·5	1425·7	890·7	507·2	147·2	82·2	19
	spp.	1·3	0·6	4·5	8·5	27·7	62·2	96·2	62·0	34·0	14·2	4·0	2·0
	α	<1	<1	2·0	3·6	11·5	17·1	25·1	13·4	8·2	4·9	1·5	<1

		1	2	3	4	5	6	7	8	9	10	11	12
1946	no.	—	—	—	—	83	310	580	1053	820	223	127	12
	spp.	—	—	—	—	20	41	77	63	34	21	7	2
	α	—	—	—	—	8·0	12	24	14	7·5	6·0	1·8	<1
1947	no.	1	0	20	171	163	371	957	960	751	283	80	297
	spp.	1	0	7	10	28	68	101	64	31	19	6	4
	α	0	0	4	2·2	12	23	28	17	6·5	4·5	1·5	<1
1948	no.	1	53	65	126	144	446	749	717	1297	161	83	64
	spp.	1	6	9	14	35	58	104	74	30	19	7	3
	α	0	1·5	3	3·2	15	17·5	30·5	21	5·2	6·0	2	1
1949	no.	6	5	71	95	122	517	881	2783	1566	97	90	56
	spp.	3	2	8	12	32	71	92	71	41	17	7	4
	α	<1	<1	2·5	3·5	13·5	21·5	26	12·5	8·0	5·8	2	1
4-yr mean	no.	2·6	19·3	52	130·7	128	411	791·7	1378·2	1108·5	191	95	107·2
	spp.	1·6	1·6	3·2	12·0	28·7	59·5	93·5	68	34	19	6·7	3·2
	α	1	1	3·2	3·0	12·1	18·5	27·1	16·1	6·8	5·6	1·8	1
8-yr mean	no.	2·1	11·9	34	86·1	142·6	551·2	1108·7	1134·4	807·8	169·1	88·6	63·1
	spp.	1·4	1·1	6·2	10·2	28·2	60·8	94·8	65	34	16·6	5·3	3·2
	α	<1	<1	2·6	3·3	11·8	17·8	26·1	14·8	7·5	5·3	1·6	<1

diminished in the less favourable climates of other months, being almost zero in winter. This decline is not just an effect of small catches. The values indicate that if uniform samples of 1000 insects were collected each month, the number of species represented would be about 95 in July compared with 19 in April and 30 in October. Species diversity in the flying population is greatest in summer.

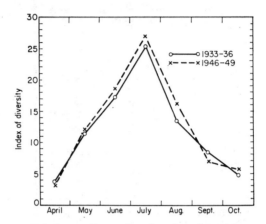

FIG. IV:32:2. Seasonal change in specific diversity of Lepidoptera at Rothamsted from captures in a light trap during two periods of four years each, 1933–6 and 1946–9. The diversities are calculated on the basis of the log series. (After Williams.)

EXERCISE 33. POPULATION DENSITY AND PARASITISM

Introduction:

Many species of animals are parasitized by other organisms. Man himself is host to a range of parasites including viruses, bacteria, Sporozoa, Mastigophora, Cestoda, Nematoda and many others internally, and mainly Arthropoda externally. Parasitic organisms affect their hosts in various ways and sometimes eventually kill them. They are thus an important factor in the regulation of animal numbers. This natural relationship is of course often disrupted in Man by therapeutic medicine (see p. 20).

In U.K. the Agromyzid leaf mining fly (*Phytomyza ilicis*) is parasitized by eight species of chalcid wasps and one species of braconid wasp. The larvae of this small fly burrow in the mesophyll of holly leaves and produce large blotches or mines beneath the leaf surface. Sometimes more than half the leaves on a tree may be attacked. The life history of the fly and its most important hymenopterous parasites, *Chrysocharis gemma*, *C. syma*, *Sphegigaster flavicornis*, and *Pleurotropis amyntas* are summarized in Fig. IV:33:1. All stages except the adults may be found within the mines at the appropriate season.

Adult leaf miners lay their eggs at the base of the midrib on the underside of the leaf in June. The larva (Fig. IV:33:2d) remains hidden in the midrib, slowly eating its way forward until in September, October or November it leaves the central tissues and enters the soft, green outer parenchyma, which it eats away below the epidermis until a large, irregular mine is formed (Fig. IV:33:2a). The mine reaches maximum size in March. Before it pupates, the larva prepares a thin triangular area on the leaf cuticle (Fig. IV: 33:2b) against which will fit a hinged emergence plate on the puparium. The

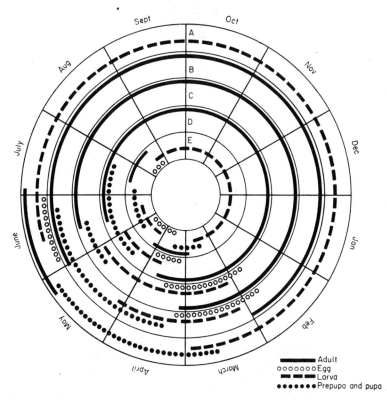

FIG. IV:33:1.   Diagram of the life history of the Holly leaf miner (*Phytomyza ilicis*) and four of its hymenopterous parasites, showing the different stages present in each month in the U.K. A—*Phytomyza ilicis*; B—*Chrysocharis gemma*; C—*Chrysocharis syma*; D—*Sphegigaster flavicornis*; E—*Pleurotropis amyntas*.

larva pupates inside the last instar skin which becomes the puparium. Its ventral surface is pressed against the epidermis and its anterior spiracles project through the attenuated area of the cuticle. The adult escapes from the puparium by pressing on the hinged plate which opens and breaks through the epidermis above it. A triangular raised flap or hole on the surface of the

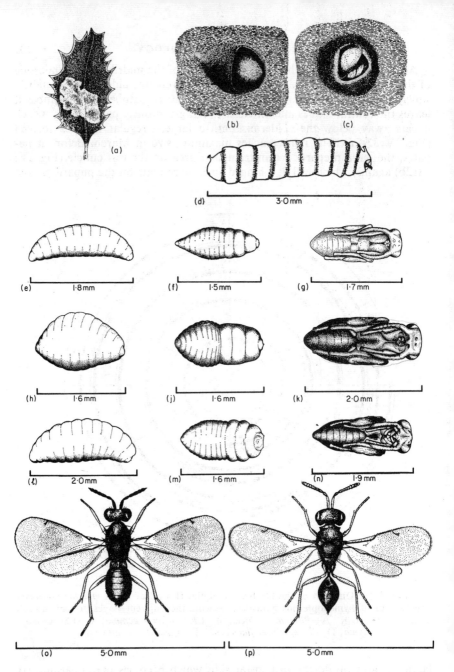

FIG. IV:33:2. Stages of *Phytomyza ilicis* and its common parasites. (After Cameron.) (a) Holly leaf mined by *Phytomyza ilicis*; (b) puparium of *P. ilicis* before emergence of adult from leaf; (c) puparium in leaf after emergence of adult; (d) larva of *P. ilicis*; (e) *Chrysocharis gemma*, mature larva; (f) *Chrysocharis gemma*, prepupa; (g) *Chrysocharis gemma*, pupa; (h) *Chrysocharis syma*, mature larva; (j) *Chrysocharis syma*, prepupa; (k) *Chrysocharis syma*, pupa; (l) *Sphegigaster flavicornis*, mature larva; (m) *Sphegigaster flavicornis*, prepupa; (n) *Sphegigaster flavicornis*, pupa; (o) *Chrysocharis gemma*, adult female; (p) *Sphegigaster flavicornis*, adult female.

leaf thus indicates that an adult fly has completed development successfully (Fig. IV:33:2c).

The most important parasite is *C. gemma* (Fig. IV:33:2o, e, f, g) which attacks the fly larvae. The adult parasite inserts a single egg through the leaf cuticle into the body cavity of a fly larva. Occasionally superparasitism occurs when two or more eggs are laid in the same larva. Attacked larvae appear flaccid and pale, dirty yellow compared with the turgid, bright, shiny, whitish-lemon, healthy larvae. The parasite larva feeds within the fly larva, and eventually kills it. It then forms a shiny jet-black pupa which *lies free inside the mine*. The adult parasite emerges from the leaf by a small, neat, round hole. The small mine contains no puparium, only the parasites' black, cast pupal skin, and is thus easily identifiable.

Sphegigaster flavicornis (Fig. IV:33:2p, l, m, n) is the second most important parasite, but it attacks the pupa of the fly. The adult parasite bores through the leaf cuticle and the tough skin of the puparium with its ovipositor. This takes about half an hour. It lays its eggs individually on the host pupa into which the parasitic larva will bore and feed and then pupate. The pupa of this parasite is black with a blueish tinge, except for antennae, wings and legs which are light, glassy brown.

The third most common parasite is *C. syma* (Fig. IV:33:2h, j, k) with a similar life history to *C. gemma* except that it attacks the fly pupa. Its pupa, *within the fly puparium*, is shiny black. All mines that have been occupied by these parasites have a small emergence hole through which the adult parasite escaped.

Pleurotropis amyntas is less common than the other parasites described but is unique among the leaf miners' parasites because it may be either a primary parasite in the pupa of the leaf miner or a secondary parasite (hyperparasite) on the miner's pupal parasite *S. flavicornis* or *C. syma*. When it behaves as a primary parasite in Spring on the dipterous pupa, its development is similar to *Sphegigaster*, but the generation of adults emerging from the fly puparium in July and August are hyperparasites and attack persisting puparia of *Phytomyza* which now contain only pupae of *S. flavicornis* or *C. syma*, all healthy flies having emerged before this date.

It is difficult to distinguish between these parasitic larvae when in the hosts' puparium, but Fig. IV:33:1 indicates the species and stage of parasite most likely to be present in mines at different times of the year. From September to early February any parasite within an unopened fly puparium will almost certainly be hyperparasitic *Pleurotropis amyntas*.

P. amyntas forms the fourth link in a parasitic food chain in which, contrary to the more familiar predatory food chain (p. 18), the successive links are smaller in size. The basic food material is synthesized by the holly leaf, *Ilex*. Within this leaf feeds the fly larva, *Phytomyza*, the second link. The larva of *Sphegigaster* forms the third link, feeding inside the pupa of *Phytomyza* and the summer generation larva of *Pleurotropis* constitutes the fourth link, feeding on the pupa of *Sphegigaster*. Thus the environment of *Pleurotropis* includes, not only the direct effect of climate on itself, but also, through its food supply, the indirect effects of climate on *Ilex*, *Phytomyza* and *Sphegigaster*.

Hypothesis:

Parasitism may cause measurable mortality in a population. Seasonal cycles of host and parasite are synchronized.

Apparatus:

Dissecting forceps and binocular microscope. Long-handled secateurs.

Site:

Any holly tree infested with *Phytomyza ilicis*: in U.K. there are few trees completely free from attack.

Time of Year:

The total mortality due to parasitism per generation of *P. ilicis* cannot be assessed until late May or June when the annual cycle is completed. This is the best time for the exercise but estimates of percentage parasitism at a given time may be made from November to June, using the life history chart (Fig. IV:33:1) to indicate which species and stages may be present in the mines. It is also instructive to sample at intervals throughout the year to follow the life-histories of the insects.

Procedure:

First estimate the proportion of leaves on the tree attacked by the leaf miner. Select twigs with about 20 leaves. Examine each leaf and record whether or not it is mined; collect mined leaves and discard the others. Continue examination until about 200 mined leaves have been collected. If 500 leaves have been examined before 200 mines have been found, estimate percentage infestation from these 500. Then continue collecting only mined leaves until 200 are available for analysis of mine contents. Try to select twigs from as wide a range of heights and aspects as possible.

In the laboratory open each mine and record the contents in the categories suggested in Table IV:33:1, with the help of the descriptions in the introduction and Fig. IV:33:2. Mines attacked by birds are torn open and may have roughly triangular beak marks on them.

Tabulate results under A, Infestation; B, *Phytomyza ilicis*; C, Parasites (Table IV:33:1). In A, express infestation as percentage of leaves with mines. In B, give a complete analysis of larval and pupal condition in *P. ilicis* and an assessment of bird damage and any other recognizable cause of mortality. In C, give the percentage parasitism, by different parasite species and stages, of both *P. ilicis* larvae and pupae separately. Give the total percentage parasitism of *P. ilicis* in larval and pupal stage.

Results:

Results of typical samples taken in early February and May are given (Table IV:33:1). The samples were taken in different years, so one is not a direct consequence of the other.

A. In February, over 60% of leaves were mined and in May nearly 40%.

B. In February, 75% of mines were occupied and 66% of these by healthy larvae; there were no healthy pupae. In May 71% of mines were occupied, but there were no healthy larvae; only 2% were occupied by healthy pupae. In February only 2% and 1% of mines were occupied by dead, but unparasitized, larvae and pupae, respectively; in May the figures were 45% and 15%. The percentage of empty mines was similar (25/29) in February and May. Of these, birds damaged between a third and a half on each occasion.

C. In February *Pleurotropis* accounted for nearly 40% of the little pupal

TABLE IV:33:1

Parasitism of *Phytomyza ilicis*

	February			May		
	Numbers	Percentages		Numbers	Percentages	
(A) *Infestation*						
Leaves with mines	216	63·5		174	39	
Leaves without mines	126	36·5		272	61	
Total leaves examined	342	100		446	100	
(B) *Phytomyza ilicis*						
Larvae parasitized	0	0		30	21	
Larvae unparasitized (hlthy.)	104	66		0	0	
Larvae unparasitized (dead)	2	2		65	45	
Pupae parasitized	8	5		6	4	
Pupae unparasitized (hlthy.)	0	0		3	2	
Pupae unparasitized (dead)	1	1		21	15	
Empty cases	42	27		19	13	
Total mines occupied	157 157	100	75	144 144	100	71
Mines damaged by birds	25	42·5		21	35	
Other empty mines	34	57·5		39	65	
Total empty mines	59 59	100	25	60 60	100	29
Total mined leaves	216	100		204	100	
(C) *Parasites*						
Chrysocharis gemma larvae	0	0		2	6·5	
Chrysocharis gemma pupae	0	0		28	93·5	
Total parasitized larvae	0 0	0	0	30 30	100	83·5
Pleurotropis amyntas larvae	3	37·5		1	16·5	
Chrysocharis syma larvae	5	62·5		5	83·5	
Total parasitized pupae	8 8	100	100	6 6	100	16·5
Total *Phytomyza ilicis* parasitized	8	100		36	100	

parasitism found. In May *Chrysocharis gemma* was the only parasite of the fly larvae and was itself mainly in the pupal stage. Parasites of the fly pupae were much the same as in February.

Conclusions:

In February most mines contained healthy larvae. Mines produced the previous year on leaves not yet shed contained empty puparia, 5% of which contained larvae of pupal parasites; 3 of these were the hyperparasite *P. amyntas* which had overwintered and 5 unusually early larvae of *C. syma* in early puparia formed after a mild winter. The presence of fly puparia and *C. syma* so early in the year illustrates how generalizations about life cycles become inaccurate in exceptional seasons.

In May, there were many fly larvae and pupae which had not been parasitized, but which had died from unknown causes. There were a few healthy pupae and fully vacated puparia from which flies had successfully emerged. The most common parasite was *C. gemma*, which had occupied 21% of the mined leaves. The combined percentage of pupal parasites was 4%. There is much variation in percentage parasitism by each species: *C. gemma* may attack most of the fly larvae, but pupal parasitism is usually much less.

Rider:

A class of students could compare different holly trees. There are often large differences in the size of infestations and the amount of parasitism between trees a few yards apart, perhaps due to the amount of sunlight falling on different trees, to local air currents influencing flight and oviposition of the tiny parasites and to small differences in texture between the leaves.

EXERCISE 34. PREDATION AND POPULATION CONTROL

Introduction:

Springtails (Collembola) and Mites (Acari) are common in soil. Under normal conditions predatory mites eat large numbers of Collembola and are one of the environmental controls maintaining the collembolan population at its normal level. The existence of such controls in an animal's environment is not usually evident, until one is removed (see pp. 20, 24).

Insecticides usually impose additional controls on insect populations; this is their purpose. However, ecological interactions are rarely simple and some insecticides are more specific than others so they can sometimes be used to remove existing natural controls. Aldrin, a chlorinated hydrocarbon, kills both springtails and mites and populations remain reduced for at least a year. DDT acts differentially, being more toxic to mites than to springtails. It can therefore be used to release the springtails from the environmental control exerted by the mites (see p. 18).

Hypothesis:

Populations released from the control exerted by predators may increase beyond their normal level.

Apparatus:

A proprietary brand of DDT dust (5% active ingredient) will give satisfactory results.

Use a sampling tool or tin to collect soil cores about 6 cm in diameter and
8 cm deep (Fig. V: 5, p. 253), polythene bags or tins to carry soil and one or
more Tullgren funnels (Fig. V: 6, p. 254) to extract the animals from it. Store
the animals in alcohol if necessary.

Site:

Four plots, each 2 m × 2 m, are required in a garden or on arable land.

Time of Year:

The experiment should extend over at least one year after the insecticide
has been applied to the soil. The best time for application is spring and
subsequent samples need to be taken at 2- or 3-month intervals.

Procedure:

Mark out four plots with paths between. Remove the vegetation and culti-
vate the soil to a depth of about 15 cm. Take three cores from each plot,
place them in a Tullgren funnel for about three days to extract the animals
and count the total number of Collembola and Acari. The springtails will
consist mostly of the small, elongate, white Onychiuridae without a springing
organ or furcula, the Isotomidae (Fig. B 31a, p. 281), and Sminthuridae
(Fig. B 31b), which live on the soil surface. All mites may be counted
together or divided into the hard, brown, sluggish Oribatidae (Fig. A 6c,
p. 265), which are phytophagous and the remainder which will consist mostly
of paler, long-legged, active predators (Fig. A 6a) whose food includes
springtails.

After the first sampling, sprinkle 3 oz (85 g) of insecticide evenly over two
of the plots and leave two as controls. (DO NOT inhale the insecticide or
contaminate the eyes.) Fork the plots to mix the insecticide thoroughly with
the soil. Repeat the sampling after 4–6 weeks and then every 3 months.
Grass and weeds may be allowed to grow on the plots after the insecticide
has been mixed.

The numbers of springtails and mites in the soil fluctuate throughout the
season, so it is best to express the results as the relative numbers in the
treated plots compared with the untreated, to eliminate the effect of these
fluctuations (Fig. IV:34:1).

Results and Discussion:

This exercise illustrates the short term effects of an insecticide on part of
the soil fauna, and the more complex long term effects which may ensue. The
initial decline in populations of both mites and Collembola was followed by
a striking rise above normal by the Collembola only. This is because
the DDT is more toxic to the mites than to the springtails, which
increased rapidly when their normal predators were suppressed.

The widely used chlorinated hydrocarbon insecticides (e.g., DDT) are highly toxic to many insects and other arthropods, and are often incorporated in pesticides or used as seed dressings. They provide crops with adequate protection against most pests in the season when applied, but some persist in the soil for years and it is these residues which may effect soil fertility through their effect on the soil fauna.

Conclusion:

When the population of predatory mites was reduced by insecticides, the population of their prey, the Collembola, increased.

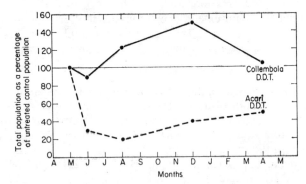

FIG. IV:34:1.   The effect of D.D.T. on populations of mites and springtails. (Redrawn after Edwards.)

Rider:

Other animal groups and insecticides may be compared in a similar way, but it must be emphasized that many of these chemicals are highly toxic to mammals, including men and *great care must be taken in their application and storage.*

EXERCISES 35–38. POPULATION DENSITY AND DISTRIBUTION OF SOIL ANIMALS

Introduction:

Animals invade each environment to the limit of their abilities. Their numbers and distribution are therefore determined by the gradients of those conditions most affecting them. The distribution of many inhabitants of the subterranean environment is affected partly by soil depth and structure, soil acidity, cultural practices and surface vegetation. Earthworm and symphylid populations are used in Exs. 35–38 to illustrate the effect of these factors on the density and distribution of soil animals (see p. 14).

Agricultural practice changes the natural vegetation, affects soil structure and thus has important effects on soil fauna and fertility. Soil fertility is closely related to the amount of organic matter incorporated into the soil, and much of the natural mixing of organic and mineral components is done by earthworms. They ingest soil, including mineral particles and organic matter, and eject the ingested material in the form of large quantities of castings which have a fine crumb structure. Some species deposit these on the surface and others below ground. The surface castings constitute a slow turnover of soil, the rate depending on the numbers, size and species present. In England the average turnover is about 11 tons per annum per acre, or 2 cm in depth every 10 years; the effect of casting within the soil is difficult to assess. Where the soil remains undisturbed for many years, the surface casting species eventually produce a superficial stone-free layer. Worm burrows enhance drainage, root penetration and aeration, and the tilth they help to create provides habitats for smaller soil fauna and micro-organisms.

EXERCISE 35. Soil Animals: pH and Population
Density of Earthworms

Hypothesis:
Some animals are intolerant of acid soils.

Apparatus:
A shallow wooden grid, 10 cm × 50 cm × 50 cm, a watering can, 40% formalin and a small spade are needed for sampling and equipment for measuring the pH of the soil, e.g. a B.D.H. colourimetric kit.

Site:
An acid soil, such as a heath, moor or coniferous woodland and a basic meadow, orchard or deciduous woodland soil.

Time of Year:
Spring to autumn.

Procedure:
Select contrasting soils and measure their pH with B.D.H. indicator. Add a few drops of indicator to a small amount of soil on a white dish, leave for a minute, tilt the dish and note the colour of the indicator. If the pH of the soils is sufficiently different, it should be detectable without further tests. For a more accurate test, add 0·5 ml of indicator to 1 ml of soil and compare the colours in a "Lovibond" comparator. The clay in suspension can be flocculated if necessary by adding barium sulphate solution. Table IV:35:1 relates pH to descriptions of soil (in an agricultural sense) in terms of acidity and for interest gives the critical pH levels below which certain crops will not grow successfully.

Chalky soils may have a pH over 8·0 and acid woodland down to about 3·5.

TABLE IV:35:1

Soil pH values and tolerant crops

pH	Description (in agricultural terms)	Crop	Critical pH
> 6·5	not acid	Red clover	5·5
6·0–6·5	slightly acid	Sugar beet	5·3
5·5–6·0	somewhat acid	Barley	5·2
5·0–5·5	acid	Wheat	5·0
4·5–5·0	very acid	Brassicae	4·9
below 4·5	extremely acid	Wild white clover	4·5
		Rye grass	4·3
		Oats, Potatoes	4·2

Extract earthworms from two to four randomly chosen areas at each site using the formalin method (see p. 218) and count and weigh the total earthworms extracted. Convert the figures into thousands and lb or kg per acre.

Results and Discussion:

On an acid peaty soil (pH 4·0–5·5) with coarse herbage and heather, estimates around 20,000–50,000 worms, equivalent to 100–200 lb per acre, were typical compared with up to half a million or more, equivalent to 500–3000 lb per acre, in a basic soil (pH 6·5–7·5). This formalin extraction method underestimates the numbers living in the upper layers of the soil (see Ex. 38) but earthworms are nevertheless strikingly more abundant in non-acid than in acid soils.

Conclusion:

Most earthworms are intolerant of acid soils. Note that acid soils with few earthworms frequently have a layer of raw humus on the surface, whereas in the presence of earthworms most undecomposed organic matter is fragmented and incorporated in the soil.

EXERCISE 36. SOIL ANIMALS: POPULATION DENSITY, SOIL TYPE AND SURFACE VEGETATION

This exercise investigates the effect of soil type and surface vegetation on the numbers of one group of animals, the Symphyla, living in soil.

The Symphyla are white, very active centipede-like animals, usually less than one centimetre long (Fig. A 9, p. 267). Do not confuse them with small white Collembola (Fig. B 31a, p. 281). They are common and widely distributed and usually feed on dead plant material, but may attack living rootlets.

Hypothesis:

The population density of soil animals depends on soil type and surface vegetation.

Apparatus:

The soil is most easily sampled with a sampling tool (Fig. V: 5, p. 253) but a tin or metal pipe about 6 cm diameter, that can be pressed or hammered into the soil to a depth of 8 cm to remove a soil core, will suffice. Polythene bags or tins are required to carry the soil, and large beakers or bowls in which the soil samples can be crumbled under water.

Site:

As wide a range of soil types and habitats as possible should be compared, from medium or heavy clays to loamy sands and fallow soil, grassland, forest litter, garden and greenhouse soils. Symphylids are likely to occur at about half the locations selected.

Time of Year:

Numbers in the surface layers of the soil are greatest in spring and autumn, the best times for sampling.

Procedure:

Collect a sample of five soil cores from each selected location. Place each core in a separate container. On a completely bare surface or one covered with vegetation it is important to take samples at random (p. 93). At sites where the ground is partly bare take two cores from the open soil and three from among the plant roots. In this way primary root feeders which are attracted to the plant roots, and saprophagous species feeding on decaying material around the roots, will both be collected and provide a representative estimate of the population. Classify the soil type at each site using the criteria and descriptions in Chapter V, p. 259. The method is subjective but with practice will give sufficiently accurate results to compare with the number of symphylids.

In the laboratory, drop the soil core into a vessel of water, carefully crumble it, stir gently and count the symphylids as they float to the surface where they can be picked off with a fine brush. For soils which produce an excessive amount of froth on the water surface, run off the floating matter into a 150 mesh sieve, wash under a stream of water to disperse the froth then refloat the debris on clean water when the symphylids may be picked off easily with a brush.

Calculate the total area of soil taken from each habitat and convert the counted number of symphylids into an estimate of millions per acre (1 acre = 4,047 sq. m).

Make two separate tables, listing the estimated numbers per acre under soil type (Table IV:36:1) and surface vegetation or cultivation (Table IV: 36:2). From these tables, make histograms to illustrate population density (Figs. IV:36:1 and IV:36:2).

Results:

Figures IV:36:1 and IV:36:2 show that symphylids were more common in various kinds of loams and less common in pure sand and heavy clay.

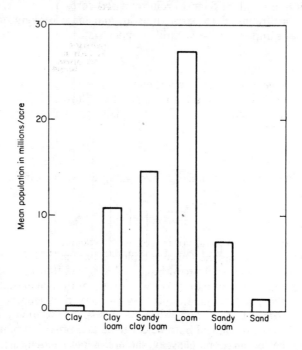

FIG. IV:36:1.   Populations density of Symphylids in different soils. (Redrawn after Edwards.)

Greenhouse soil was especially well populated and well-cultivated soil generally contained high populations. Some grassland and forest soils had lower populations.

Conclusions and Discussion:

Symphylids are most suited by cultivated loams and light clay soils. These soils often contain a high proportion of organic matter for food which is distributed more evenly in cultivated soils, especially greenhouse and garden soils, than in undisturbed ones. Clay loams hold more moisture than sandy soils which contain fewer symphylids because they are often dry and perhaps because the sand grains scratch their delicate cuticle causing the animals to lose water excessively. Few occur in very heavy clays probably because there are insufficient crevices.

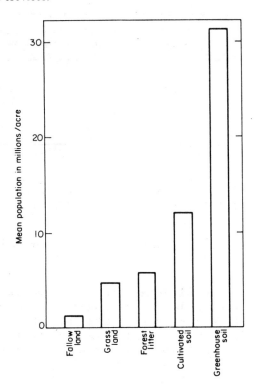

FIG. IV:36:2. Population density of Symphylids in soils under different cultivation and vegetation. (Redrawn after Edwards.)

TABLE IV:36:1

Populations of symphylids in different soil types

Soil type	No. of sites	Population mean	Range (millions per acre)
Clay	3	0·5	0·1–1·3
Clay loam	6 ·	10·8	1·4–21·3
Sandy clay loam	7	14·6	4·2–19:7
Loam	10	27·2	4·6–88·2
Sandy loam	5	7·3	3·3–10·8
Sand	2	1·2	0·7–1·7

TABLE IV:36:2

Population of symphylids in soils with different vegetation cover

Habitat	No. of sites	Population mean	Range (millions per acre)
Fallow land	3	1·3	0·5–1·7
Grassland	5	4·8	1·8–7·2
Forest litter	3	5·8	4·7–6·9
Cultivated soil	7	12·1	4·2–28·5
Greenhouse soil	7	31·4	5·8–88·2

EXERCISE 37. SOIL ANIMALS: POPULATION DENSITY OF
EARTHWORMS AND SOIL CULTIVATION

Hypothesis:

Cultural practices on surface vegetation create a characteristic soil fauna.

Sites:

As wide a variety of surface vegetation as possible, especially those super-imposed by human activity on the original flora, e.g. plantations, orchards, lawns and farmland. Whenever possible sample beneath different vegetation growing on the same soil type to make direct comparisons, e.g. an old ever-green shrubbery and a well-kept vegetable patch in neighbouring gardens.

Time of Year:

Spring and autumn.

Procedure:

Sample as many differently managed sites as possible for earthworms using the formalin method (see p. 218). Take two samples from each site. Try to trace the history of the site for at least the last 5 yr. Tabulate results under major vegetative type; trees, herbs, bare soil, Table IV:37:1 (columns 1, 3, 5) and record number of worms per sq. m (columns 2, 4, 6). Arrange the her-baceous (grassland) column in sites of increasing cropping intensity, i.e. the extent to which organic matter is removed from the land at harvest.

Results:

Table IV:37:1 shows that the woodland had a wide range of worm popu-lations from 0 to 500 per sq. m. Grassland had a similar range, with de-creasing population as less organic material was returned to the soil. Soil disturbance in arable land decreased populations.

Conclusions and Discussion:

Different cultural practices greatly affect the total number and species of earthworms present in the soil, probably because of the degree of disturbance caused, the different fertility levels maintained and the amount of organic matter available.

Vigorous cultivation of arable land reduces the numbers of worms, but if the soil moisture remains high they may recover within a year. In arable apple orchards the leaves falling in autumn provide additional organic matter; they are dragged into burrows, especially by *Lumbricus terrestris*. Undisturbed soils in grassed-down orchards where the grass is mown and left as a mulch on the surface, have a greater organic matter content and a much greater earthworm population than soils in arable orchards. Large numbers of earthworms may occur in deciduous woodland where there is also much organic matter. Unfertilized grass cut for hay supports a smaller population than similar grassland that is grazed by cattle or sheep because their dung and urine return to the soil. The dung itself is eaten by the worms as well as stimulating leaf and root development in the sward.

A good sporting turf consists mainly of fine-leaved grasses, e.g. *Festuca rubra* and *F. ovina*, growing through a spongy raw humus and is free of worm casts. This type of sward is encouraged by the use of ammonium sulphate which breaks down to provide nitrogen as a plant nutrient, and sulphuric acid which lowers the pH of the soil and encourages fine-leaved fescues but discourages earthworms. Coniferous woodland soils are also acid because of the slow breakdown of needles and accumulation of organic acids, and also because they tend to be grown on originally acid subsoils that can be used for little else.

Numbers of many other animals, e.g. Collembola and dipterous larvae, gradually increase when soils are cultivated, especially when organic manures are used. Initial cultivation may temporarily reduce their numbers, because they are disturbed and may be exposed to unfavourable weather and predators, but usually the populations recover and numbers eventually exceed those present before the land was cultivated. The animals are essential for the first stages of the decomposition of organic matter which is eventually converted

TABLE IV:37:1

Approximate numbers of earthworms in sites under different cultures

Column 1 Woodland	2 No./sq. m	3 Grassland	4 No./sq. m	5 Arable	6 No./sq. m
Mixed deciduous	80–500	Orchard	400–700	Orchard	100–140
		Pasture	350–450		
		Meadow	160–240	Field crop	70–150
Douglas Fir	0–10	Lawn	0–30		

into soluble plant nutrients by fungi and bacteria; the nutrients stimulate plant growth, crops yield more and finally produce more rotting vegetation on which the animals can feed.

Cultivations at different seasons have different effects on the numbers of animals present. Earthworms, myriapods and wireworms are most abundant in spring and autumn, but numbers, especially of earthworms, decline when there are early frosts after autumn ploughing. Spring ploughing reduces those carabid beetles which hibernate as adults, but larval hibernators are less affected. Crops which occupy fields during winter encourage a rich fauna and flora on the surface of the soil, but summer crops impoverish the surface populations because fields are disturbed in spring just when animals are beginning to re-colonize them after the winter.

EXERCISE 38. SOIL ANIMALS: DISTRIBUTION AND DEPTH OF EARTHWORM POPULATIONS

Hypothesis:

Different species live at different depths below the soil surface.

Apparatus:

A shallow wooden grid 10 cm × 50 cm × 50 cm, watering can, 40% formalin, spade, and containers in which to collect earthworms.

Sites:

Most garden, grassland or deciduous woodland soils that are fairly deep and with a subsoil that is not pure sand, clay or rock.

Time of Year:

Spring to autumn, but the exercise may be less successful in mid-summer.

Procedure:

Hand sorting: Remove 0·1 or 0·2 sq. m of soil to a depth of 20 cm. Crumble the soil and collect the worms by hand. Separate the worms into two groups, mature *Lumbricus terrestris* and others; mature *L. terrestris* are 90–300 mm long and 6–9 mm diameter, strongly pigmented, brown-red or violet dorsally, yellowish ventrally with a prominent orange-red, saddle-shaped clitellum (see Fig. A 12, p. 267) on segments 31–37 or 32–37. The body is cylindrical but the tail is flattened dorso-ventrally. Partly grown specimens are similar but without a clitellum. Very young ones are difficult to distinguish but their contribution to the total weight of populations of this species is small and they may be included with the other worms. Weigh the worms fresh or store in 5% formalin until convenient. (The weight after preservation in formalin is about 25% less than fresh weight, so a correction must be made for this.)

Formalin method: Add 25 ml of 40% formalin to 1 gal of water and apply half the solution to the soil within the quadrat. When the worms cease to appear at the surface, usually after about 20 min, pour in the remaining solution. Collect the worms for another 20 min, divide into groups as above and weigh them.

Tabulate (Table IV:38:1) under adult *L. terrestris* and other worms (column 1), numbers extracted by formalin and hand sorting (columns 2, 3), weight extracted by formalin and hand sorting (columns 4, 5).

TABLE IV:38:1

Numbers and weight of earthworms per 0·25 sq. m extracted by formalin and by hand sorting the top 20 cm of soil

Column 1	2	3	4	5
	No. per 0·25 sq. m		Weight (g)* per 0.25 sq. m	
	Formalin	Hand sorting	Formalin	Hand sorting
Adult *L. terrestris*	16	5	46	4
Other worms	88	175	14	25

*1 ton = 1016 kg

Results:

Both methods produce fewer adult *L. terrestris* than the total of other species but formalin is relatively more effective for *L. terrestris* adults than for the others. This shows up even more clearly in terms of weight, because the other worms are smaller and weigh less. Formalin extracted 3 times the weight of adult *L. terrestris* as other worms whereas digging out the top 20 cm extracted 6 times the weight of other worms as adult *L. terrestris*.

Conclusions:

Different species live in different layers in the soil; some, e.g. *L. terrestris*, live in well developed burrow systems going down 2–3 m. The formalin quickly runs down these deep, vertical burrows and extracts this species from the soil more efficiently than the smaller surface-dwelling species, where burrow systems are shallow, narrow and mostly horizontal and into which the formalin does not penetrate so easily. Hand sorting is a more efficient method of extracting these surface dwellers.

Rider:

Combine the numbers of *L. terrestris* collected by the formalin method and the numbers of other worms collected by hand sorting and calculate the total numbers and weight of worms per acre (1 acre = 4,047 sq. m). This may be up to 3 million, equivalent to a fresh weight of up to 3000 lb per acre.

EXERCISE 39. Distribution Gradients

Introduction:

Each species of animal has its own characteristic pattern of distribution (Exs. 41, 42) which is superimposed on the basic geographical gradients created by physical features of geology and climate. These may act directly (Exs. 35, 36) or indirectly through the biological environment; in particular the distribution of animals depends on that of plants (Ex. 43) (see Plate 3).

Geographical gradients may be slight and almost imperceptible over large areas, or confusingly patchy. At the edges of watercourses, on the seashore, down the sides of valleys, chalk pits, clay pits, even across a well-worn footpath in a playing field, garden or park, environmental gradients are steeper than usual. They can be clearly seen in the zonation of plants and the resultant zonation of animals. Within zones where the habitat is similar, different species may live together as communities.

Hypothesis:

The distribution and abundance of animals is the result of ecological gradients created by the physical geography of their habitat.

Apparatus:

This exercise will have to be adapted to the available sites, and the apparatus will vary accordingly. Plastic clothes lines marked at 2 m intervals, measuring poles painted in alternate 15 cm bands of red and white, spirit level, sweep nets, water nets, tubes, rubber boots, pH indicator, thermometer and exposure meter may be required.

Sites:

Any area with a well-marked linear change in physiography, e.g. a canal.

Time of Year:

Any time. The exercise becomes more instructive as more information accrues, especially in relation to seasonal changes.

Procedure:

(1) *Vegetation maps:* Select a typical 15 m length of canal (Fig. IV:39:1a) and lay out a straight base line along the towpath a few feet away from the bank. From this line stretch graduated plastic clothes lines at right angles across the canal at 3 m intervals. Decide how much information to show on the maps; whether to represent communities, or individual plants or a combination of both depends on the size of the area and the size and spacing of individual plants. Examine each 3 m-wide strip in turn. Map the vegetation on graph paper at a reduced scale of 1:60 (Fig. IV:39:1b). On canal banks, the boundaries of the main communities fall into well-marked zones (aquatic, reed swamps and marsh) so it should be possible to plot to an accuracy of ± 15 cm, and to record the location of clumps of conspicuous individuals among the normal dominant plants. Note also the width and steepness of the banks and the position of the water's edge. At one end of the mapped area take a transverse profile of the shoreline and canal bottom. Measure the bank profiles above the water with a spirit level and vertical rules, noting

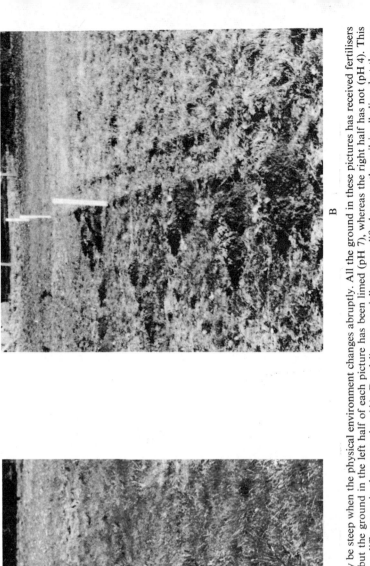

A

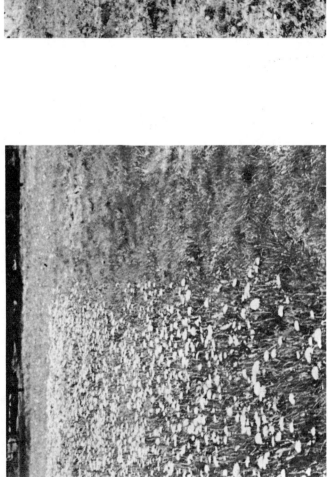

B

PLATE 3. Gradients of plants and animals may be steep when the physical environment changes abruptly. All the ground in these pictures has received fertilisers containing N, K, Na and Mg for many years but the ground in the left half of each picture has been limed (pH 7), whereas the right half has not (pH 4). This difference in the physical environment produces differences in the vegetation (A). Dandelions, especially, are prolific because the soil is alkaline, but they are absent from the more acid soil. The alkaline soil containing greater amounts of rotting organic matter supports many more earthworms (see Exs 35, 36). These in turn are eaten by moles whose presence is revealed by the distribution of mole hills (B). (Photo: by courtesy of Joan M. Thurston, Rothamsted.)
Facing page 220.

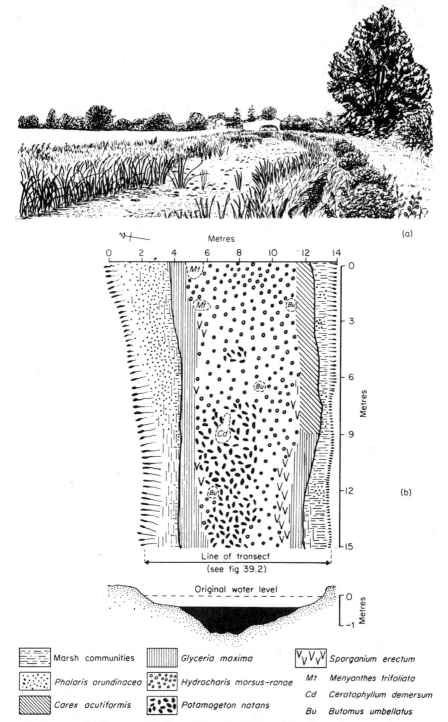

Metres

(a)

(b)

Line of transect
(see fig 39.2)

Original water level

	Marsh communities		Glyceria maxima	Vᵥ Vᵥ Vᵥ	Sparganium erectum
Phalaris arundinacea		Hydrocharis morsus-ranae	Mt	Menyanthes trifoliata
Carex acutiformis		Potamogeton natans	Cd	Ceratophyllum demersum
				Bu	Butomus umbellatus

FIG. IV:39:1. (a) Drawing of part of the Redwith/Maesbury Marsh reach of the Shropshire Union Canal, England. Disuse and low water level have enabled floating plants to spread and reedswamp to develop (after Sinker). (b) Vegetation map of the same reach; the path is on the right-hand side. The transverse profile (vertical exaggeration × 2) is beneath the map; the blackened part represents water (after Twigg).

the differences in height at 25 cm intervals; take depth soundings at the same intervals across the channel.

On the final map, communities are best shown by differential shading, and individual plants by letters or symbols. The banks may be indicated by hachures. Rule scale lines along both axes. Draw the profile below the map with a vertical exaggeration ($\times 2$) to emphasize its shape. (Fig. IV:39:1b).

(2) *Transects:* It is often most instructive to study zonations and successions in plant communities along a line perpendicular to the zones. Various types of "transects" may be suitable in different habitats but one of the most useful is a "ladder transect" which records quantitative and qualitative data and shows how changes in communities and individual species are associated with variations in habitat.

Relative assessments are difficult to make for species of different stature and growth form. An assessment of the proportion of an area covered by a given plant may give very different results to an assessment of the number of plants present. Possible cover/abundance and grouping scales are given below.

Cover/Abundance Scale:

1 = very scant (covering less than $\frac{1}{20}$ of the ground)
2 = covering $\frac{1}{20}$ to $\frac{1}{4}$ of the ground surface
3 = covering $\frac{1}{4}$ to $\frac{1}{2}$ of the ground surface
4 = covering $\frac{1}{2}$ to $\frac{3}{4}$ of the ground surface
5 = covering $\frac{3}{4}$ to $\frac{4}{4}$ of the ground surface.

Grouping scale:

1 = growing singly
2 = grouped or tufted
3 = small patches, or cushions
4 = small colonies, or extensive patches
5 = pure populations.

A more rapid subjective estimate may be made by dividing the plants into the categories abundant, frequent, occasional and rare. This gives a satisfactory general impression but tends to make small plants (e.g. *Lemna minor*) conspicuous by comparison with larger ones (e.g. *Sparganium erectum*); it ignores any vertical layering.

Plot the results on squared paper as in Fig. IV:39:2. First draw the profile at the bottom of the page, then list the recorded species in an order which conveys the sequence of changes in the vegetation. Plot the frequency variations for each species as histograms along a line across the profile so that the frequency in a particular square appears above the appropriate part of the profile. Add the boundaries of the main zones to the diagram.

This presentation offers the maximum amount of information in a minimum

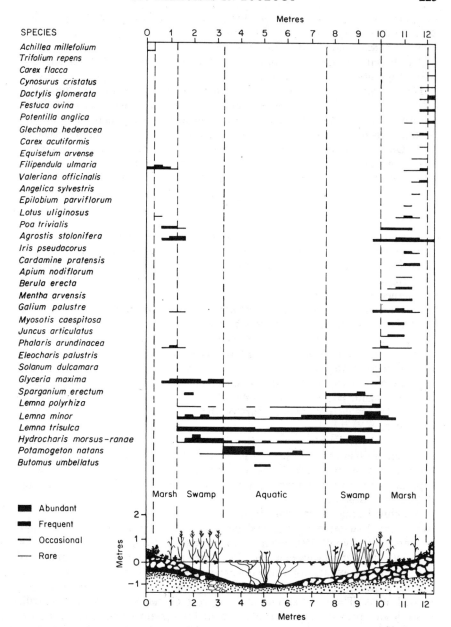

Fig. IV:39:2. "Ladder transect" diagram across the Redwith/Maesbury Marsh reach. Species frequencies are plotted in the form of histograms for comparison with the profile drawn below in which the dominants of each zone are represented diagrammatically (after Twigg).

of space and remains easily readable. Other variables such as temperature, pH, light intensity and speed of the water flow may be plotted beneath the diagram if seasonal changes are to be measured.

Additional notes on the zonation and grouping of the vegetation, and habitat factors, should be made.

The association between vegetation and fauna. Most freshwater animals are inconspicuous and active and are difficult to sample quantitatively. Comparative population counts between different localities or microhabitats may be made using standardized sampling methods based on volume, area or time. Qualitative observations, particularly of the habitats and habits of the larger animals, are more convenient and show the association between the vegetation and animal populations, and the mode of life and method of locomotion of different species.

Results:

Vegetation: In the stretch of canal illustrated, three zones, aquatic, swamp and marsh were recognizable from the middle to either bank. *Hydrocharis morsus-ranae* and *Lemna minor* were co-dominant over most of the aquatic zone frequently mixed with *L. polyrhiza*; *L. trisulca* formed a continuous submerged tier. There was a small, pure stand of *Ceratophyllum demersum* among an area of *Potamogeton natans*. The reed swamp dominants grew mostly in pure stands with *Glyceria maxima* in a broad, conspicuous belt between the discontinuous fringe of *Sparganium erectum* and the shore. In one stretch *Carex acutiformis* replaced *Glyceria*, and *Equisetum fluviatile* and *Rumex hydrolapathum* were occasional associates. The marsh zone was delineated by a narrow strip of *Berula erecta* at the water line and thence became very mixed, with *Phalaris arundinacea* common. Mixed grasses grew on the dry, calcicolous upper part of the bank with clumps of alder and hawthorn.

Flow was negligible except after heavy rain and the stretch was well illuminated except for the shade cast by the plants themselves. Slumped soil formed most of the shore-line except where the original canal walls were exposed, and from the reed swamp to the centre there was an increasing thickness of organic mud.

Animals: In the aquatic zone were animals that require open, but sheltered water, the "swimmers" and "skaters" such as pond skaters or Water Striders (Fig. IV:39:3B), swimming caddis (Fig. IV:39:3N) and whirligig (Fig. IV:39:3A) and water beetles. The Water Spider (Fig. IV:39:3D) prefers still water and plants for attachment; water mites (Fig. IV:39:3E, F) and Phantom Midge larvae (*Chaoborus* sp.) occur in the most stagnant and shaded areas. The leafy shoots of Hornwort *C. demersum*, with their stiff, spiky leaves provide a favourable habitat for certain Caddis. *Cyrnus* (Fig. IV:39:3M) spins a silken tent and *Agraylea* attaches the ends of its fine-grained case, like a hammock, to the leaves. The more active snails, flatworms (Fig. IV:39:3T) and leeches (Fig. IV:39:3S) also occurred.

In the swamp zone, the increased shelter produces more stagnation and

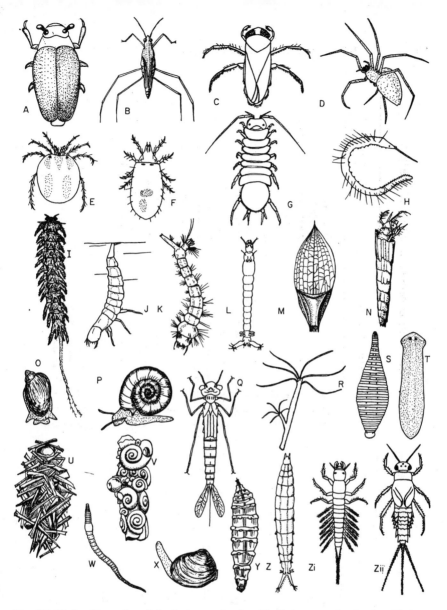

Fig. IV:39:3. Examples of freshwater animals with their approximate size. Skaters, A and B; Swimmers, C–N; Crawlers and Clingers, O–V; Burrowers, W–Z$_{11}$.
A. Whirligig beetle, *Gyrinus* sp., 3–7 mm; B. Pond skater or Water Strider, *Gerris* sp., 8 mm; C. Water boatman, *Corixa* sp. 18 mm; D. Water spider, *Argyroneta aquatica* 10 mm; E. Water mites, *Hygrobates* sp., 1·5 mm; F. Water mites, *Megapus* sp., 1 mm; G. Water louse, *Ascellus aquaticus*, 20 mm; H. Oligochaet worm, *Stylaria lacustris*, 15 mm; I. Beetle larva, *Haliplus* sp., 6 mm; J. Beetle larva, *Dytiscus* sp., 40 mm; K. Mosquito larva, *Culex* sp., 40 mm.; L. Midge larva, *Chironomus* sp., 17 mm; M. Tent caddis larva, *Cyrnus flavidus* fixed to a leaf; N. Swimming caddis larva, *Triaenodes* sp., 15 mm; O. Snails, *Limnaea* sp., 9 mm; P. Snails, *Planorbis* sp., 20 mm; Q. Dragonfly nymph, *Coenagrion puellum*, 29 mm; R. *Hydra* sp., 16 mm; S. Leech, *Glossosiphonia* sp., 44 mm; T. Flatworm, *Planaria lugubris*, 17 mm; U. Crawling caddis larva, *Limnophilus rhombicus*, 20 mm; V. Crawling caddis larva, *Limnophilus flavicornis*, 25 mm; W. Lumbricid worm, *Lumbricillus* sp., 25 mm; X. Pea shell snails, *Pisidium* sp., 3 mm; Y. Fly larva, *Tabanus* sp., 20 mm; Z. Crane fly larva, *Tipula* sp., 20 mm; Z$_1$. Alder fly larva, *Sialis* sp., 22 mm; Z$_{11}$. Mayfly larva, *Ephemerella* sp., 14 mm.

TABLE IV:39:1

Detailed list of species inhabiting different parts of a freshwater environment in Central England

Skaters	Swimmers	Crawlers and Clingers	Burrowers
Insects	Worms	Hydroids	Worms
Gerris	Stylaria	Hydra	Eiseniella
Hydrometa stagnorum			Lumbricillus variegatus
Velia caprai	Arachnids	Sponges	Tubifex
Gyrinus marinus	Argyroneta aquatica	Spongilla	
	Hydrarachna		Snails
	Hygrobates	Flatworms	Anodonta cygnea
	Megapus	Bdellocephala punctata	Pisidium
		Dendrocoelum lacteum	Sphaerium
	Crustaceans	Planaria lugubris	
	Eucrangonyx gracilis	Polycelis nigra	Insects
			Caenis horaria (nymphs)
	Insects	Leeches	Ephemera vulgata (nymphs)
	Corixa	Erpodella octoculata	Sialis lutaria (larvae)
	Ilyocoris cimidoides	Erpodella testacea	Molanna angustata (larvae)
	Notonecta	Glossiphonia complanata	Chironomus plumosus (larvae)
	Triaenodes (larvae)	Glossiphonia heteroclita	Tipula (larvae)
	Agabus bipustulatus	Haemopis sanguisuga	
	Dytiscus marginalis	Helobdella stagnalis	
	Haliplus lineolatus	Hemiclepsis marginata	
	Hyphydrus ovatus	Pisicola geometra	
	Laccophilus hyalinus	Theromyzon tessulatum	
	Noterus capricornis		
	Noterus clavicornis	Snails	
	Platambus maculatus	Ancylus lacustris	
	Anopheles (larvae)	Bithynia tentaculata	
		Hydrobia jenkinsii	

Chironomus sp. (larvae other than *C. plumosus*
Culex (larvae)
Dixa (larvae)
Forcipomyia (larvae)

Limnaea auricularia
Limnaea pereger
Limnaea stagnalis
Physa fontinalis
Planorbis albus
Planorbis carinatus
Planorbis complanatus
Planorbis corneus
Planorbis crista
Planorbis laevis
Planorbis vortex
Viviparus viviparus

Crustaceans
Asellus aquaticus

Insects
Aeshna grandis (nymphs)
Coenagrion puellum (nymphs)
Nepa cinerea
Agraylea (larvae)
Anabolia nervosa (larvae)
Cyrnus (larvae)
Limnophilus extricatus (larvae)
Limnophilus flavicornis (larvae)
Limnophilus rhombicus (larvae)
Phryganea (larvae)
Polycentropus (larvae)
Cataclysta lemnata (larvae)
Nymphula nymphaeata (larvae)
Anacaena limata
Hydroporus (*Graptodytes*) *pictus*
Tabanus (larvae)

oxygen deficiency. The animal population in the water here was low, but there were many animals on the plants themselves, mostly "crawlers" e.g snails (Fig. IV:39:3O, P), leeches and flatworms, especially on the Branched Bur-reed (*Sparganium erectum*). The cavities at the bases of the leaf sheaths collect food on which detrital feeders like the Water Louse (*Ascellus aquaticus*) (Fig. IV:39:3G) may feed, and tabanids (Fig. IV:39:3Y) and chironomid larvae (Fig. IV:39:3L) occurred within the broken aerenchyma of old leaves. Fewer animals occurred in the water among the dense *Glyceria*. Its dead leaves rot quickly and the leaf sheaths enfold the stem tightly so there is little shelter at the base of the plant. On the aerial parts, the snails *Succinea* and *Zonitoides nitidus* were found.

The soil-free roots of alder and grasses under the bank had a distinctive population including beetles (*Anacaena limata* and *Hydroposus pictus*) and a number of caddis [*Analolia, Phryganea, Polycentropus* and *Limnophilus* spp. (Fig. IV:39:3U, V)] most of which use the roots for case-making material as well as for attachment.

In the silt and organic mud on the bottom, search for "burrowers" or bottom "crawlers", e.g. the mayfly *Caenis horaria*, the caddis *Molanna angustata*, peashell snails (Fig. IV:39:3X) and worms. A more comprehensive list of invertebrates likely to be found in various types of canal, including their habitats and mode of progression is given in Table IV:39:1. Make up your own table.

Conclusions:

The zonation of plants is self-evident and the zonation of animals conforms with it. Many of the physical factors underlying this zonation can be elucidated, and explained in terms of the topography, geology, soil type and moisture distribution at the site, and seasonal temperatures. A more complete ecological background to the plants and animals may be accumulated by several separate surveys for each of these features.

EXERCISE 40. POPULATION DENSITY AND AVAILABLE SITES

Introduction:

Many species of thrips (Fig. B 10a, p. 273) overwinter as adults. One of the most common species, particularly in agricultural districts in S. England, is *Limothrips cerealium*. This species feeds on grasses and cereals and the females migrate to overwintering sites in August and September and emerge again in spring. Large numbers are often found hibernating in most unlikely places. Indoors they squeeze behind picture frames and skirting boards and outdoors they occur in all manner of crevices among litter, old birds' nests, old cocoons and particularly in bark. At this time of year they are strongly sensitive to close contact with their surroundings. The size of suitable crevices is critical and they prefer them to be between 0·1 mm and 0·5 mm wide so that both dorsal and ventral body surfaces are in contact with the walls of the crevice.

The size of the overwintering population in bark depends on the number of suitable crevices it contains and hence the population density differs on different trees (see p. 14).

Hypothesis:

Population density is limited by the number of available places to live.

Apparatus:

A 2·5 cm (1 in.) chisel, mallet, penknife, small polythene bags or envelopes, and a white tray.

Site:

Tree trunks at least 20 cm in diameter, preferable within 1 km of cereal fields. Try to select a loose, flaky bark (e.g. pine) to contrast with a moderately compact bark (e.g. horse chestnut) and a very compact one (e.g. poplar).

Time of Year:

October to March.

Procedure:

First obtain permission to cut the bark.

If samples are taken carefully as described, they should not harm the tree, but if old trees, soon to be felled, can be found, use them. Take 2·5 cm square samples by making four cuts with the chisel and removing the bark to about half its depth. Be careful not to cut down to the cambium layer joining the bark to the wood. Paint over the cut surface with linseed oil, pruning paint or brown plastic plaint.

Take ten sample units from each species of tree, scattered at random over several trees if possible. Collect each unit in a separate plastic bag or envelope.

In the laboratory split the layers of bark with a penknife over a white tray and tap the slivers sharply to dislodge all thrips. Count the living and shrivelled, dead specimens that fall out of each sample. Compute the mean numbers per 6·25 sq. cm (1 sq. in.), alive and dead, for each tree species (Table IV:40:1).

Estimate the height of the main trunk for representative trees of each species and measure the girth at the highest point you can reach from a standing position. Estimate the bark surface area per tree (height × girth in sq. m.). Compute the total thrips population per tree.

Results:

From Table IV:40:1; on Austrian Pine (*Pinus laricio* var. *nigricans*), Horse Chestnut (*Aesculus hippocastanum*) and Lombardy Poplar (*Populus italica*), respectively, the mean over-wintering population was 13·1, 3·6 and 0·9 per 6·25 sq. cm. Because of the difference in sizes of the tree the total populations per tree differed less, being 566, 173 and 52 thousand, respectively.

The numbers found dead were unrelated to the numbers found alive, being 8·8, 46·2 and 5·3 for Pine, Chestnut and Poplar, respectively.

Conclusions:

No statistical test was necessary to show that the numbers, both of live and dead thrips differed greatly on different trees. Such a test may be necessary where differences are less (see p. 103). Ten times as many living thrips over-wintered on the Austrian Pine as on the Lombardy Poplar, in spite of the fact that the pine trees were smaller.

The number of dead thrips bore no obvious relation to the number living. It seems evident that this is not a measure of mortality in the current years' population. This can be confirmed by estimating numbers in the population at the beginning and the end of winter when it will be found that mortality is small. The dead thrips have accumulated over several or many years, depending upon the age of the trees and the rate at which bark is sloughed off. Austrian Pine loses scales of bark frequently once it is about 20 years old, so dead thrips accumulate most rapidly on young trees; Horse Chestnut only loses bark when it is older, and Poplar hardly at all.

TABLE IV:40:1

Numbers of thrips per 6·25 sq. cm (1 sq. in.) of bark

	Austrian Pine		Horse Chestnut		Lombardy Poplar	
	Alive	Dead	Alive	Dead	Alive	Dead
	14	5	0	59	0	2
	4	13	11	26	1	11
	1	4	1	30	0	5
	11	14	7	34	1	9
	13	8	4	36	3	6
	30	11	1	33	1	3
	41	24	1	59	0	9
	7	4	7	33	1	2
	3	1	2	60	2	6
	7	4	2	92	0	0
$\sum$	131	88	36	462	9	53
$\bar{N}$	13·1	8·8	3·6	46·2	0·9	5·3
Height (m)	18		15		20	
Girth (m)	1·5		2·0		1·8	
Surface area (sq. m.)	27		30		36	
Total live pop. per tree (thousands)	566		173		52	

EXERCISES 41 AND 42. COMPETITION AND DISTRIBUTION

Introduction:

Great numbers of small disc or button-shaped galls, sometimes known as spangle galls, often develop on the lower surface of oak leaves in autumn (Fig. IV:41:1). The insects which produce them are larvae of the agamic generations of 4 species of the gall wasp genus *Neuroterus* (*Cynipidae*) and their life histories in S. England are summarized in Fig. IV:41:2.

The spangle galls fall to the ground in September and October before the leaves fall and the fully grown Cynipid larvae overwinter in the galls. Larvae in galls which remain attached to dead leaves on the tree throughout the winter, dry up and die. Parthenogenetic females emerge from the spangle galls on the ground in spring and lay eggs on the oak buds, with the possible exception of *N. fumipennis* which may lay in expanded leaves. The resulting galls, which appear quite different from the spangle galls, contain immature stages of the sexual generations. Adults of both sexes emerge from these galls whilst they are still attached to the trees in summer and the fertilized females lay in the veins of the oak leaves. The larvae emerging from these eggs induce the development of spangle galls.

Young trees are particularly subject to heavy infestations, though infestations are erratic from district to district and from season to season. The galls may occupy a high proportion, occasionally up to 80%, of the leaf surface. In autumn, a small area of leaf surrounding each mature gall dries and turns brown and the gall falls to the ground. Younger galls, encompassed by these dried up areas surrounding a mature gall, die. Thus there is *intraspecific* competition for space to avoid arrested development. In areas where more than one species of *Neuroterus* occur, there is likely to be *interspecific* competition for space at such high densities (see p. 19).

Apparatus:

A hand lens and long-handled secateurs are useful but not essential.

Site:

The distributions are most clearly separated, and sampling is easiest, on young oaks (*Quercus robur*) about 10 years old, and 3 m high. Results obtained from such trees may then be compared with results from mature trees.

Time of Year:

Leaves should be collected from mid-August to late September before the galls begin to fall.

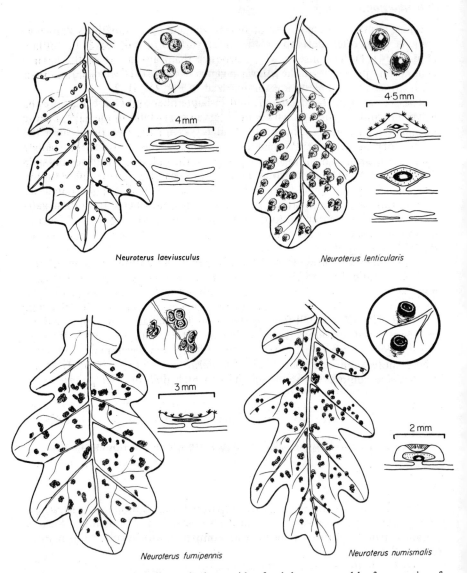

Neuroterus laeviusculus

4 mm

Neuroterus lenticularis

4·5 mm

Neuroterus fumipennis

3 mm

Neuroterus numismalis

2 mm

FIG. IV:41:1.   Spangle galls on the lower side of oak leaves caused by four species of gall wasps (*Neuroterus* spp.). At the side of each leaf are enlarged drawings of galls (inset) and typical cross-sections.

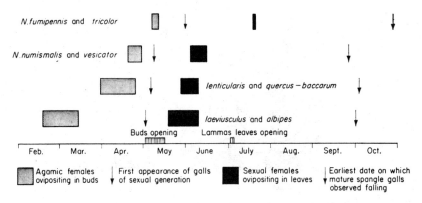

FIG. IV:41:2. The timing of the life cycles of *Neuroterus* species in Wytham Wood, S. England, 1958 (after Askew).

EXERCISE 41. COMPETITION AND DISTRIBUTION: SPANGLE GALLS ON YOUNG OAK TREES

Hypothesis:

The distribution of similar species is often complementary; their habitats overlap but do not coincide.

Procedure:

The susceptibility of different trees to attack varies due to differences between dates of leaf opening and probably physiological variation between trees. It is better therefore to take samples from more than one tree if possible.

From young trees bearing large numbers of, for example, *numismalis*, *lenticularis* and *laeviusculus* galls, collect 50 leaves approximately 50 cm from the trunk at successive intervals, 10 cm apart, from the ground up to 180 cm. Count the numbers of each species at each level, calculate the percentage representation at successive heights and plot the results as a histogram (Fig. IV:41:3).

Collect a further series of samples from two to four horizontal branches at the same height above ground, say 100 cm, but at successive 10 cm intervals from the base of the branch to its tip. Treat the results as described above (Fig. IV:41:4).

Results:

The histograms show that *numismalis* galls predominate at the tops and towards the periphery of the tree, *lenticularis* at the middle and *laeviusculus* towards the bottom and near the bole. Each species predominates in one particular zone, though their total distributions overlap completely.

Conclusion:

See Exercise 42, p. 236.

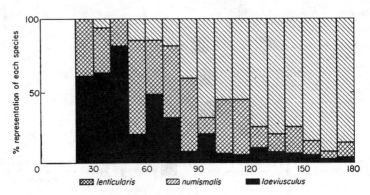

FIG. IV:41:3. Percentage histogram for the distribution of galls of *Neuroterus* species at different heights in cm, measured at 50 cm from the trunk (after Askew).

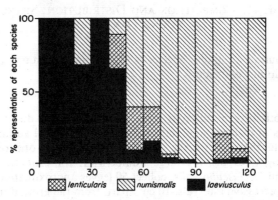

FIG. IV:41:4. Percentage histogram for the distribution of galls of *Neuroterus* species at different distances in cm from the trunk measured at a height of 100 cm (after Askew).

EXERCISE 42. COMPETITION AND DISTRIBUTION: SPANGLE GALLS
ON OAK LEAVES

Hypothesis:
See Ex. 41.

Procedure:
Collect five uninfested oak leaves and divide each into ten areas of equal depth from petiole to tip. Number these divisions 1–10, then measure their areas in mm². Calculate the mean area of the five replicates for each division (Fig. IV:42:1).

Find leaves on which each of the three common gall species occurs alone and collect sufficient to produce a total of at least 200 galls of each species.

Leaf division	Mean area mm²	Gall distribution		
		num.	lent.	laev.
10	225	263	28	10
9	450	655	114	33
8	550	786	193	51
7	600	721	238	75
6	575	550	213	72
5	575	380	174	91
4	475	220	132	105
3	300.	72	90	90
2	225	24	29	55
1	150	1	10	31

FIG. IV:42:1.   The distribution of spangle galls of *Neuroterus numismalis, N. lenticularis,* and *N. laeviusculus* on leaves divided into ten areas (after Askew).

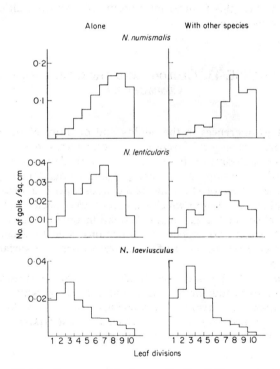

FIG. IV:42:2.   Distribution per unit leaf area of galls of *Neuroterus* spp. on leaves on which only one species was present (left-hand diagrams) and on leaves on which more than one species occurred (after Askew).

Divide the leaves into ten divisions and calculate for each species the number per sq. cm of leaf area in each division. Repeat this for each gall species using leaves on which two or more of the species are present. Present the results on adjacent histograms (Fig. IV:42:2).

Results:

Whether there was one or more species of gall on a leaf, the spatial zonation was similar and thus independent of the presence of other species. Note that *numismalis* galls occur almost up to the leaf margins whereas *laeviusculus* galls occur nearer to the midrib. If *lenticularis* predominates at a site, it may be concentrated towards the leaf apices rather than at the centre of leaves.

Conclusions (*Exs.* 41 *and* 42):

N. numismalis forms galls at the top of the tree and at the apex and edge of leaves, *N. laeviusculus* galls occur near the bases of the tree and leaves, *N. lenticularis* produces galls near the middle of the tree and the middle of leaves. The distributions of the three species are dissimilar on young trees and we may assume that this reduces interspecific competition, although this is very difficult to measure.

EXERCISE 43. GEOLOGY AND THE COMPOSITION OF COMMUNITIES

Introduction:

There are big differences in the quality and quantity of the invertebrate fauna on different geological formations. A calcareous rock like chalk produces a shallow, well-drained basic soil, called a rendzina, which supports a rich and varied vegetation and a correspondingly abundant fauna. In contrast, sandstones often produce an acidic podsol with a characterstically uniform heathland vegetation which supports fewer species and numbers of animals. The effect of the geological formation on the fauna, through its influence on the soil and vegetation, is demonstrated by a comparison of invertebrates living on the soil surface in areas where the surface rocks are different (see p. 7).

The distribution of Man is also partly determined by the surface rocks. For instance, dry sands producing heathland and heavy clays with poor drainage resulting in marshes, are avoided for dwellings. Thin soils on limestones and granite moors are unsuitable for crops which are most productive on alluvial deposits, silts and loams on plains and in valleys.

Hypothesis:

The composition of the fauna in a community is influenced by the underlying rock.

Apparatus:
Grids, 30 cm square, hand fork and trowel, garden shears, specimen tubes, forceps and pooter, a flora and soil pH indicator.

Sites:
Comparison may be made between any two widely different soil types. This example compares 5 sq. m of chalk downland and greensand heathland in S.E. England. The fauna on clay areas, peat and silt fens, limestone valleys or gritstone moors also provide interesting contrasts with each other. The exercise can be adapted to compare the fauna of litter in different types of woodland and hedgerow.

Time of Year:
Any time between late spring and autumn is suitable, though most individuals will be found in summer. This example was done in July.

Procedure:
It is best to work in pairs and the scale of the exercise depends on the number of students participating. Try to cover a total of 2–5 sq. m of a typical piece of habitat with the quadrats. Take the same area at each site.

TABLE IV:43:1

List of vegetation at two sites

Chalk Downland	Greensand Heath
Linum catharticum (Purging flax)	*Agrostis* sp. (Bent-Grass)
Trifolium dubium (Lesser yellow trefoil)	*Ulex minor* (Dwarf Furze)
Lotus corniculatus (Bird's foot trefoil)	*Calluna vulgaris* (Ling)
Poterium sanguisorba (Salad Burnet)	*Erica cinerea* (Bell Heath)
Euphrasia sp.	*Deschampsia flexuosa* (Wavyhair grass)
Origanum vulgare (Marjoram)	*Cladonia pyxidata* (Cup lichen)
Plantago lanceolata (Ribwort)	
Galium verum (Lady's Bedstraw)	
Leontodon sp. (Hawkbit)	
Hieracium pilosella (Mouse-ear Hawkweed)	
Cladium mariscus (Sedge)	
Festuca ovina (Sheep's Fescue)	
Avena sp. (Wild Oat)	
Holcus lanatus (Yorkshire Fog)	
Hypnum sp. (Moss)	
No. of species: 15	6

No. of species common to both sites: 0

Lay out the quadrats (p. 96) and identify the most common plants within it. Cut and remove the tall vegetation if necessary. Collect all animals from the soil surface, then overturn small stones and scrape to about 1 cm depth to find any remaining creatures. Avoid colonies of animals like ants which require specialized sampling techniques. Make notes on the general condition of the vegetation and ground cover. Determine the pH of the soil (see Ex. 35). Count and determine the samples to orders. List species of plants present, marking those common to both sites (Table IV:43:1). Tabulate sample under taxon (column 1), total on chalk (column 2), total on heath (column 3) and ratio of total on chalk to total on heath obtained from $\log (N + 1)$, as explained in Chapter III (p. 68), to form a sequence in column 4 (Table IV:43:2).

Results:

The greensand heath was more acid (pH 5·4) than the chalk down (pH 7·4). From Table IV:43:1 there was a more varied flora on the chalk down (15

TABLE IV:43:2

Number of individuals from different taxa on chalk down and greensand heath with the ratio of one to the other

Column 1	2	3	4
Taxon	Chalk	Heath	Ratio Chalk/Heath
Mollusca	147	0	148
Annelida	46	0	47
Diplopoda	37	0	38
Isopoda	218	8	25
Chilopoda	38	1	19
Dermaptera	12	0	13
Collembola	25	3	8·3
Orthoptera	5	0	6·0
Diptera	21	15	1·4
Opiliones	15	11	1·3
Thysanoptera	12	10	1·2
Coleoptera	49	62	0·79
Araneida	73	104	0·71
Acari	9	17	0·55
Lepidoptera	2	11	0·25
Psocoptera	0	3	0·25
Hemiptera	4	74	0·066
Total	713	319	

species) than the greensand heath (6 species) and no species were common to both sites. The vegetation was also more luxuriant on the down which had a matt of moss on the soil surface, whereas there were bare patches of soil on the heath (from notes taken).

Table IV:43:2 shows that there were twice as many animals on chalk as on heath and numbers of Mollusca, Annelida, Diplopoda, Isopoda, Chilopoda, Dermaptera, Collembola and Orthoptera on chalk were greater than on heath. Within limits $\overset{x}{\div} 2$ (see p. 75), Diptera, Opiliones, Thysanoptera, Coleoptera, Araneida and Acari were equally common at both sites. Lepidoptera, Psocoptera and Hemiptera were more common on the heath.

Conclusions:

The different pH, vegetation, soil and drainage on the two types of underlying rock produced differences in the animal populations there. Isopoda, Chilopoda and Diplopoda have highly permeable integuments and can only survive in high humidities. Gasteropoda, Diplopoda, Chilopoda and Isopoda thrive where there is adequate calcium carbonate. Both these conditions are fulfilled on chalk but not on heathland. The more primitive insects, especially Collembola and Dermaptera, were also fewer on heath than on chalk. More highly evolved insects and arachnids were more versatile, occurring equally commonly in both sites. The sample for Psocoptera was small and meaningless. The large number of Hemiptera on greensand heath were mainly heather bugs *Kleidocerys truncatulus*.

EXERCISE 44. Community Structure

Introduction:

There is an underlying structure to communities, implying inter-relationships between species and often referred to loosely as the "balance of nature".

In any large sample of individuals from a natural community with many species, there are more species with only one individual than with two; more with two than with three; more with three than four, and so on (Plate 4). At the other end of the scale there are several species with many individuals, a few species that are very common and always one that is exceedingly common and whose numbers predominate. This dominant species may often contain a quarter, or even occasionally more than half, of all the individuals in the sample (see p. 23).

Hypothesis:

The way in which the individuals are distributed amongst the species in a natural community follows a regular sequence which can be described by the logarithmic series.

Apparatus:

A light trap (Chapter V, p. 250) and means of identifying Macrolepidoptera, i.e. South (1961), Holland (1913) (Appendix N) or a reference collection.

Sites:

Anywhere outdoors. The trap will obtain a bigger catch if it is not in competition with other bright lights, but this is not essential.

Time of Year:

Throughout the year. Better results will be obtained by continuous trapping, but one week per month will suffice.

Procedure:

The practical work is the same for all the exercises using light traps and is given in Chapter V (p. 250). It is best to operate such traps as a routine as far as possible because the nights on which insects will fly are not easy to forecast and a whole year's catches will enable several exercises to be done on the same material.

Most exercises can be done with Macrolepidoptera which are the easiest group to identify to species. If the trap is started in January, very few insects will be caught at first and this makes identification easier. Most museums have a reference collection of moths which will be helpful when identities are doubtful, but a reference collection can be built up over a period of two or three years which will cover nearly all future catches. For this purpose the insects should be pinned and set with wings displayed to facilitate comparison and this should be done also with specimens difficult to identify (Oldroyd, 1958) (see Appendix N).

Tabulate the catch daily by species and summarize each month to give totals of each species, the total catch and the number of species. Summarize again at the end of the year, keeping the families separate (Table IV:44:1, 2, 3).

Use a whole year's catch or the longest series of catches available. Count up the number of species with one individual only, the number with two, with three, etc. and make a frequency table (Table IV:44:4). Sum the total number of individuals and the total number of species. Calculate α and x and the expected number of species with 1, 2, 3, etc. individuals from the log series (see Chapter III, p. 107). Make a frequency diagram and plot the expected values on it (Fig. III:35).

Results:

The frequency distribution is a hollow curve; with no zero value for the x axis, since, when there are no individuals per species, there is no catch and y cannot be measured. The highest value of y is 37 when x is 1 and next is $y = 22$ when $x = 2$; there are 4 points at $y = 11-12$, a long series from $y = 6$ at $x = 7$ down to $y = 2$ at $x = 48$, then only singletons up to $x = 1799$, with one exception (2 with 64). The fitted values follow the same course as far as $x = 30$, beyond which a few values can be computed at intervals to show that the curve continues to fall.

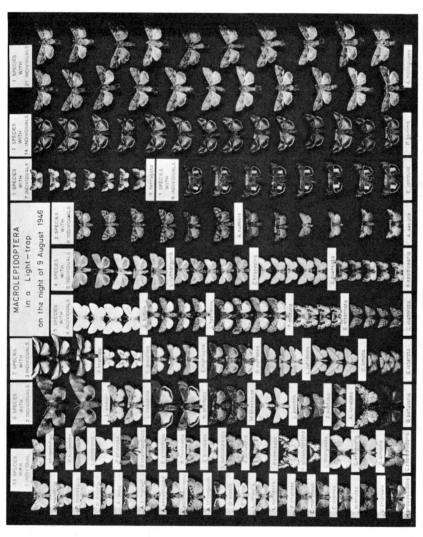

PLATE 4. A large catch of moths in a light trap on a single night at Rothamsted Experimental Station, England, showing the logarithmic series of number of individuals per species; there are many species with few individuals and few species with many individuals (see Chapter III, p. 107).

Facing page 240.

TABLE IV:44:1

Daily species record of macrolepidoptera from Burleigh School, Hatfield, England, May, 1964

	1	2	3	4	5	6	7	8	9	10	11	12	13	14	15	16	17	18	19	20	21	22	23	24	25	26	27	28	29	30	31	Σ
Noctuidae																																
Orthosia gothica		1									1																					2
Orthosia gracilis					1							1		1																		3
Geometridae																																
Lycia hirtaria	1		1										1																			3
Ligdia adustata														1																		1
Gonodontis bidentata													1																1			2
Menophra abruptaria																					1											1
Opisthograptis luteolata																	1			1						1			2	2	2	9
Ectropis biundulata																							1									1
Xanthorhoe ferrugata																									2		1				1	4
Xanthorhoe fluctuata																		1														1
Eupithecia vulgata																1	1		1		2		1	3	4	1	2		3	3	1	23
Notodontidae																																
Lophopteryx capucina																		1														1
Hepialidae																											1			2	1	4
Hepialus lupulina																																
Arctiidae																																
Cycnia mendica															2	1	1	1														5
Drepanidae																																
Cilix glaucata																									1	1						2
Total	1	1	1		1						1	1	2	2	2	2	3	3	1	1	3		2	3	7	3	4		6	7	5	62

TABLE:IV:44:2

Part of a monthly species summary of macrolepidoptera from
Burleigh School, Hatfield, England, in 1964

	Jan.	Feb.	Mar.	Apr.	May	June	July	Aug.	Sept.	Oct.	Nov.	Dec.	Total
Thyatiridae													
Polyploca ridens				1									1
Habrosyne pyritoides						1	3						4
Notodontidae													
Lophopteryx capucina					1			2					3
Hepialidae													
Hepialus lupulina					4	3							7
Arctiidae													
Cycnia mendica					5								5
Spilosoma lutea						3	3						6
Drepanidae													
Cilix glaucata					2		2						4
Drepana falcataria							1						1
Lasiocampidae													
Malacosoma neustria							16						16
Individuals	0	0	0	1	12	7	25	2	0	0	0	0	47
Species	0	0	0	1	4	3	5	1	0	0	0	0	9

TABLE IV:44:3

Annual summary of families of Macrolepidoptera from Burleigh School, Hatfield, England

| | 1964 | | | | | | | | | | | | |
	Jan.	Feb.	Mar.	Apr.	May	June	July	Aug.	Sept.	Oct.	Nov.	Dec.	Total
Sphingidae	0	0	0	0	0	0	0	0	0	0	0	0	0
Endromidae	0	0	0	0	0	0	0	0	0	0	0	0	0
Saturnidae	0	0	0	0	0	0	0	0	0	0	0	0	0
Notodontidae	0	0	0	1	0	0	0	2	0	0	0	0	3
Thyatiridae	0	0	0	0	1	1	3	0	0	0	0	0	5
Drepanidae	0	0	0	0	2	3	0	0	0	0	0	0	5
Lymantriidae	0	0	0	0	0	0	0	0	0	0	0	0	0
Lasiocampidae	0	0	0	0	0	0	16	0	0	0	0	0	16
Nolidae	0	0	0	0	0	0	0	0	0	0	0	0	0
Lithosiidae	0	0	0	0	0	0	0	0	0	0	0	0	0
Arctiidae	0	0	0	0	5	3	3	0	0	0	0	0	11
Callimorphidae	0	0	0	0	0	0	0	0	0	0	0	0	0
Limacodidae	0	0	0	0	0	0	0	0	0	0	0	0	0
Zygaenidae	0	0	0	0	0	0	0	0	0	0	0	0	0
Sesiidae	0	0	0	0	0	0	0	0	0	0	0	0	0
Hepialidae	0	0	0	0	4	0	3	0	0	0	0	0	7
Noctuidae	0	0	0	4	5	28	42	54	19	5	0	0	157
Geometridae	0	0	3	7	45	14	54	51	8	2	2	2	188
Total no. of individuals	0	0	3	12	62	49	121	107	27	7	2	2	392
Total no. of species	0	0	2	6	15	18	32	26	15	5	2	1	80
Index of diversity α	0	0	<1	2	6	10	13	11	12	1	∞	<1	30

TABLE IV:44:4

Frequency distribution of individuals per species in a light trap sample of Macrolepidoptera collected at Rothamsted Experimental Station, England, during 1935

Individuals per sp.	No. of species Obs.	Log series	Individuals per sp.	No. of species Obs.	Log series
1	37	38·0	51	1	
2	22	18·9	52	1	
3	12	12·5	53	1	
4	12	9·3	58	1	
5	11	7·4	61	1	
6	11	6·1	64	2	
7	6	5·2	69	1	
8	4	4·5	73	1	
9	3	4·0	75	1	0·3
10	5	3·6	83	1	
11	2	3·3	87	1	
12	4	3·0	88	1	
13	2	2·7	105	1	
14	3		115	1	
15	2	2·3	131	1	
16	2		139	1	
17	4		173	1	
18	2		200	1	0·06
19	0		223	1	
20	4	1·7	232	1	
21	4		294	1	
22	1		323	1	
23	1		603	1	
24	0		1799	1	
25	1	1·3			
28	2				
29	2				
30	0	1·1			
33	2		Total individuals (N)	= 6814	
34	2				
38	1		Total no. of species (S) =	197	
39	1				
40	3	0·8	α =	38	
42	2				
48	2		x =	0·994	

Conclusions:

The computed curve describes the frequency distribution quite closely despite the slight scatter of points about it. For practical purposes, the distribution of individuals within species, in the sample of Macrolepidoptera caught in the light trap, conformed to a logarithmic series.

EXERCISE 45. COMMUNITY DIVERSITY

Introduction:

"Diversity" is a measure of the amount of variability in the species composition of a community such as the trees in a forest, the flora of a field, the birds in a wood, or the moths in an area sampled by a light trap. A conifer plantation comprizing only one species of tree has no diversity. Even a natural coniferous forest usually shows little diversity. Growing crops have no diversity if they are kept free of weeds. In contrast a piece of grazed chalk downland usually has a very diverse flora and one of Man's main impacts on the flora and fauna is to affect its diversity (see p. 24).

Insects are often used to demonstrate "diversity" because they are easy to sample and are so profuse that sampling is unlikely to damage the fauna. Since it is necessary to separate communities into populations of species to investigate diversity, even though it is not essential to be able to name each species, moths, most of which are easily identified, are convenient insects to use.

Hypothesis:

The species composition of a community is a measurable character, "diversity". Man's activities usually diminish diversity.

Apparatus:

As Ex. 44.

Site and Time of Year:

As for Ex. 44.

Procedure:

Calculate the index of diversity as described in Chapter III (p. 107) from the total number of individuals (N) and the total number of species (S). Try to operate the light trap for a whole year to get a good estimate of α.

Take out every eighth night's catch and recount the number of species and the number of individuals for this one-eighth of a year sample. Repeat for other one-eighth year samples, beginning on successive days. This will give eight separate estimates for N, the total number of individuals, and S, the total species. Take the mean of the eight estimates of N and S. Read off α from Appendix K, p. 372. Compare α for the one-eighth year sample with α from the whole year's sample (Table IV:45:1).

If possible, compare α from different sites, either by using more than one trap, or by co-operating with other people in parallel exercises. The diversity of the moth population can be related to the physiography of the site which can be assessed using an Ordnance Survey map (25 in. to the mile). On this map, locate the site of the trap and using it as centre, describe a circle of 2·5 in. radius. Within this circle trace all the detail and, using different shading, distinguish: (a) buildings, surfaced roads and areas of concrete; (b) water;. (c) arable fields; (d) cut grass or lawns; (e) grazed grass or pasture; (f) gardens, rough land with shrubs, hedgerows, roadside verges, allotments,

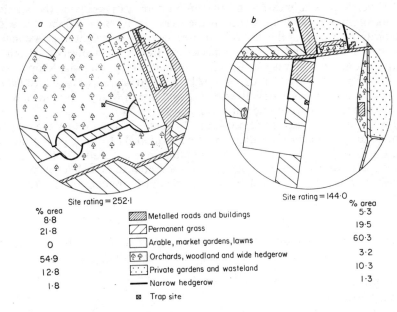

Site rating = 252·1

% area		% area
8·8	▨▨ Metalled roads and buildings	5·3
21·8	◿ Permanent grass	19·5
0	☐ Arable, market gardens, lawns	60·3
54·9	🜋🜋 Orchards, woodland and wide hedgerow	3·2
12·8	⠿ Private gardens and wasteland	10·3
1·8	▬ Narrow hedgerow	1·3
	⊠ Trap site	

Site rating = 144·0

FIG. IV:45:1.   Maps of 20-acre sites centred on a light trap; (a) has a high physiographic score and (b) a low score.

orchards and woodland (Fig. IV:45:1). Measure the total area under each heading. Convert each to a percentage of the total area of the circle. Allot a physiographic score to each type of land as follows: 0 for categories (a) and (b); 1 for categories (c) and (d); 2 for category (e); 3 for category (f). Multiply the score by the percentage area and sum (Table IV:45:2). Plot α against the total score for the site (Fig. IV:45:2).

Results:

From Table IV:45:1: This table includes one-eighth year samples, mean samples from several years, many for a whole year and also 4-year and 8-year samples. In spite of the very great difference in total number of insects in the sample, the estimate of α differs only slightly. This slight difference is because the time taken to collect the samples increases with each increase in sample size, hence a slightly greater community is sampled over 8 years than over 1; the communities are not static in time or place. Allowing for this, the estimate of diversity is remarkably constant, even though at one extreme there are 32,853 individuals in 285 species and at the other, 250 individuals in 67·7 species (this is a mean value, hence the fraction of a species). In other words, diversity is a property of a community, not dependent upon the sample size, nor necessarily on population density; compare 1935 and 1936.

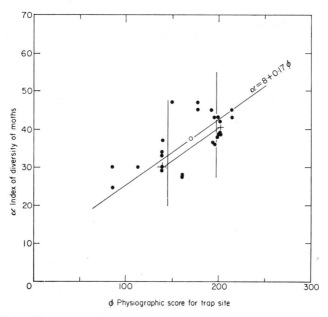

ϕ Physiographic score for trap site

FIG. IV:45:2. The relationship between the Index of Diversity (α) of a moth population and physiographic site score (ϕ) for a number of sites in different years. The significant relationship between α and ϕ is defined by the regression line α = 8 + 0·17 ϕ, established graphically in Chapter III.

Figure IV:45:2 shows results from many different sites, not all available in towns, hence co-operation with observers in the country would add variety and interest to the exercise. When a Rothamsted light trap is used, the results can be added to Fig. IV:45:2. Ultraviolet lamps sample from a somewhat different population of moths whose diversity is slightly higher.

The regression line in Fig. IV:45:2 is drawn through the computed mean for all values and is parallel to the line joining the medial crosses (see p. 80). It shows a relationship that is significant when tested by the quadrant method.

Conclusion:

The figure shows that diversity is greatly affected by vegetation and land use in the surrounding 20 acres, even when the total moth population is not much less. The major differences in physiography are those imposed by Man, buildings and cultivation, hence Man has an important influence on the community diversity of moths.

TABLE IV:45:1

Catches in Lepidoptera in a light trap at Rothamsted Experimental Station, showing the relation of the diversity to the length of the trapping period

Year	Individuals	Species	Diversity
One-eighth of a year, average			
1933	440	89·6	34·0
1934	409	86·8	33·7
1935	855	107·1	32·2
1936	250	67·7	30·9
Whole year			
1933	3454	173	38·4
1934	3276	168	37·4
1935	6530	191	36·8
1936	1961	154	39·2
1946	3124	156	35·1
1947	3786	184	40·2
1948	3771	187	41·1
1949	6107	191	37·5
4 years			
1933–36	15,221	234	39·7
1946–49	16,972	254	42·9
All 8 years	32,853	285	43·8

TABLE IV:45:2

Example of method for deriving the site score of an area within the
trapping range of a light trap (Site a, see Fig. IV:45:1)

Land type	Suggested physiographic rating	% area occupied	% area × rating
Metalled roads and buildings	0	8·8	0
Arable, market gardens and lawns	1	0	0
Permanent grass	2	21·8	43·6
Orchards, woodland, wide hedgerow	3	54·9	164·7
Private gardens, wasteland	3	12·8	38·4
Narrow hedgerow	3	1·8	5·4
Site score			Total 252·1

CHAPTER V

Ecological Apparatus and Techniques

COLLECTING

Each piece of collecting apparatus used in the exercises is described briefly below, but the sampling procedures themselves are either described in Chapter III or in each exercise separately, because the same apparatus may be used or sited differently, depending on the animals sought.

TRAPS

Water Trap Water traps catch insects that fly in daylight. Shallow tins of any shape are suitable providing that they are 5–8 cm deep and about 0·1 sq. m in area. Groups of students may use larger traps but the larger the trap, the greater the catch and the more time required for sorting. The inside of the traps should be painted; white is the most generally attractive colour; yellow will also attract many species, especially aphids. Half fill traps with water and add a few drops of detergent so that insects sink and cannot drift to the edges and escape. Remove insects with forceps or a pipette.

Light Trap (Fig. V:1) Light traps are used to catch crepuscular and night-flying insects. The simple "Rothamsted" model, which uses a 200 watt bulb with a tungsten element, catches sufficient insects for ecological studies without reducing local populations. The insects collect in the killing jar (a "Kilner" preserving jar) screwed beneath the light. The inner sides of the jar are coated with plaster of paris soaked in tetrachlorethane. The fumes quickly kill the insects and the catch is collected daily by replacing the full jar with an empty one.

Pitfall Trap (Fig. V:2) Any shallow jar sunk into the ground with its opening at soil level will serve as a pitfall trap and catch animals running on the surface of the ground. Ensure that earth is packed against the outer rim of the opening after sinking the jar. Cover the opening with a piece of wood or bark, raised on small stones, to prevent rain entering. Captured specimens may be removed with forceps without disturbing the trap. Pitfall traps usually catch most animals at night.

Sticky Traps Cards, boards, jars or lengths of large diameter piping each with total surface area of 0·1–0·2 sq. m may be smeared with transparent or pale brown tree-banding grease and used as sticky traps. It is better to smear the grease on to transparent acetate sheeting and clip this on to the solid structure with paper clips for support. Traps which are more convenient to handle and erect, but which catch fewer insects, may be made from glass boiling tubes (15–20 cm long × 2·5–4 cm diameter) inverted over the top of canes stuck in the ground. Cover the tubes with grease except for 5 cm at the closed end; leave this clean to hold. Insects may be removed from the acetate sheet or from the tubes in or out of doors, using either forceps or a small brush dipped in solvent, e.g. xylol, benzene or petroleum ether. To clean traps completely, soak the grease in paraffin, then wipe it off. Do not use dark coloured greases as these obscure the catch.

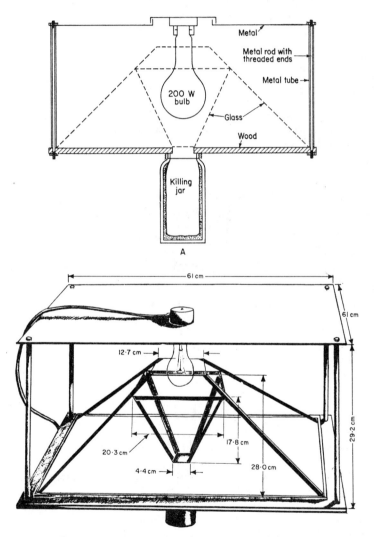

Fig. V:1.   "Rothamsted" light trap. (a) Cross section of trap. (b) Drawing of assembled trap showing essential dimensions for construction.

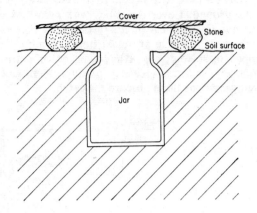

FIG. V:2.   Pitfall trap. One pound size honey jars make useful traps. Avoid large containers or frogs and hedgehogs may enter and eat the catch.

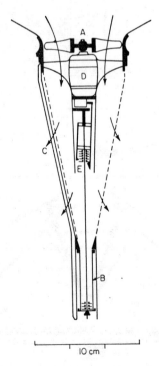

FIG. V:3.   Suction trap. Arrows indicate direction of air flow. A, fan inlet; B, collecting tube; C, filter net inside iron framework; D, fan motor; E, disc-dropping device.

Suction Traps Suction traps sample the aerial population of flying insects and are equally effective by day or night. A fan sucks into the trap a constant volume of air per hour. The air escapes through a metal gauze cone, but the animals are retained and fall into a collecting jar containing alcohol. More complicated models (Fig. V:3) segregate the catch into pre-selected periods (usually of 1 hr) throughout the day. No exercises necessitating suction traps are included in Chapter IV because they are expensive, but they fulfil all the functions of the other aerial traps, and have additional uses, so are described here.

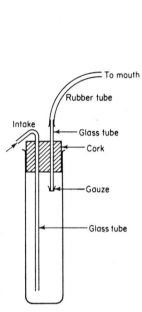

FIG. V:4. Pooter.

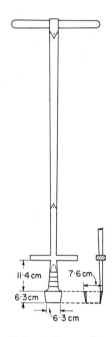

FIG. V:5. Soil sampling tool. The graduations indicate the depth of the soil core taken.

Nets For sweeping vegetation use strong, white canvas nets preferably with a rigid triangular or stirrup-shaped frame. Kite nets, for catching flying insects or specimens resting on the tips of vegetation, should be strong, but with a loose transparent weave and preferably with a circular frame. The bag of a kite net must be long enough to fold over and close the inlet when the frame is held horizontally, and thus retain the catch inside the net.

Pooters Small, living animals are most easily removed from nets or transferred from tube to tube with a pooter (Fig. V:4). To use this effectively, suck air into it in short vigorous bursts while holding the intake as close as possible to the animal. It is useful to have a number of collecting tubes and spare stoppers available in the field to interchange with the pooter head.

SOIL SAMPLES

Remove soil cores to the required depth using a soil sampling tool (Fig. V:5) or a metal pipe or tin of about 6 cm diameter that can be hammered into the soil.

EXTRACTION FROM SOIL AND LITTER

Extract animals from soil or leaf litter in a Tullgren funnel (Fig. V:6). Heat and light from the bulb drive the animals into a collecting tube containing alcohol. Leave samples in the funnel for about 3 days, until the litter is completely dry.

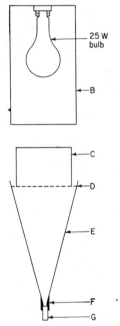

Fig. V:6.   Tullgren funnel. The 25-watt bulb is mounted in the metal cylinder (B). A litter or soil sample is placed in the smaller cylinder (C) on the sieve (D) which has ten meshes per inch. As the sample dries, the animals fall down the funnel (E) into a collecting tube containing 70% alcohol. The tube is held on to the funnel by a rubber collar (F). Overall length 67 cm.

LABELLING

Animals should be put into tubes or pill boxes as collected, with clearly written labels giving the date, number of the experiment, details of habitat (e.g. in meadow on dandelion flower), method of collection (e.g. sweeping), collector's name, and any other relevant information. *All* tubes and boxes in which living animals are placed should contain pieces of blotting paper or filter paper to which the animals can cling to prevent them shaking and to absorb condensation on the sides of the container.

KILLING AND PRESERVATION

KILLING BOTTLES

Kill insects to be stored dry, in a killing bottle. A safe model can be made with a screw-top or corked jar, or tube. Pour into it a layer of plaster of paris about 2 cm deep and allow to set and dry. Pour a few drops of ethyl acetate or tetrachlorethane on to the plaster and the fumes from it will kill most insects quickly. Ethyl acetate is the least dangerous and least objectionable chemical to use. Replenish the killing agent when its smell disappears from the bottle. When a pooter is used to collect animals from a net, it is useful for one of the spare specimen tubes to contain plaster and killing agent. *Never* suck a pooter attached to a killing bottle.

PRESERVATION

Specimens should be identified fresh where possible, especially slugs, earthworms, roundworms, spiders, bugs and midges (Chironomidae and Cecidomyiidae). When this is impossible, Table V:1 gives suitable quick methods for killing and preserving specimens. Complete methods of preservation and setting, for permanent storage and for detailed examination beyond the scope of this book, are given by Smart (1954), Wagstaffe and Fidler (1957) and Oldroyd (1958) (see Appendix N, p. 380).

Insects collected in water traps and on sticky traps are best preserved in 80% alcohol. In most of the exercises described, detailed identification is unnecessary and all insects can be transferred directly to 80% alcohol. However, if perfectly preserved specimens are required, only hard-bodied animals should be placed directly into 80% alcohol and soft-bodied animals should be immersed for a few hours in 30%, then a few hours in 50% before final transfer to 80% alcohol.

PRESERVATIVES

Alcohol, a general preservative "Industrial" methylated spirit (95%) diluted to about 80%, i.e. 1 volume of water to every 8 volumes of the industrial spirit, is the cheapest and most useful general preservative. It is wise to add a little glycerol (about 3%) to this preserving fluid, so that if corks and stoppers crack or fall out, specimens do not dry irrevocably and can be re-moistened.

Formalin, for preserving earthworms Formalin is usually purchased as a 40% solution of formaldehyde. Dilute 5 parts of the concentrated fluid with

TABLE V:1

Killing and preserving methods

Scientific name	Common name	Killing	Preserving
COELENTERATA *Hydrozoa* Hydrida	Hydra	Allow to extend in a drop of fresh water. With a pipette squirt hot (60°C) Bouin's fluid over the specimen, from foot to tip	80% alcohol
PORIFERA	Sponges	Wash in water and drop into 95% alcohol	95% alcohol
PLATY- HELMINTHES *Turbellarea*	Flatworms	Allow to extend in water. Flood with 10% formalin, leave for 15 min and wash in 50% alcohol	80% alcohol
ANNELIDA *Chaetopoda* Oligochaeta (Terrestrial)	Earthworms Enchytraeid worms	Wash off soil. Drop into 5% formalin or hot water	5% formalin
Hirudinea	Leeches	Allow to extend in 5–10% solution of chloral hydrate. When inert, straighten on to a glass slide and flood with 5% formalin for a few minutes	Compress between two slides, bind these together with thread and store in 80% alcohol
ARTHROPODA *Malacostraca* Isopoda	Woodlice Sowbug Hog slater	Drop into a mixture of 100 parts alcohol and 4 parts glycerol; leave for 10–14 days	80% alcohol
Chilopoda *Diplopoda*	Centipedes Millipedes	Drop into 50% alcohol or kill with ethyl acetate fumes	80% alcohol or store dry
Insecta Collembola	Springtails	Drop into 80% alcohol	80% alcohol
Orthoptera	Grasshoppers Katydids Crickets Mantids Cockroaches Walking sticks	Ethyl acetate fumes	Dry. Remove gut from large specimens

Scientific name	Common name	Killing	Preserving
Dermaptera	Earwigs	Ethyl acetate fumes or collect into 80% alcohol	Dry or in 80% alcohol
Plecoptera	Stone flies	Drop into 80% alcohol	80% alcohol
Psocoptera	Book lice Bark lice	Drop into 80% alcohol	80% alcohol
Ephemeroptera	Mayflies	Drop into 80% alcohol	80% alcohol
Odonata	Dragonflies	Ethyl acetate fumes or drop into 80% alcohol	Dry or in 80% alcohol
Thysanoptera	Thrips	Drop into 80% alcohol plus drop of ethyl acetate	80% alcohol
Hemiptera Heteroptera	Shield bugs Stink bugs Lace bugs Assassin bugs Plant bugs Pond skaters Water striders	Ethyl acetate fumes	Dry
Hemiptera Homoptera	Frog-hoppers Spittle insects Leaf-hoppers Cicadas Tree-hoppers	Ethyl acetate fumes or collect into 80% alcohol	Dry or 80% alcohol
	Whitefles Psyllids Aphids	Collect into 95% alcohol	95% alcohol
Neuroptera Mecoptera Trichoptera	Lacewings Scorpion flies Caddis flies	Ethyl acetate fumes or collect in 80% alcohol	Dry or 80% alcohol
Lepidoptera	Moths Butterflies	Ethyl acetate or tetrachlorethane fumes	Dry
Coleoptera	Beetles	Ethyl acetate fumes or immerse large, non-hairy species in boiling water	Dry or 80% alcohol
Strepsiptera	Stylops	Drop into 80% alcohol	80% alcohol
Hymenoptera	Bees, Wasps, Ants, Sawflies	Ethyl acetate fumes, or non-hairy species in 80% alcohol	Dry or 80% alcohol

TABLE V: 1—*continued*

Scientific name	Common name	Killing	Preserving
Diptera	Two-winged flies	Ethyl acetate fumes or collect in 80% alcohol	Dry or 80% alcohol
Araneida	Spiders Harvestmen Harvest-spiders	Collect into 80% alcohol	80% alcohol
Acarina	Mites	Collect into 80% alcohol	80% alcohol Clear and examine in 50–100% warm lactic acid, then return to alcohol for storage
MOLLUSCA *Gastropoda*	Slugs	Drown in boiled water	Fresh killed specimens are most easily identified
	Snails	Drown in boiled water	48 hr in 50% alcohol, then to 80% alcohol
	(Shells)	Drown as above and remove the soft part of the animal with a needle or flexible wire	Dry with a cloth or in dry sand, and store dry

Collect and preserve all larvae and immature insects in 80% alcohol or in van Emden's fluid. The larval cuticle should be pierced with a pin before immersion.

95 parts of water for general use. (This dilution is sometimes referred to as "5%" formalin.)

Van Emden's Fluid, for preserving larvae for future examination of external features

Glacial acetic acid	2 parts
Distilled water	30 parts
Formaldehyde, 40%	5 parts
Alcohol (95% industrial methylated spirit)	15 parts

Bouin's Fixative, for preserving animals for future histological examination

Picric acid, a saturated aqueous solution	30 parts
Formaldehyde, 40%	10 parts
Glacial acetic acid	2 parts

SOIL TEXTURE

To determine soil texture first moisten some soil until it reaches its maximum stickiness or plasticity. Then work a small amount of moistened soil between fingers and thumb. It is important to break down the soil thoroughly to eliminate the effects of structure and consistency. The method is an attempt to assess the relative proportions of coarse sand, fine sand, silt and clay in the soil and to categorize it according to the proportion of these groups present. This is a subjective method but with practice it will give sufficiently accurate results to compare with the number of animals in soil. Coarse sand consists of grains between 0·2 mm and 2 mm diameter. These are large enough to grate against each other and can be detected individually by touch and sight. In fine sand (0·05–0·2 mm) the grating is less obvious, though individual grains can be detected but not easily distinguished by touch or sight. Silt feels smooth and soapy, but only very slightly sticky, whereas clay is characteristically sticky. Dry clays may require a lot of water and manipulation before they develop maximum stickiness.

Soils can be classified into types using the criteria given below.

DESCRIPTION OF SOIL TEXTURES

Sand Soil consisting mostly of coarse and fine sand and containing so little clay that it is loose when dry and not sticky at all when wet. When rubbed it leaves no film on the fingers.

Sandy Loam Soil in which the sand fraction is still quite obvious, which moulds readily when sufficiently moist but in most cases does not stick appreciably to the fingers. Threads do not form easily.

Loam Soil in which the fractions are so blended that it moulds readily when sufficiently moist and sticks to the fingers to some extent. It can with difficulty be moulded into threads but will not bend into a small ring.

Silt Loam Soil that is moderately plastic without being very sticky and in which the smooth soapy feel of the silt is the main feature.

Clay Loam The soil is distinctly sticky when sufficiently moist and the presence of sand fractions can only be detected with care.

Silt Soil in which the smooth, soapy feel of silt is dominant.

Sandy Clay The soil is plastic and sticky when moistened sufficiently, but the sand fraction is still an obvious feature. Clay and sand are dominant and the intermediate grades of silt and very fine sand are less apparent.

Clay The soil is plastic and sticky when moistened sufficiently and gives a

polished surface on rubbing. When moist the soil can be rolled into threads
With care a small proportion of sand can be detected.
Silty Clay   Soil which is composed almost entirely of very fine material but
in which the smooth soapy feel of the silt fraction modifies the stickiness of
the clay.

SOIL ACIDITY

Simple kits for testing soil acidity can be purchased complete with
instructions on their use.

TEMPERATURE MEASUREMENT USING THERMOCOUPLES

If the two ends of a piece of wire are attached to the terminals of a galvanometer, there is no voltage difference between the ends of the wire so no current
flows through the circuit. However, if this wire is cut, the two cut ends
joined to the ends of a wire of different metal and the two junctions thus
produced are maintained at different temperatures, an electromotive force
(E.M.F.) is created in the circuit. An electric current, which is indicated by
the galvanometer, then flows.

This type of electric circuit, called a thermocouple, can be used to measure
temperature. The actual points of contact between the wires can be kept
extremely small, so thermocouples are useful for measuring temperature in
small and awkwardly placed sites, for example, inside the bodies of small
animals.

MATERIALS

Approximately 32 S.W.G. insulated copper wire Constantan (= Eureka,
or Contra) wire, and a sensitive, low resistance galvanometer. "Cambridge"
spot or "Cambridge" Pot galvanometers are both suitable.

CONSTRUCTION

Remove the insulation from the end 0·5 cm of two 60 cm pieces of copper
wire and from the ends of a 30 cm piece of "Constantan" wire. Twist the
wires together to form short probes about 2 cm long and solder the bared
ends. Attach the free ends of the copper wires to the positive and negative
terminals of the galvanometer.

If the soldered junctions of a thermocouple prepared in this way are
simultaneously immersed in two beakers of water at different temperatures,
an E.M.F. will be generated and current flow will register on the galvanometer.

The reproducible feature of a thermocouple output is the E.M.F. and this
increases almost linearly with difference in temperatures between the two
junctions. Thus if the temperature of one junction is known (melting ice at
0°C), or measured (using, for example, an ordinary mercury-in-glass thermometer), the galvanometer can be calibrated to read the temperature of the
other junction by noting the deflection at a number of temperatures within

the desired range, e.g. if a temperature difference of $\theta°C$ between the two junctions produced a reading of x milliamps then a difference of $1°C$ would give a reading of x/θ m amps, or 1 m amp $\equiv \theta/x°C$.

Since the galvanometer is a current-measuring device and not primarily a voltage-measuring instrument, the reading is only proportional to the voltage (E.M.F.) provided that the resistance of the circuit is constant (Ohm's Law). Thus, if the wires are stretched and the joints are faulty, the calibration will be upset. Similarly, if several thermocouples are to be used with the same galvanometer (e.g. for readings in rapid succession from a number of points) care must be taken to ensure that the resistances of the couples are as similar as possible.

A thermocouple calibrated in this way can be used to measure an unknown temperature simply by observing the current in the circuit when one junction (the "reference" junction) is controlled or measured as in the calibration experiment and the other (the "experimental" junction) is buried in the environment or body whose temperature is required.

As the E.M.F. × temperature difference curve is not exactly linear, an error will be introduced unless the reference temperature in the experiment is the same as the reference temperature at which the thermocouple was calibrated. Conduction of heat is very efficient in metals, particularly in copper, and unless the junction and a reasonable length of leads are within the temperature zone being measured, there is also a danger that heat will be conducted to or from the "outside" through the wires themselves, giving a false result. If melting ice in a vacuum flask is used as the reference temperature, add some water to the flask to ensure that the temperature of the reference junction is not below $0°C$, support the junction in the floating ice at the top of the flask and shake the flask regularly.

CHAPTER VI

Keys to Common Land Invertebrates

With a little practice these keys permit rapid identification to Class, Order or Family of most animals likely to be caught in tens or hundreds, during the exercises suggested in this book. To make the keys easy to follow, difficult taxonomic groups and some uncommon animals, which are likely to occur only in ones and twos, are omitted. However, a few rare, but spectacular species, like Stag beetles, are included, because even single specimens of such noticeable animals attract attention. Parasites on mammals and birds, such as lice, fleas, bed bugs, ticks and mites, tape-worms and roundworms, most aquatic animals and animals living in stored products or in buildings are omitted.

As most exercises suggested use insects, this Class is keyed in greatest detail. Where more precise identification is required or where the fauna is somewhat different, as in parts of the U.S., separate keys are given in the exercises themselves, or reference made to another suitable published key. Moths and butterflies (Lepidoptera) are identified most easily from coloured pictures, so keys to this order are omitted. A good flora is also necessary for some exercises.

It is best to identify specimens whilst fresh, but the keys are also suitable for pickled or dried animals (see p. 256). Special difficulties with pickled specimens are mentioned. Outstanding characters, visible to the naked eye or with a × 10 hand lens, are used wherever possible. For very small animals, or where it is essential to see detailed structures, the minimum magnification necessary to see these easily is given, e.g. × 60. Once identified correctly, specimens of some orders may be kept as a reference collection and used for subsequent identification.

The characteristic appearance of most families is illustrated with drawings of a typical animal or parts of animals. The drawings are arranged on the page opposite to the relevant text where possible. Full use *must* be made of these figures and the size ranges given beneath them to identify specimens correctly. Measurements should be taken from the front of the head to the tip of the abdomen, excluding any additional appendages. The positions of especially important features, or those difficult to see, are indicated by arrows. Specialized habitats occupied, or characteristic movements made, e.g. hopping, are also mentioned.

Under the numbers on the left-hand side of these pages are listed two or more choices. After deciding which description best fits the animal you are examining, read off the number on the right and find this number lower down on the left. Repeat this procedure until the name of the animal follows the description selected. Colloquial words are used frequently followed by technical terms in brackets. Separate keys have a different code letter and the numbering of figures in each starts at 1. The sequence of keys is:

KEY TO MAJOR GROUPS OF LAND INVERTEBRATES
(Except Orders of Insects)

1. With segmented legs.   2

—Without legs   7

2. Three pairs of legs or if 2 pairs, with brightly coloured wings. Body usually in 3 fairly distinct regions, head, thorax and abdomen (Fig. A,1a, b)   .   .   .   .   class INSECTA

—Three pairs of legs, body elongated and distinctly segmented. (Immature specimens see p. 225, Fig. J)   .   .   .   .   class INSECTA

—Three pairs of legs, but body short and indistinctly segmented. No wings (Fig. A2)   .   .   .   .   order ACARI
(Immature specimens)

—Four or more pairs of legs, rarely 2 pairs. No wings. Body in 1 or 2 regions   3

3. Four, 2 or apparently 5 pairs of legs. Body usually rounded .   .
.   .   .   .   class ARACHNIDA...4

—More than 5 pairs of legs. Body usually elongated   .   .   .   5

4. A pair of large palps with pincers on the head, resembling lobster's claws (Fig. A3)   .   .   .   .   .   .   order CHELONETHI
(False scorpions)

—Body > 1 mm, with 1 region, no waist. Long palps resembling short legs, true legs with the second pair the longest (Fig. A4). order OPILIONES
(Harvestmen or harvest spiders)

—Body in 2 regions, distinct waist. Second pair of legs rarely longest (Fig. A5, a). Silk glands near the tip of the abdomen (Fig. A5,b)
.   .   .   .   ·   order ARANEIDA
(Spiders)

—Body in 1 main region with a small additional region sometimes visible. Round or slightly elongated with 4 pairs of legs, or sausage-shape with 2 pairs of legs (Fig. A6,a, b, c, d)   .   .   order ACARI
(Mites)

continued

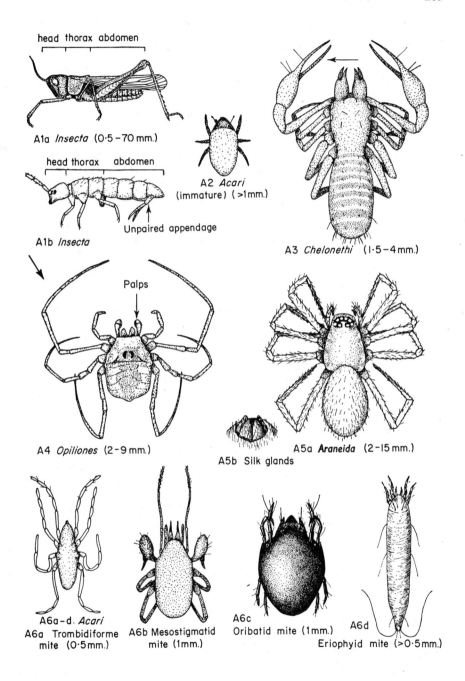

head thorax abdomen

A1a *Insecta* (0·5–70 mm.)

head thorax abdomen

Unpaired appendage

A1b *Insecta*

A2 *Acari* (immature) (>1mm.)

A3 *Chelonethi* (1·5–4 mm.)

Palps

A4 *Opiliones* (2–9 mm.)

A5b Silk glands

A5a **Araneida** (2–15 mm.)

A6a–d. *Acari*
A6a Trombidiforme mite (0·5mm.)

A6b Mesostigmatid mite (1mm.)

A6c Oribatid mite (1mm.)

A6d Eriophyid mite (>0·5mm.)

5. Body no more than 3 × longer than wide. Seven pairs of legs. Segments resemble plates of armour. Black, steely-blue or grey (Fig. A7)   .
.   .   .   .   class MALACOSTRACA
order ISOPODA
(Woodlice, Sowbugs)

—Body very elongated, more than 3 × longer than wide .   .   .   6

6. 1–2 mm. Nine pairs of functional legs. Very delicate branched antennae (× 100). Body with 12 segments. Whitish. In soil and litter (Fig. A8)   .   .   .   .   class PAUROPODA
(Pauropods)

—1–8 mm. Long, unbranched beaded antennae. Seven to 12 pairs of legs. Body with 15–22 segments. Whitish. In soil and litter (Fig. A9)   .
.   .   .   .   class SYMPHYLA
(Symphylids)

—Body flattened dorsoventrally with 1 pair of legs per segment. Poison "jaws" at front of head. In soil and litter (Fig. A10,a, b)   .
.   .   .   .   class CHILOPODA
(Centipedes)

—Body usually cylindrical, occasionally flattened dorsoventrally, but always with 2 pairs of legs on most segments. In soil and litter (Fig. A11, a, b, c)   .   .   .   .   class DIPLOPODA
(Millipedes)

7. Body pointed at both ends and composed of many distinct, ring-like segments (Fig. A12)   .   .   .   .   .   class CHAETOPODA...8
order OLIGOCHAETA

—Body with a large, blunt sucking pad at one end and a small pad at the other. Composed of more than 14 ring-like segments (Fig. A13)
class HIRUDINEA
(Leeches)

—Body with less than 14 ring-like segments, often with a hardened head capsule or dark hooks at one end   .   .   .   class INSECTA*

—Body unsegmented   9

* Immature specimens. See p. 225, Fig. IV:39; 3, Y, Z.

continued

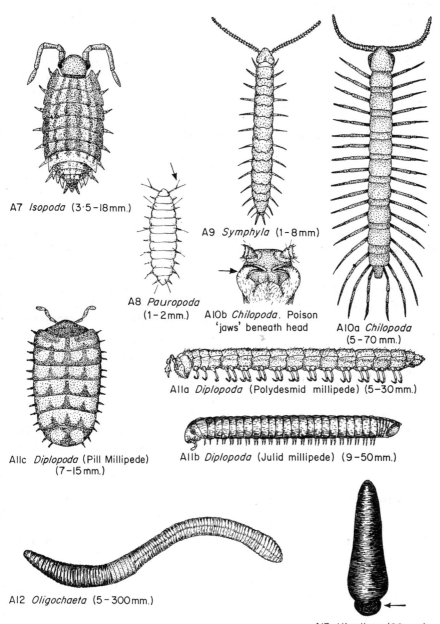

A7 *Isopoda* (3·5–18mm.)

A9 *Symphyla* (1–8mm)

A8 *Pauropoda* (1–2mm.)

A10b *Chilopoda*. Poison 'jaws' beneath head

A10a *Chilopoda* (5–70 mm.)

A11a *Diplopoda* (Polydesmid millipede) (5–30mm.)

A11c *Diplopoda* (Pill Millipede) (7–15mm.)

A11b *Diplopoda* (Julid millipede) (9–50mm.)

A12 *Oligochaeta* (5–300mm.)

A13 *Hirudinea* (20mm.)

8. White or pale pink, < 25 mm, bristles along sides arranged as Fig.
   A14,a (×15)   .   .   .   .   .   .   family ENCHYTRAEIDAE
                                                       (Enchytraeid worms)

—Pink, red or purple, > 25 mm. Bristles along sides arranged as
   Fig. A14,b (×15)   .   .   .   .   .   family LUMBRICIDAE
                                                       (Earthworms)

9. Flat or dome-shaped scales, sedentary. On twigs or beneath leaves
   (Fig. A15)   .   .   .   .   class INSECTA
                                                       order HEMIPTERA
                                                       (Scale insects)*

—Elongated or cyst-like. Usually in or on soil   10

10. Golden-brown, cyst-like > 1·5 mm (Fig. A16)   .   family LUMBRICIDAE
                                                       (Earthworm cocoons)

—Transparent, pale amber or translucent white, oval > 1·5 mm in
   clusters, strings or single   .   .   .   .
                    .   .   .   Invertebrate eggs—probably of slugs, snails or centipedes

—White. Thin and cylindrical, < 1·0 mm or 40–200 mm (Fig. A17,a),
   or small white or brown spherical or lemon shaped cysts, < 1·5
   mm with at least 1 small protrusion (Fig. A17,b, c)   .   class NEMATODA
                                                       (Roundworms and eelworms)

—Brown or black. Extremely long, up to 300 mm. Like a piece of horse-
   hair (Fig. A18)   .   .   .   .   .   .   class NEMATOMORPHA
                                                       (Threadworms or hairworms)

—With tentacles; slimy   .   .   .   .   class GASTROPODA...11

—Characters other than above   .   .   .   .   .   OTHER TAXA

11. With an external shell (Fig. A19)   .   .   .   .   .   (Snails)

—Without an external shell (Fig. A20)   .   .   .   .   (Slugs)

* The segmented body is completely hidden by the scale.

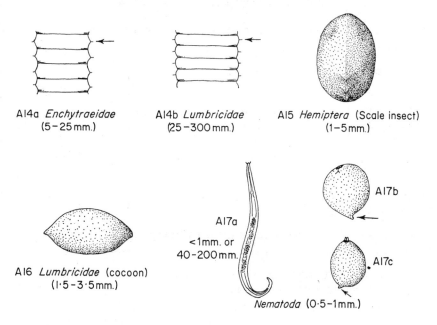

A14a *Enchytraeidae* (5-25 mm.)

A14b *Lumbricidae* (25-300 mm.)

A15 *Hemiptera* (Scale insect) (1-5mm.)

A16 *Lumbricidae* (cocoon) (1·5-3·5mm.)

A17a

<1mm. or 40-200mm.

A17b

A17c

Nematoda (0·5-1mm.)

A18 *Nematomorpha* (100-300mm.)

A19 *Gastropoda* (Snail) (Shell height 2-35mm.)

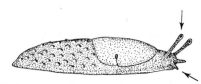

A20 *Gastropoda* Slug (up to 140mm.)

KEY TO ORDERS OF INSECTA

1. With complete (Fig. B1,a, b, c) or partially developed wings (Fig. B1,d, e, f) 2

—No trace of wings 17

2. Two pairs of wings. N.B. examine carefully, lower pair may be hidden by hard or leathery upper pair 3

—One pair of wings 16

3. Both pairs of wings covered with dense scales or hairs, or with mealy, white, powdery wax 4

—Both pairs of wings membranous, mostly transparent and of similar texture 6

—Forewings hard or leathery, mostly opaque and of different texture from the membranous hind wings 11

4. Wings completely covered in flattened scales. Often brightly coloured, wing expanse > 5 mm (Fig. B2,a, b) . . . LEPIDOPTERA
(Moths and butterflies)

—Wings covered in mealy, white, powdery wax (yellow in alcohol), wing expanse about 3 mm. Sucking mouthparts (Fig. B3,a, b)
HEMIPTERA–HOMOPTERA
(Whiteflies)

—Wings only partially clothed in hairs 5

5. Fore and hind wings of similar shape with many small cross veins and veinlets, especially around the edges, giving a lace-like appearance. Short hairs on wings. Brown (Fig. B4) . . NEUROPTERA
(Lacewings)

—Hind wings usually broader than forewings, few cross veins. Margins of wings often with long hairs. Brown or black. Usually near water (Fig. B5,a, b) . . . . . . TRICHOPTERA
(Caddis flies)

continued

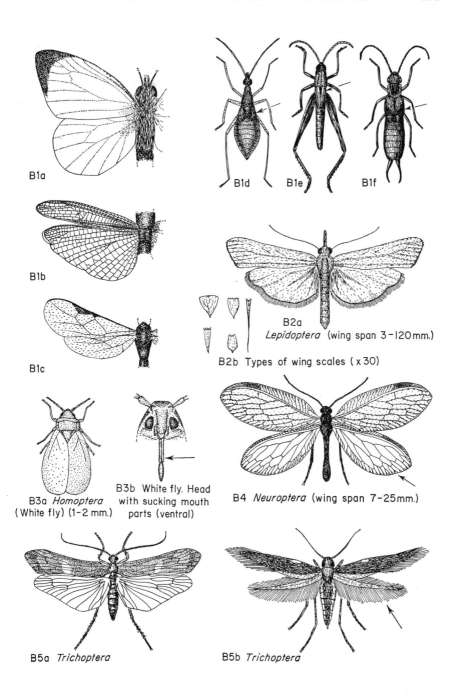

B1a

B1d   B1e   B1f

B1b

B1c

B2a
Lepidoptera (wing span 3-120mm.)

B2b Types of wing scales (x30)

B3a *Homoptera*   B3b White fly. Head   B4 *Neuroptera* (wing span 7-25mm.)
(White fly) (1-2 mm.)   with sucking mouth
    parts (ventral)

B5a *Trichoptera*   B5b *Trichoptera*

6. Abdomen terminated with 2 or 3 long, straight filaments (cerci).
Usually near water (Fig. B6)   .   .   .   .   EPHEMEROPTERA
(Mayflies)

—Abdomen without any, or with one long terminal appendage   .   7

7. Long thin bodies, overall length > 25 mm, often brightly coloured.
Bulging eyes. Wings with complex pattern of small cross veins and
a conspicuous opaque spot near each tip. Usually near water (Fig.
B7,a, b)   .   .   .   .   ODONATA
(Dragonflies)

—Bodies unlike Fig. B7 or overall length < 25 mm.   .   .   .   8

8. Wings with many dark, well-defined blotches. Head produced
downwards into a short snout. Tip of abdomen in ♂♂ curved upwards and forwards (Fig. B8,a, b)   .   .   .   .   MECOPTERA
(Scorpion flies)

—Wings with many small cross veins and veinlets, especially around
the edges giving a lace-like appearance. Body green, yellowy-brown
or grey (Fig. B9)   .   .   .   .   NEUROPTERA
(Lacewings)

—Wings clear, or with diffuse tinted areas, or with 1 opaque spot near
the tip of the forewings   9

9. < 3 mm. Wings strap-like and fringed with long hairs. Occasionally
2 dark bands traversing each forewing. Body black, dark brown or
yellow (Fig. B10,a, b)   .   .   .   .   .   .   THYSANOPTERA
(Thrips)

—Margins of wings bare or very slightly hairy   10

continued

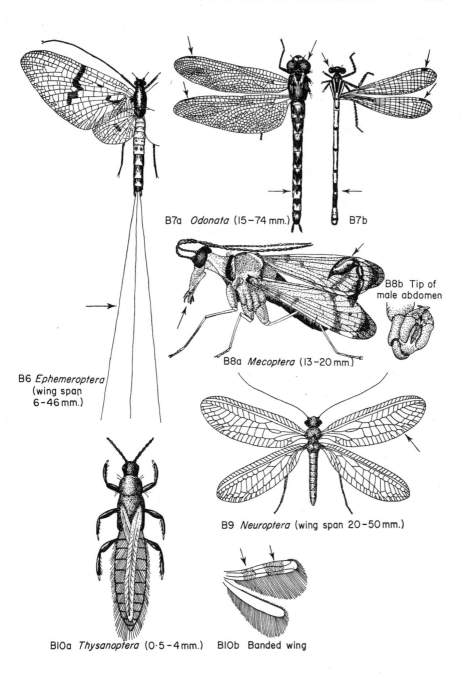

B7a *Odonata* (15–74 mm.) B7b

B8b Tip of male abdomen

B8a *Mecoptera* (13–20 mm.)

B6 *Ephemeroptera* (wing span 6–46 mm.)

B9 *Neuroptera* (wing span 20–50 mm.)

B10a *Thysanoptera* (0·5–4 mm.) B10b Banded wing

10. Dull grey or brown. Wings held roof-like with venation as in Fig.
B11. Antennae with 12 or more segments. Biting mouthparts.
Often found on bark   .   .   .   . PSOCOPTERA
(Psocids, barklice, booklice)

—Antennae with less than 10 segments. Venation as in Fig. B12,a.
Piercing and sucking mouthparts (Fig. B12,b). Often with tubular
outgrowths on the sides of the abdomen (Fig. B12,c, d) . HEMIPTERA–
HOMOPTERA
(Plant lice, aphids)

—With a set of characters as follows:
either (i) densely furry (as a bumble bee) (Fig. B13,a) ⎫
or   (ii) with a pronounced waist between thorax and │
abdomen and an opaque spot near the tip │
of the forewings (Fig. B13,b);   │ HYMENOPTERA
or   (iii) with a waist, few veins but elbowed antennae ⎬ (Bees, wasps,
· (Fig. B13,c);   │ ants and
or   (iv) dorso-ventrally flattened, no waist, opaque │ sawflies)
spot on leading edge of forewings (Fig. │
B13,d)   ⎭

11. Forewings short and most of abdomen exposed 12

—Forewings long and all or most of abdomen covered .   .   . 14

12. Abdomen terminated by a pair of hardened, curved pincers (Fig.
B14,a, b)   .   .   .   . DERMAPTERA
(Earwigs)

—Abdomen without terminal appendages or with short straight ones 13

continued

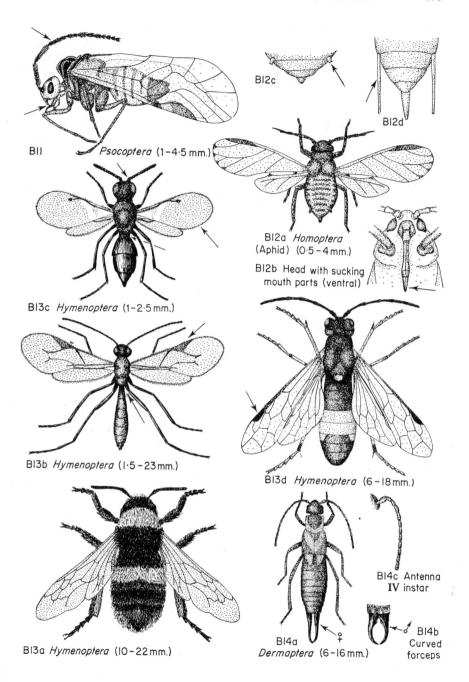

B12c

B12d

B11   *Psocoptera* (1–4·5 mm.)

B12a *Homoptera*
(Aphid) (0·5 – 4 mm.)

B12b Head with sucking
mouth parts (ventral)

B13c *Hymenoptera* (1–2·5 mm.)

B13b *Hymenoptera* (1·5 – 23 mm.)

B13d *Hymenoptera* (6 – 18 mm.)

B14c Antenna
IV instar

B14b
Curved
forceps

B14a   ♀
Dermaptera (6 – 16 mm.)

B13a *Hymenoptera* (10 – 22 mm.)

13. Hind legs much longer and thicker than other legs (Fig. B15,a).
Biting mouthparts (Fig. B15,b) . . . . ORTHOPTERA
(Grasshoppers, crickets)

—Hind legs slightly longer and usually thicker than other legs (Fig.
B16,a). Piercing and sucking mouthparts (Fig. B16,b) . HEMIPTERA

—All legs similar. Biting mouthparts (Fig. B17) . . COLEOPTERA
(Staphylinid beetles)

14. The basal two-thirds of the forewings stiffer than the terminal one-
third which is often membranous. When wings are folded the mem-
branous areas overlap to form a diamond shaped patch (Fig. B18,a).
Piercing and sucking mouthparts (Fig. B18,b, c) . . HEMIPTERA–
HETEROPTERA
(Plant bugs)

—Forewings of uniform consistency 15

15. Hind legs about twice as long as, and thicker than, other legs.
Biting mouthparts (Fig. B15,b). Wings mostly held roof-like over
abdomen (Fig. B19), occasionally flattened . . ORTHOPTERA
(Grasshoppers, crickets, katydids)

—Hind legs slightly longer and thicker than other legs. Piercing and
sucking mouthparts (Fig. B16,b). Wings held roof-like over abdomen
(Fig. B20) . . . . . . HEMIPTERA–HOMOPTERA
(Plant Hoppers)

—All legs similar. Forewings often hard and held flat or domed over
the abdomen (Fig. B21,a, b) . . . . . COLEOPTERA
(Beetles)

continued

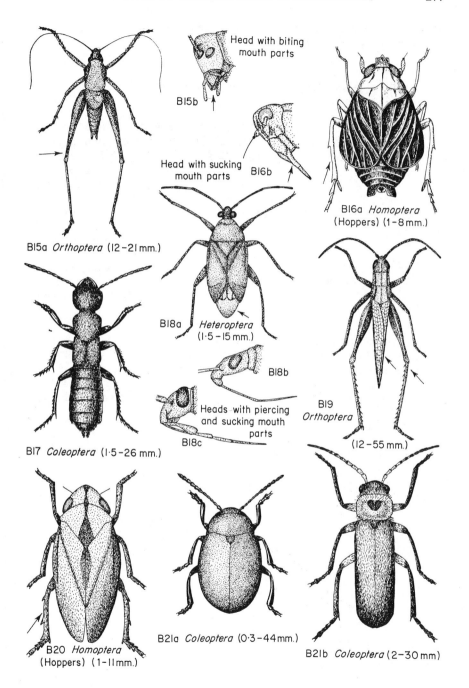

B15b Head with biting mouth parts

B15a *Orthoptera* (12–21mm.)

Head with sucking mouth parts B16b

B16a *Homoptera* (Hoppers) (1–8mm.)

B18a *Heteroptera* (1·5–15 mm.)

B17 *Coleoptera* (1·5–26 mm.)

B18b

B18c Heads with piercing and sucking mouth parts

B19 *Orthoptera*

(12–55 mm.)

B20 *Homoptera* (Hoppers) (1–11mm.)

B21a *Coleoptera* (0·3–44mm.)

B21b *Coleoptera* (2–30 mm)

16. Forewings membranous, hind wings reduced to small, clubbed
structures (halteres) (Fig. B22,a, b)　　.　　.　　.　　.　　DIPTERA
(Two-winged flies)
Very common and almost all insects with 1 pair of wings
will be flies. They can be distinguished from other, rarer,
2-winged insects as follows:

　—Forewings membranous, hind wings minute and flattened. Last
division of leg (tarsus) with 1 segment and 1 claw. Compound eyes
like small blackberries. Rare (Fig. B23)　　.　　.　HEMIPTERA–
HOMOPTERA
(Coccid males)

　—Forewings reduced and club-like, hind wings membranous. Rare
(Fig. B24)　.　　.　　.　　.　　.　　.　　.　　.　STREPSIPTERA
(Stylopid males)

　—Forewings hard, fused together to form an immovable cover to the
abdomen (Fig. B25)　.　　.　　.　　.　　.　　.　　.　COLEOPTERA
(A few ground beetles and weevils)

17. Scale-like, sedentary and adpressed on bark or leaves (Fig. B26,a),
or sluggish and covered in particles or strands of mealy white wax
(Fig. B26,b). Legs minute or invisible. Common in hothouses　　.
.　　.　　.　　.　　.　　.　　.　　.　　.　HEMIPTERA–HOMOPTERA
(Scale insects and mealy bugs)

　—Legs clearly visible　.　　.　　.　　.　　.　　.　　.　　.　　.　18

　—Characters other than above　.　　.　　.　　.　　.　Other orders

continued

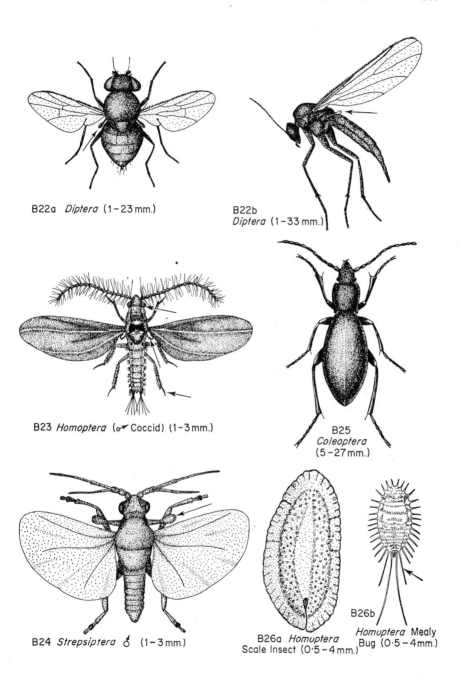

B22a *Diptera* (1–23 mm.)

B22b
Diptera (1–33 mm.)

B23 *Homoptera* (♂ Coccid) (1–3 mm.)

B25
Coleoptera
(5–27 mm.)

B24 *Strepsiptera* ♂ (1–3 mm.)

B26a *Homoptera*
Scale Insect (0·5–4 mm.)

B26b
Homoptera Mealy
Bug (0·5–4 mm.)

18. Three long, segmented appendages at or near the tip of the abdomen
(Fig. B27). Around hearths, ovens and on beaches . THYSANURA
(Bristle-tails, silverfish)

—Two segmented appendages at the tip of the abdomen (Fig. B28). In
soil . . . . DIPLURA

—Tip of abdomen bare or with a short, single straight appendage . 19

19. Distinct waist between thorax and abdomen. Antennae elbowed
(Fig. B29) . . . . HYMENOPTERA
(Ants)

—No waist, antennae usually straight 20

20. Each dorsal edge of the abdomen with a small, unsegmented, tubular
or barrel-shaped outgrowth (cornicle) (Fig. B30) . HEMIPTERA–
HOMOPTERA
(Aphids)

—Stubby or elongated and sometimes with a forked appendage be-
neath the abdomen. Elongated (Fig. B31,a) or globular (Fig. B31,b).
White, grey or black. In the soil or beneath rotting bark . COLLEMBOLA
(Springtails)

—Abdomen without appendages, orange-brown or black (Fig. B32)
THYSANOPTERA
(Thrips)

—Opaque, yellowish, often with a conspicuous median white stripe
internally. No compound eyes or antennae (though the forelegs may
resemble antennae). In litter and soil (Fig. B33) . . PROTURA

—Characters other than above . . . . . Other orders

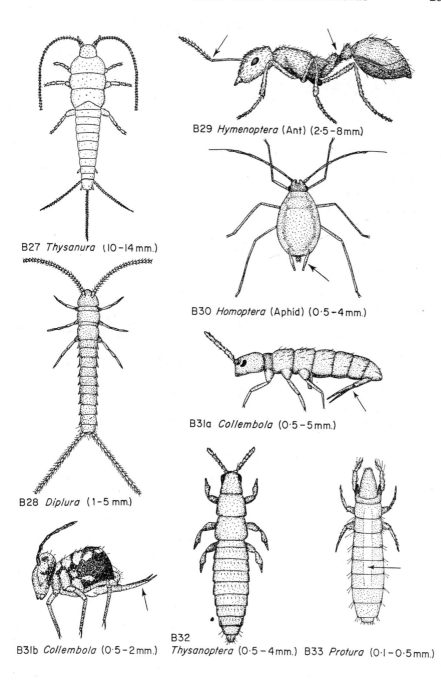

B27 *Thysanura* (10–14 mm.)

B28 *Diplura* (1–5 mm.)

B31b *Collembola* (0·5–2 mm.)

B29 *Hymenoptera* (Ant) (2·5–8 mm.)

B30 *Homoptera* (Aphid) (0·5–4 mm.)

B31a *Collembola* (0·5–5 mm.)

B32 *Thysanoptera* (0·5–4 mm.)   B33 *Protura* (0·1–0·5 mm.)

KEY TO FAMILIES OF HEMIPTERA

(Hoppers, Plant lice and Bugs with elongated piercing mouthparts)

1. Forewings with firm, uniform texture, often coloured, held roof-like over body. Vigorous hoppers 2

—Wingless, or with delicate white or transparent wings. Walk slowly 3

—Flattened; very long legs. Skate on water surfaces . . . 4

—Flattened; tips of forewings usually membranous and overlapping; forewings sometimes short. Run quickly on the ground or on vegetation 5

2. Brown. Thorax dark brown, enlarged and thorn-like (Fig. C1,a, b). On broom and in hedgerows . . . . . MEMBRACIDAE
(Treehoppers)

—Brown or fawn, or with large red spots. Streamlined shape. Hind legs with a *cluster* of spines on tip of the last long segment (tibia) (Fig. C2,a, b). On grasses, herbaceous plants, shrubs and trees .
. . . . CERCOPIDAE
(Froghoppers, spittle insects)

—Green, brown, yellow, straw or speckled. Hind legs with a *row* of spines along the outer edge of the last long segment (tibia) (Fig. C3,a, b, c). On grasses herbaceous plants, shrubs and trees . JASSIDAE
(Leafhoppers)

—Brown, or fawn, with grey or black shading. Hind legs with a scythe-shaped spur at the tip of the last long segment (tibia). Wings either covering abdomen or stumpy (Fig. C4,a, b, c). Among long herbage . . . . DELPHACIDAE

continued

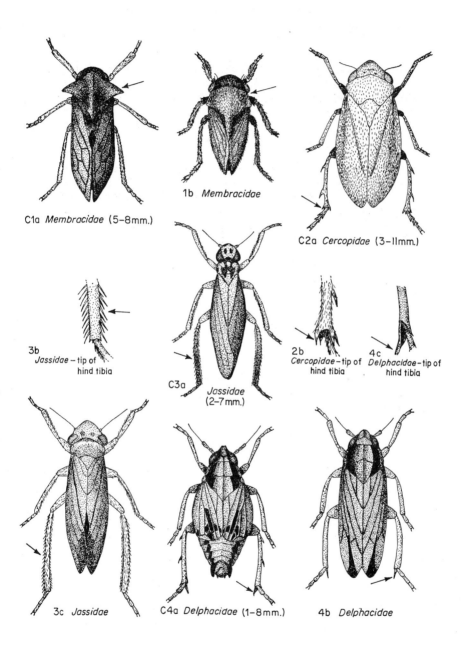

C1a *Membracidae* (5–8mm.)

1b *Membracidae*

C2a *Cercopidae* (3–11mm.)

3b
Jassidae – tip of
hind tibia

C3a *Jassidae*
(2–7mm.)

2b
Cercopidae – tip of
hind tibia

4c
Delphacidae – tip of
hind tibia

3c *Jassidae*

C4a *Delphacidae* (1–8mm.)

4b *Delphacidae*

3. White or grey, thorax and wings powdery (Fig. C5). On Brassicas
   and in glasshouses   .   .   .   . ALEYRODIDAE
   (Whiteflies)

—Green, yellow, pink, grey or black. Small tubular or barrel-shaped
   outgrowths (cornicles) on the abdomen. Wingless, or winged (Fig.
   C6,a, b, c, d, e). On many grasses, herbaceous plants, shrubs and
   trees   .   .   .   . APHIDIDAE
   (Aphids, greenflies, blackflies)

4. Dark brown or black above, dense silvery hairs beneath. Very
   elongated, wings either stumpy or cover most of abdomen. The first
   long segment of the hind legs (femur) reaches beyond the end of the
   body (Fig. C7)   .   .   .   . GERRIDAE
   (Pond skaters, water striders)

5. Brown or black; stumpy heads with large prominent compound eyes
   (Fig. C8). Run rapidly on mud or sand in wet places   . SALDIDAE
   (Shore bugs)

continued

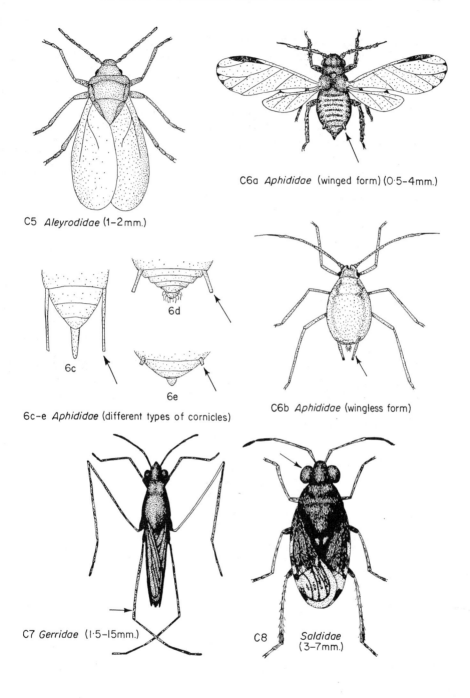

C6a *Aphididae* (winged form) (0·5-4mm.)

C5 *Aleyrodidae* (1-2 mm.)

6d

6c

6e

6c-e *Aphididae* (different types of cornicles)

C6b *Aphididae* (wingless form)

C7 *Gerridae* (1·5-15mm.)

C8 *Saldidae* (3-7mm.)

5. (continued)

—Brown or grey. Upper surface, especially the forewings covered with raised, lace-like patterns (Fig. C9). On herbaceous plants, and shrubs . . . . TINGIDAE
(Lace bugs)

—Green, brown or fawn, occasionally black and white. Back covered by a large shield about half as long as the forewings; often widest across the thorax (Fig. C10,a, b). On herbaceous plants, shrubs and trees . . . . PENTATOMIDAE
(Shield bugs, stink bugs)

—Fawn or brown with grey or black shading. Elongated. Long, thin antennae, curved piercing mouthparts (proboscis) (Fig. C11,c). Forewings short or cover the abdomen (Fig. C11,a, b). Raptorial forelegs. Among long grass NABIDAE
(Damsel bugs)

—Black, dark brown and cream. Conical head with 2 simple eyes and straight 3-segmented piercing mouthparts (proboscis) (Fig. C12,a, b, c). Forewing notched midway along leading edge. On flowers and low herbage . . . . ANTHOCORIDAE
(Flower bugs)

—Green, yellow, brown, fawn, occasionally black or speckled. No simple eyes and straight 4-segmented piercing mouthparts (proboscis) (Fig. C13,a, b, c). Forewing notched beyond middle of leading edge; veins on membrane recurved. Common on many herbaceous plants, shrubs and trees . . . . MIRIDAE
(Capsid bugs, leaf bugs, plant bugs)

—Characters other than above . . . . . Other families

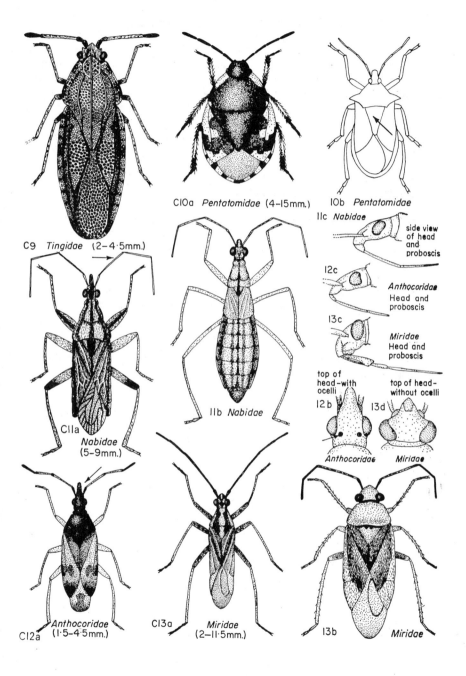

C9 *Tingidae* (2–4·5mm.)

C10a *Pentatomidae* (4–15mm.)

10b *Pentatomidae*

11c *Nabidae*

side view of head and proboscis

12c *Anthocoridae* Head and proboscis

13c *Miridae* Head and proboscis

11b *Nabidae*

11a *Nabidae* (5–9mm.)

top of head – with ocelli

12b *Anthocoridae*

top of head – without ocelli

13d *Miridae*

12a *Anthocoridae* (1·5–4·5mm.)

C13a *Miridae* (2–11·5mm.)

13b *Miridae*

KEY TO FAMILIES OF DIPTERA
(Two-winged Flies)

It is impossible to key out all common families of flies using simple, obvious characters. Many families of Acalypterates are therefore omitted from this key, and common flies which do not key out probably belong to this group.

1. Long, antennae with 6 or more segments ($\times$ 30) (Fig. D1,a, b). Legs delicate, often much longer than the body   2

—Black. Stout antennae with 7–12 segments ($\times$ 30) (Fig. D2,a, b). Legs wholly or partly thickened (Fig. D2,c)   9

—Stumpy, 3-segmented antennae often almost invisible to naked eye, terminal segment with 1 long bristle or swollen and ringed ($\times$ 30) (Fig. D3,a–e). Legs about as long as the body   10

2. Wing *membrane* and *margins* coated with dense hairs ($\times$ 60) when fresh; venation similar to Fig. D4,a or b; minute flies; 3 mm.   .   3

—Wing *veins* covered with dense scales ($\times$ 30) (Fig. D5) .   .   .   4

—Wing membrane and veins almost bare, margins slightly hairy ($\times$ 30) (Fig. D6,a, b)   5

continued

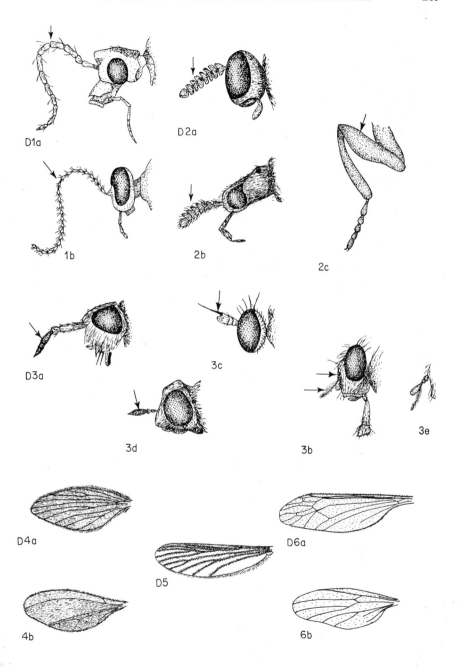

D1a

D2a

1b

2b

2c

D3a

3c

3d

3b

3e

D4a

D5

D6a

4b

6b

3. Grey or black. Body and wings densely covered with fine hairs ($\times$ 30), wings often pointed with many veins, held roof-like over the body (Fig. D7) . . . . PSYCHODIDAE
(Moth flies, owl midges)

—Body orange, yellow or pale brown. Wings with very fine hairs, few distinct veins, held vertically when at rest ($\times$ 60) (Fig. D8,a). Antennae with bead-like segments adorned with whirls of hairs ($\times$ 60) (Fig. D8,b). Mouthparts indistinct and soft . . CECIDOMYIIDAE*
(Gall midges, gall gnats)

—Grey, brown or black. Venation as in Fig. D9,a ($\times$ 30). Mouthparts conspicuous, rigid and adapted for piercing ($\times$ 60) (Fig. D9,b) .
. . . . CERATOPOGONIDAE
(Biting midges, no-see-ums)

4. Ten veins reach the wing margin. Biting mouthparts and plumose antennae in ♂♂ (Fig. D10,a). Rest with hind legs raised (Fig. D10,b)
. . . . CULICIDAE
(Mosquitoes)

* The more primitive sub-families (e.g. Lestremiinae) are difficult to separate. They may be dark brown and have more definite venation (Fig. D8,c).

continued

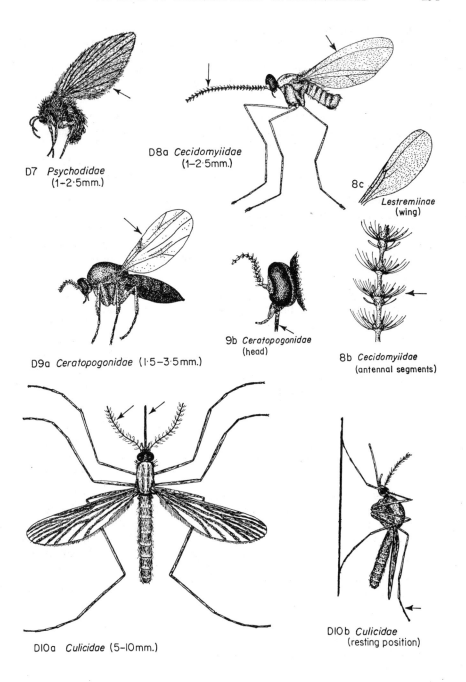

D7 *Psychodidae*
(1–2·5mm.)

D8a *Cecidomyiidae*
(1–2·5mm.)

8c
Lestremiinae
(wing)

D9a *Ceratopogonidae* (1·5–3·5mm.)

9b *Ceratopogonidae*
(head)

8b *Cecidomyiidae*
(antennal segments)

D10a *Culicidae* (5–10mm.)

D10b *Culicidae*
(resting position)

5. Antennae very hairy (Fig. D11,c) plumed (Fig. D11,a) or with bead-like segments adorned with whorls of hairs (× 60) (Fig. D8b)   .   6

—Antennae bare or with few hairs, thread-like, each segment of almost uniform diameter (Fig. D12,a, 13,a, 14,a)   7

6. Green, grey, yellow, brown or black. Delicate and gnat-like. Humped thorax partly shields the head. Usually 6 veins reaching wing margins (× 30) (Fig. D11,a). Rests with fore-legs raised (Fig. D11,b) .   .
.   .   .   .   CHIRONOMIDAE
(Midges or harlequin flies)

—Orange, yellow or pale brown. Wings with few distinct veins. Antennae with bead-like segments adorned with hairs (× 60) (Fig. D8,a)   .   .   .   .   CECIDOMYIIDAE
(Gall midges—pickled specimens and species with almost bare wings)

7. "Hunch-backed". Broad wings with 6–8 veins reaching the margin (× 30). Pale brown or black. Leg segments nearest the thorax (coxae) elongated and the last long segment (tibia) spurred (× 30) (Fig. D12,a).   .   .   .   .   MYCETOPHILIDAE*
(Fungus gnats, mushroom gnats)

—Narrow wings, with at least 10 veins, reaching the margin. Slender bodies and long, brittle legs   8

8. Brown, grey or yellow. ♂♂ with blunt tip to abdomen (Fig. D13,a) ♀♀ with pointed tip (Fig. D13,b). No simple eyes (ocelli) behind the antennae (× 60) (Fig. D13,c). Absent in winter   .   .   TIPULIDAE
(Crane flies, daddy longlegs)

—Dull grey (Fig. D14,a) with 3 simple eyes (ocelli) just behind the antennae (× 60) (Fig. D14,b), and with a short bent vein reaching the hind margin of the wing near its base. Common in spring, autumn and winter   .   .   .   .   .   .   TRICHOCERIDAE
(Winter gnats)

—Thorax shiny black, with yellow and black markings on abdomen, yellow legs   .   .   .   .   PTYCHOPTERIDAE
(Phantom crane flies)

* The primitive Sciarinae (Root gnats or dark-winged fungus gnats) are dark with venation as in Fig. D12,b. The more highly evolved Mycetophilinae are pale, sometimes with spotted wings and venation as in Fig. D12,a.

continued

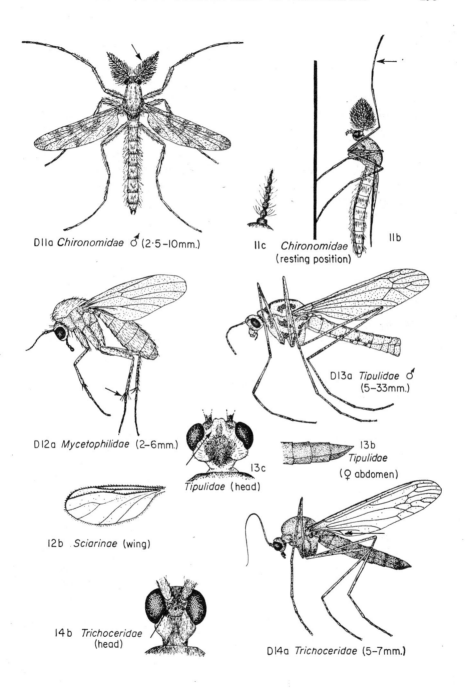

D11a *Chironomidae* ♂ (2·5–10mm.)

11c *Chironomidae* (resting position)

11b

D12a *Mycetophilidae* (2–6mm.)

D13a *Tipulidae* ♂ (5–33mm.)

13b *Tipulidae* (♀ abdomen)

13c *Tipulidae* (head)

12b *Sciarinae* (wing)

14b *Trichoceridae* (head)

D14a *Trichoceridae* (5–7mm.)

9. Black. < 3·0 mm. Wings with anterior veins very strong and posterior veins feint (× 60) (Fig. D15). Occur in great numbers around decaying organic matter   .   .   .   .   .   SCATOPSIDAE
(Refuse flies)

—Black. 3·5–12 mm. Hairy. First long segment (femur) of the forelegs thickened (Fig. D16,a, b)   .   .   .   .   BIBIONIDAE
(St. Mark's flies, fever flies)

10. Wings with false vein (Fig. D17,a). Black or brown with yellow bands or spots, or brown and slightly furry (Fig. D17,b, c, d). Hovering flight   .   .   .   .   SYRPHIDAE
(Hover flies, flower flies)

—Wings without false vein   11

11. Wings distinctly patterned, and their hind margins not bordered by a vein (Fig. D18,a, b, c, d)   12

—Wings plain or tinted brown. If patterned, with a vein extending right round the wing margin as Fig. D27,a (p. 299)   .   .   .   13

12. Body densely furry, like a bumble bee. Mouthparts often elongated (Fig. D19)   .   .   .   .   BOMBYLIIDAE
(Beeflies)

—Body with few hairs, not furry. Wings often waved gently as flies rest .   .   .   .   either
TRYPETIDAE
(Fruit flies)
OTITIDAE
(Pictured-wing flies)
PALLOPTERIDAE
SCIOMYZIDAE
(Marsh flies)
OPOMYZIDAE
AGROMYZIDAE
(Leaf miners)

continued

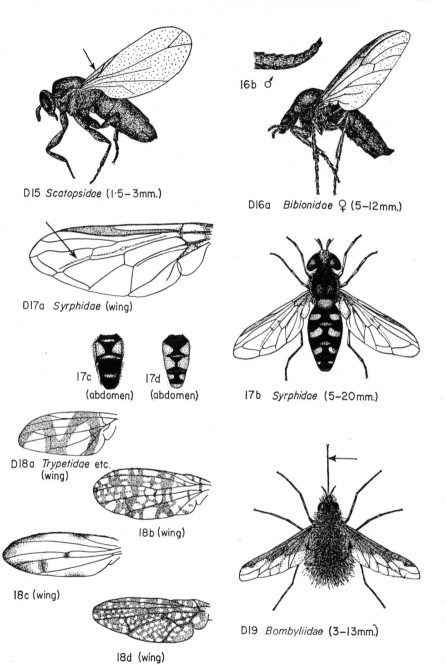

D15 *Scatopsidae* (1·5–3mm.)

16b ♂

D16a   *Bibionidae* ♀ (5–12mm.)

D17a *Syrphidae* (wing)

17c
(abdomen)

17d
(abdomen)

17b   *Syrphidae* (5–20mm.)

D18a *Trypetidae* etc.
(wing)

18b (wing)

18c (wing)

D19 *Bombyliidae* (3–13mm.)

18d (wing)

13. Abdomen and/or thorax shiny blue, green or chequered black and
grey. > 3 mm 14

—Body dull; grey, brown, yellow or black or densely furry, or slightly
shiny on dorsal part of abdomen. If wholly shiny < 3 mm . . 15

14. Black or yellow spherical head. Abdomen tapered basally to produce a constriction between abdomen and thorax. Often with pigmented spot near the wing tip on the leading edge. Wave wings gently
as flies rest. Often on rotting vegetation (Fig. D20) . . SEPSIDAE

—Flattened body, green, black or yellow. Third antennal segment
ringed (× 60) (Fig. D21,b). Pronounced discal cell on wings which
may be partially tinted brown (Fig. D21,a) . . STRATIOMYIIDAE
(Soldier flies)

—Abdomen and/or thorax burnished green or blue. Venation as Fig.
D22,a, b. Long legs and ♂♂ may have prominent genitalia at tip of
abdomen (Fig. D22,c) . . . . . DOLICHOPODIDAE
(Long-headed flies)

—Thorax and abdomen wholly bright metallic green or blue. Venation
as Fig. D23 . . . . CALLIPHORIDAE*
(Blue bottles, green bottles)

15. Whole body or underside of abdomen densely furry, hairs partly
obscuring the cuticle, especially ventrally 16

—Body sparsely hairy or bare 17

16. Like a small bumble bee; mouthparts often elongated (Fig. D19)
(p. 295) . . . . BOMBYLIIDAE
(Bee flies)

—Body with golden hairs; size and shape as a housefly (Fig. D24). On
dung . . . . CORDILURIDAE
(Muck flies)

17. Slender, tapering, conical abdomen, sometimes curled upwards in
♂♂, covered with fine hairs (Fig. D25,a, b) . . RHAGIONIDAE
(Snipe flies)

—Abdomen blunt or if tapering, curls downwards 18

* A few species of Muscidae and Tachinidae could key out to this family, and a few
chequered grey Calliphoridae may be overlooked. Their correct determination is
difficult.

continued

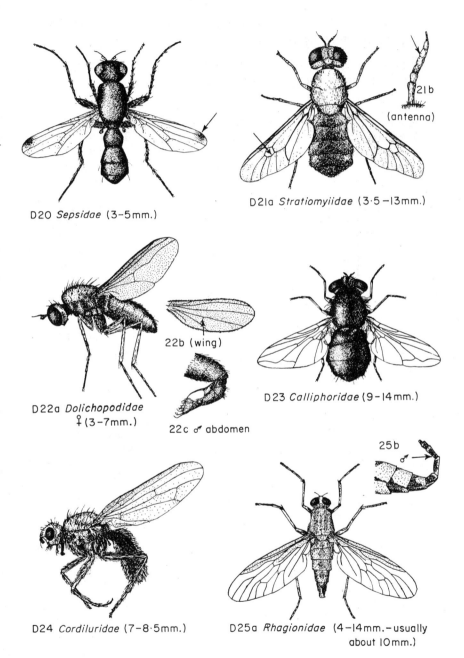

D20 *Sepsidae* (3-5mm.)

D21a *Stratiomyiidae* (3·5 –13mm.)

21b
(antenna)

22b (wing)

D22a *Dolichopodidae*
♀(3-7mm.)

22c ♂ abdomen

D23 *Calliphoridae* (9-14mm.)

D24 *Cordiluridae* (7-8·5mm.)

D25a *Rhagionidae* (4-14mm.-usually
about 10mm.)

25b

18. Mouthparts rigid for piercing ($\times$ 30) (Fig. D26,a, b, c, d). Venation similar to Fig. D27,a or D28,a 19

—Mouthparts inconspicuous, or fleshy and flabby ($\times$ 30) or with a large sucking pad on the tip (Fig. D26,e) 20

19. Robust, flattened flies. Head at least as wide as thorax (Fig. D27,a). Eyes in living or freshly killed specimens iridescent. Third antennal segment ringed ($\times$ 60) (Fig. D27,b). A vein always extends right around the wing ($\times$ 30). Always fly solitarily . . TABANIDAE
(Horse flies, gad flies, clegs)

—Abdomen not flattened. Long or short mouthparts (Fig. D28,a, D26,a, b). Third antennal segment ringed and pointed ($\times$ 60) (Fig. D28,b). Small species often fly in swarms . . . . EMPIDAE
(Dance flies)
ASILIDAE
(Robber flies, assassin flies)

20. Wing tips pointed (Fig. D29). Main veins almost parallel. . .
LONCHOPTERIDAE
(Pointed-winged flies)

—Wing tips blunt 21

21. Black or grey, with "hunchbacked" appearance. Wings with thick bristly foreveins near to the thorax and usually 3, occasionally 4 delicate veins traversing the wing membrane, without cross veins ($\times$ 30) (Fig. D30) . . . . PHORIDAE
(Humpbacked flies)

—Wings with cross veins 22

continued

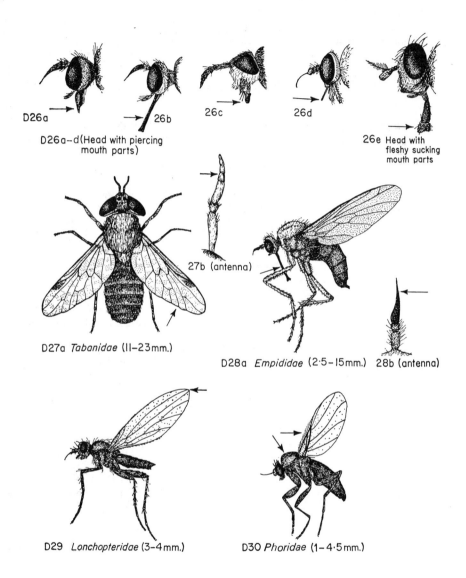

D26a

→ 26b

26c

26d

D26a-d(Head with piercing
mouth parts)

26e Head with
fleshy sucking
mouth parts

27b (antenna)

D27a *Tabanidae* (II-23mm.)

D28a *Empididae* (2·5-I5mm.) 28b (antenna)

D29 *Lonchopteridae* (3-4mm.)

D30 *Phoridae* (1-4·5mm.)

22. Triangular shape often visible between eyes. Abdomen and thorax bright shiny black or black and yellow. (In pickled specimens abdomen may have brown stripes.) Last long vein on each wing with a slight irregularity. < 3 mm (Fig. D31) . . . . CHLOROPIDAE
(Grass or stem flies)

—Body dull or slightly shiny. Usually > 3 mm 23

23. Dull grey, often with indistinct stripes on the thorax. Elongated mouthparts with a fleshy pad on the tip. Venation as in Fig. D32. Around buildings, flowers and rotting organic matter . MUSCIDAE
(Houseflies, face flies, stable flies, root flies)

—Mouthparts thick and fleshy. First short segment of hind leg (1st tarsal segment) very broad (×30) (Fig. D33,a, b). Posterior wing veins do not reach hind margin of wing. On manure and decaying organic matter. Often jump. . . . . . . SPHAEROCERIDAE

—Mouthparts small and delicate. First short segment of hind leg (1st tarsal segment) long and slender (Fig. D34,a). Long feathered antennal bristle (arista) with a forked tip (×60) (Fig. D34,b) . . . . DROSOPHILIDAE
(Fruit flies, vinegar flies)

—Mouth opening wide and face strongly arched (Fig. D35,a). Venation as in Fig. D35,b. Usually on vegetation or ground bordering fresh or salt water . . . . EPHYDRIDAE
(Shore flies)

—Characters other than above . . . . . Other families

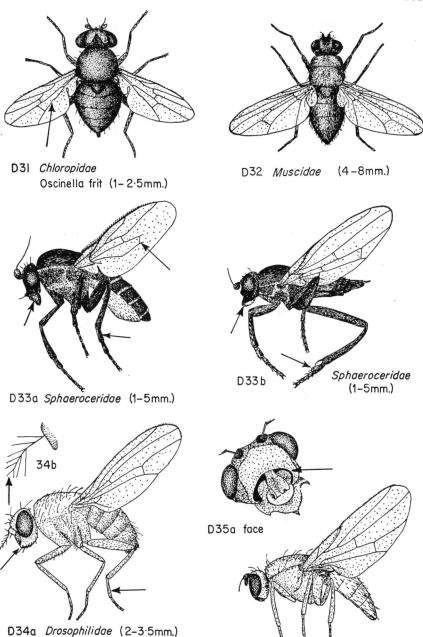

D3I *Chloropidae*
Oscinella frit (1- 2·5mm.)

D32 *Muscidae*   (4-8mm.)

D33a *Sphaeroceridae* (1-5mm.)

D33b   *Sphaeroceridae*
(1-5mm.)

34b

D34a *Drosophilidae* (2-3·5mm.)

D35a face

D35b *Ephydridae* (I·5-5mm.)

KEY TO FAMILIES OF COLEOPTERA
(Beetles—with pigmented wing-cases and biting mouthparts)

1. Water beetles. Hind legs flattened or oar-like for swimming, claws reduced (Fig. E1,a, b, c) 2

—Terrestrial beetles. Hind legs not oar-like, claws usually conspicuous (Fig. E2,a, b, c, d) 4

2. Body completely covered by wing cases (elytra). Swim beneath the water surface 3

—Tip of body protruding beyond wing cases (elytra). Steely-blue metallic lustre. Whirl in swarms on water surfaces (Fig. E3) GYRINIDAE
(Whirlygig beetles)

3. Filamentous antennae (Fig. E4) . . . . DYTISCIDAE
(Predaceous diving beetles)

—Clubbed antennae, plus a pair of filamentous palps (Fig. E5). In water or on damp land . . . . HYDROPHILIDAE
(Water scavenger beetles)

4. Wing cases (elytra) very short, at least 4 abdominal segments exposed. Antennae filamentous (Fig. E6,a, b) . . STAPHYLINIDAE
(Rove beetles)

—Wing cases covering most or all of the abdomen, or if 4 abdominal segments exposed, then antennae boldly clubbed (Fig. E9,b) (p. 305). 5

5. Wing cases (elytra) with distinct yellow, red or black markings visible to naked eye 6

—Wing cases uniformly coloured, or if multicoloured, boundaries of coloured areas indistinct 10

6. Wing cases (elytra) parallel-sided, flexible and papery (Fig. E7) .
. . . . CANTHARIDAE
(Soldier beetles)

—Wing cases hard 7

continued

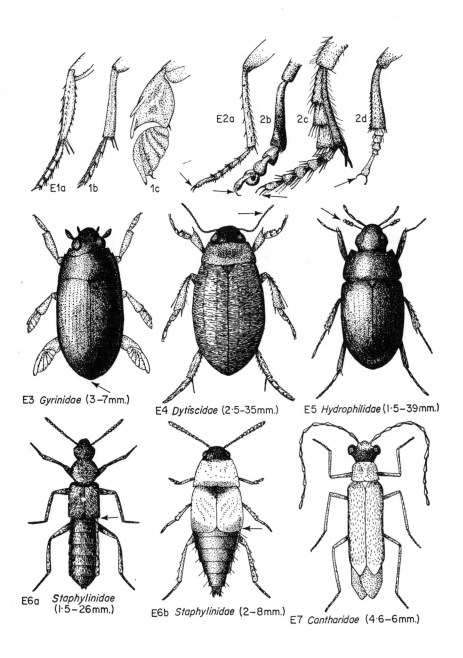

E3 *Gyrinidae* (3–7mm.)

E4 *Dytiscidae* (2·5–35mm.)

E5 *Hydrophilidae* (1·5–39mm.)

E6a   *Staphylinidae*
          (1·5–26mm.)

E6b *Staphylinidae* (2–8mm.)

E7 *Cantharidae* (4·6–6mm.)

7. Antennae threadlike (Fig. E8,a, b, c) 8

—Antennae clubbed (Fig. E9,a, b, c, d) 9

—Apparently 2 pairs of antennae, 1 threadlike and 1 clubbed (Fig. E5) (p. 303) . . . . HYDROPHILIDAE

8. Yellow stripes on each black wing case (elytron). Thick hind legs (Fig. E10) . . . . . CHRYSOMELIDAE (HALTICINAE) (Flea beetles)

—Irregular black spots on yellow wing cases . . CHRYSOMELIDAE (other sub-families) (Leaf beetles)

—Yellow spots and/or stripes on green or brown wing cases. Head held horizontally (Fig. E21,a) (p. 311) and jaws (mandibles) with long, pointed, overlapping tips (Fig. E11,a, b) . . CICINDELIDAE (Tiger beetles)

—Yellow stripes on black background or black blotches on a yellow background. Head held vertically (Fig. E20,a) (p. 311). Long antennae (Fig. E12,a, b) . . . . . . CERAMBYCIDAE (Longhorn beetles)

continued

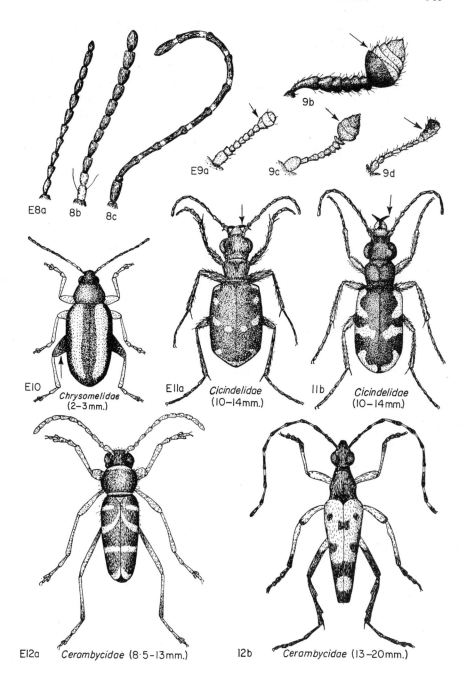

E8a   8b   8c

9b

E9a   9c   9d

E10   *Chrysomelidae*
       (2–3mm.)

E11a   *Cicindelidae*
        (10–14mm.)

11b   *Cicindelidae*
       (10–14mm.)

E12a   *Cerambycidae* (8·5–13mm.)

12b   *Cerambycidae* (13–20mm.)

9. < 10 mm, more than half as wide as long, with 2–22 regular or
irregular spots (Fig. E13,a, b, c, d, e, f)   .   .   .   COCCINELLIDAE
(Ladybirds, lady beetles)

— < 10 mm, half as wide as long, with 4 yellow spots (Fig. E14)
MYCETOPHAGIDAE
(Hairy fungus beetles)

— > 10 mm, yellowy-orange and black (Fig. E15,a, b). Abdominal
tip may be exposed .   .   .   .   SILPHIDAE
(Carrion beetles)

continued

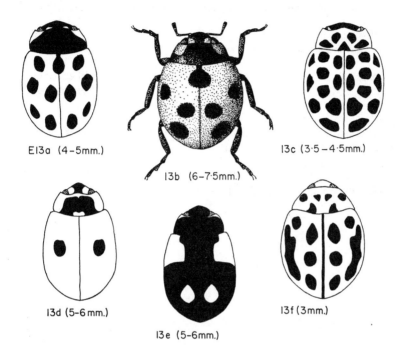

E13a (4-5mm.)

13b (6-7·5mm.)

13c (3·5 – 4·5mm.)

13d (5-6 mm.)

13e (5-6mm.)

13f (3mm.)

E13a-f *Coccinellidae* (3-7·5mm.)

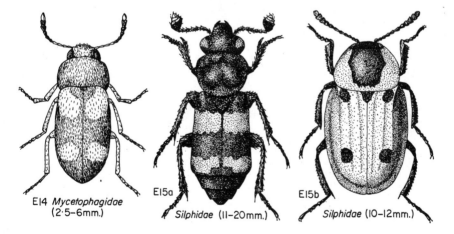

E14 *Mycetophagidae*
(2·5-6mm.)

E15a

Silphidae (11-20mm.)

E15b

Silphidae (10-12mm.)

10. Antennae uniformly threadlike or feathery (Fig. E16,a, b, c, d) or
with slight, *gradual* thickening towards the tip (× 30) (Fig. E16,e).   11

—Antennae boldly clubbed (Fig. E17,a, b, c, d, e) or last 3 segments
only markedly dilated (Fig. E17,f)   19

—Apparently 2 pairs of antennae, 1 threadlike and 1 clubbed (Fig.
E5) (p. 303)   .   .   .   .   HYDROPHILIDAE

11. In silhouette, distinct division between head and thorax, *and* thorax
and abdomen (Fig. E18,a, b, c)   12

—In silhouette, division between either head and thorax, or thorax and
abdomen, inseparable (Fig. E19,a, b, c, d)   14

continued

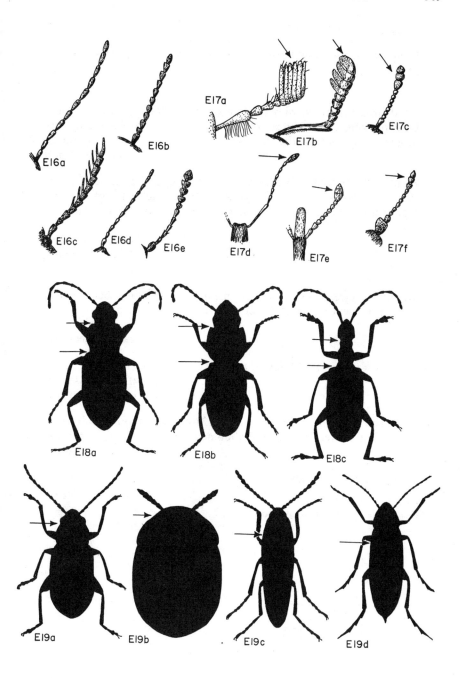

E16a

E16b

E16c

E16d

E16e

E17a

E17b

E17c

E17d

E17e

E17f

E18a

E18b

E18c

E19a

E19b

E19c

E19d

12. Wing cases (elytra) hard 13

—Wing cases soft and flexible in fresh specimens, papery in dry
specimens   ⸳.   17

13. Head held vertically, with jaws (mandibles) beneath (Fig. E20,a).
Base of wing cases (elytra) about twice as wide as back edge of
thorax. Wood-borers (Fig. E20,b)   .   .   .   . CERAMBYCIDAE
(Longhorn beetles)

—Head held horizontally with jaws in front (Fig. E21,a). Base of wing
cases as wide as, or only slightly wider than, back edge of thorax.
Uniformly metallic black, brown, with green or violet sheen. Ground
dwellers (Fig. E21,b, c)   .   .   .   .   .   . CARABIDAE
(Ground beetles, rain beetles)
(except *Amara* spp.)

14. Stumpy, convex, shiny, usually metallic body. One bell-shaped
(tarsal) segment near the end of each leg (Fig. E22,a, b) .   .   .
CHRYSOMELIDAE
(Flea* and leaf beetles)

—Stumpy, dull metallic body, shaped like a tortoise (Fig. E23)
CHRYSOMELIDAE (CASSIDINAE)
(Tortoise beetles, helmet beetles)

—Flattened oval body; similar to Fig. E21,b, but without the distinct
separation between thorax and abdomen Fig. E21, d   . CARABIDAE
(*Amara* spp.)

—Long and narrow body   15

* Flea beetles have broadened hind femora.

continued

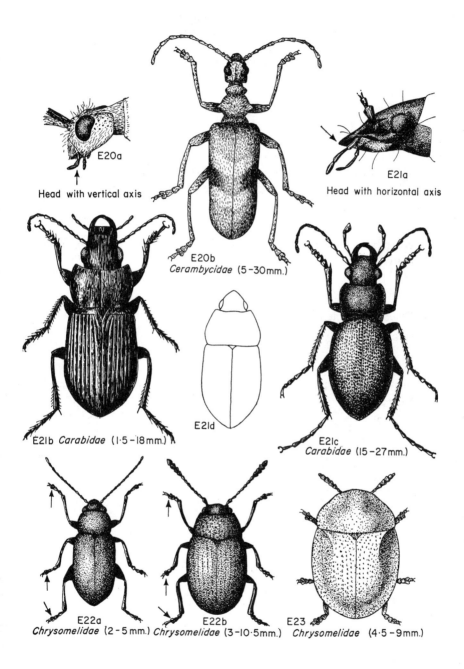

E20a

Head with vertical axis

E21a

Head with horizontal axis

E20b
Cerambycidae (5-30mm.)

E21d

E21b *Carabidae* (1·5-18mm.)

E21c
Carabidae (15-27mm.)

E22a
Chrysomelidae (2-5mm.)

E22b
Chrysomelidae (3-10·5mm.)

E23
Chrysomelidae (4·5-9mm.)

15. Wing cases (elytra) strap-like and only partly meet down the middle of the back. Live in wasps' nests (Fig. E24) . . RHIPIPHORIDAE

—Wing cases meet down the middle of the back 16

16. Wing cases (elytra) flexible 17

—Wing cases (elytra) rigid 18

17. Wing cases (elytra) parallel-sided and papery, covered with velvety hairs. Legs loosely jointed. Common on flowers (Fig. E25) . . . . CANTHARIDAE
(Soldier beetles)

—Wing cases bright red, widest near the tip, feathery antennae (Fig. E26) . . . . PYROCHROIDAE
(Cardinal beetles)

18. Head constricted strongly behind the eyes. Common on flowers (Fig. E27) . . . . MORDELLIDAE
(Flower beetles, tumbling flower beetles)

—No constriction between head and thorax, which has acute points at its hind angles. Boat shaped (Fig. E28). Jump when placed on their backs . . . . ELATERIDAE
(Click beetles)

continued

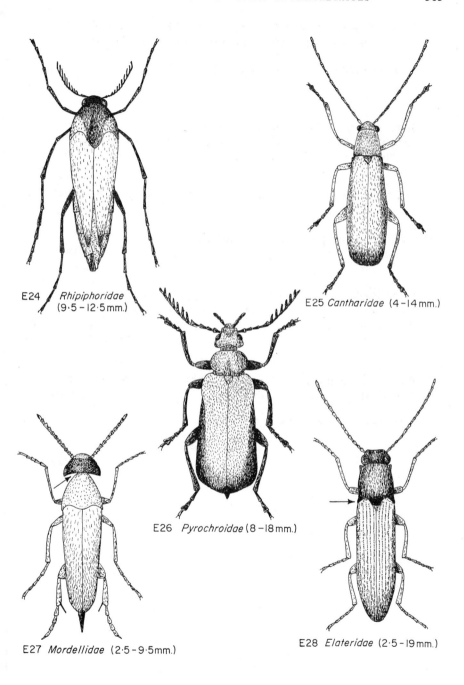

E24   *Rhipiphoridae*
(9·5 – 12·5 mm.)

E25 *Cantharidae* (4 – 14 mm.)

E26   *Pyrochroidae* (8 – 18 mm.)

E27 *Mordellidae* (2·5 – 9·5 mm.)

E28 *Elateridae* (2·5 – 19 mm.)

19. Front of head extended to form a snout (rostrum) extending far beyond the eyes (check this in side view) (Fig. E29,a, b, c, d).   .   20

—Front of head rounded extending only slightly beyond the eyes, no snout   22

20. Snout (rostrum) stubby and ventral, antennae widely spaced with angular bend when extended. Thorax bulbous. Wing cases (elytra) terminate abruptly (Fig. E30) .   .   .   .   .   .   SCOLYTIDAE
(Bark beetles)

—Snout extending forward or curved beneath head, wing cases rounded   21

21. Snout long and pointed. Pear-shaped, metallic black, violet or red. Antennae straight (Fig. E31) .   .   .   .   .   .   APIONIDAE
(Seed weevils)

—Snout (rostrum) stumpy or long. Antennae with angular bend (pull the tip to ascertain this) and partly slotted into a groove (scrobes) (Fig. E32,a, b, c)   .   .   .   .   .   .   CURCULIONIDAE
(Weevils, snout beetles)

22. Wing cases (elytra) densely covered with fine hairs or flattened scales (Fig. E33)   .   .   .   .   DERMESTIDAE

—Wing cases bare, or if hairy > 15 mm   23

continued

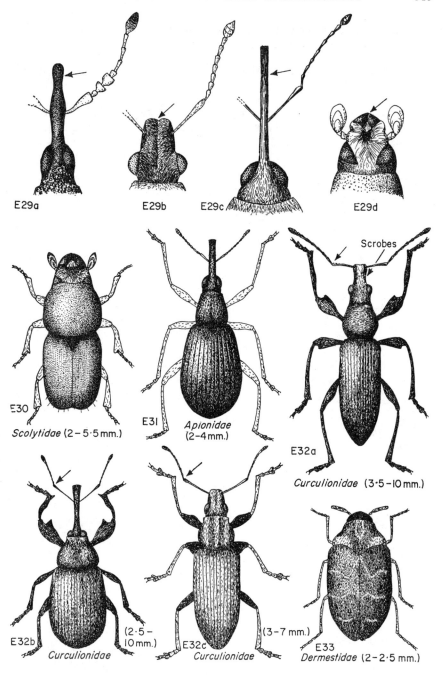

E29a

E29b   E29c   E29d

Scrobes

E30
Scolytidae (2 – 5·5 mm.)

E31   *Apionidae*
(2–4mm.)

E32a

Curculionidae (3·5–10 mm.)

E32b   (2·5–
10mm.)   E32c   (3–7 mm.)
Curculionidae   *Curculionidae*   E33
Dermestidae (2–2·5 mm.)

23. Antennal club asymmetrical, formed from separate or densely com-
pacted plates (Fig. E34,a, b, c) 24

—Antennal club symmetrical (Fig. E17,c, f) (p. 309)   .   .   . 25

24. Black or deep chestnut. Large forward-protruding jaws (mandibles),
or a horn on the head. Plates of antennal club clearly separated (Fig.
E35,a, b, c)   .   .   .   . LUCANIDAE
(Stag beetles, horn beetle)

—Black, buff, brown or metallic green. Plates of antennal club compact. Many species in dung (Fig. E36,a, b)   .   . SCARABAEIDAE
(Chafer beetles, dung beetles)

continued

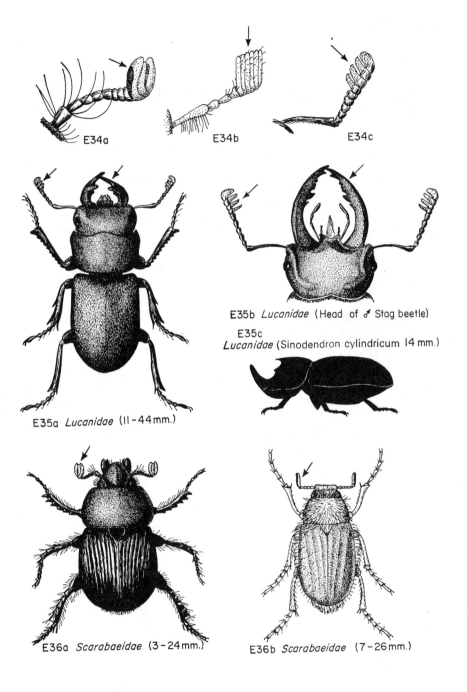

E34a

E34b

E34c

E35b *Lucanidae* (Head of ♂ Stag beetle)

E35c
Lucanidae (Sinodendron cylindricum 14 mm.)

E35a *Lucanidae* (11-44mm.)

E36a *Scarabaeidae* (3-24mm.)

E36b *Scarabaeidae* (7-26mm.)

25. > 4 mm 26

—< 3·5 mm   27

26. Black. Flattened, usually with 3 ridges on each wing case (elytron)
(Fig. E37)   ·   .   .   .   .   ·   .   .   SILPHIDAE
(Carrion beetles)

—Metallic green, violet, black or brown. Highly convex, no distinct
ridges on wing cases (elytra)   .   .   .   .   . CHRYSOMELIDAE
(Leaf beetles, some large species. See Fig. E22,b, p. 311)

27. Minute black beetles < 1·5 mm with feathery hind wings (Fig. E38)
.   .   .   . PTILIIDAE

—Beetles usually with angular thorax shaped as in Fig. E39,a; thorax
always narrower at its base than the front of the wing cases (elytra)
(Fig. E39,b, c)   .   .   .   . LATHRIDIIDAE

—Beetles usually with rounded thorax shaped as in Fig. E40,a; thorax
always with its base almost as wide as the front of the wing cases
(elytra)   .   .   .   . CRYPTOPHAGIDAE

—Beetles with slightly shortened wing cases (elytra) and tip of abdomen visible (Fig. E41). Common on flowers   .   . NITIDULIDAE
(Pollen beetles)

—Characters other than above   .   .   .   .   . Other families

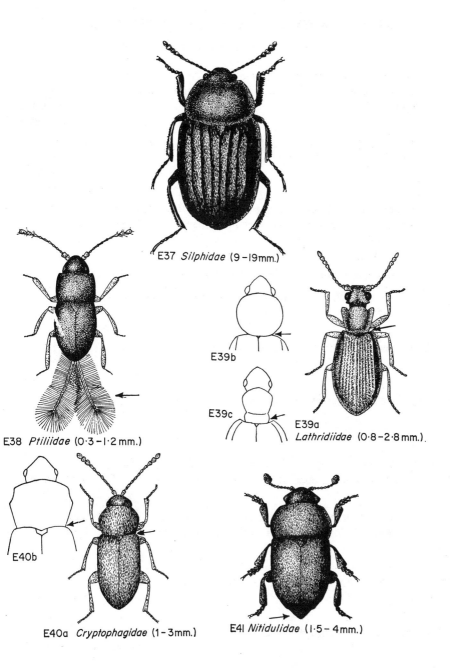

E37 *Silphidae* (9 –19mm.)

E39b

E39c

E39a
Lathridiidae (0·8–2·8mm.)

E38 *Ptiliidae* (0·3–1·2 mm.)

E40b

E40a *Cryptophagidae* (1–3mm.)

E41 *Nitidulidae* (1·5– 4mm.)

KEY TO SUPERFAMILIES OF HYMENOPTERA
(Bees, Wasps, Ants and Sawflies)

1. Densely furry, with none of the underlying cuticle visible. Black or
brown, or with 1 to 3 yellow, buff or orange stripes. Hind tibiae with
spurs (Fig. F1)   .   .   .   . APOIDEA
(Bumble bees)

—Moderately furry, with some of the underlying cuticle visible. Uniformly brown. Hind legs with tibiae flattened and broad, without
spurs (Fig. F2)   .   .   .   . APOIDEA
(Honey bees)

—Slightly furry or bare. Either uniformly black brown, metallic blue
or green, or with at least 4 yellow stripes or black spots on the
abdomen   2

2. Abdomen cylindrical with yellow stripes or a broad brown band . 3

—Abdomen without yellow stripes, or with 1 yellow stripe, or broad
and flat 4

3. Wings entirely transparent. Hairs not feathered (× 60) (Fig. F3,a, b)
.   .   .   . VESPOIDEA
(Wasps)

—Wings smokey near the tip. Hairs feathered (× 60) (Fig. F4,a, b)   .
.   .   .   . APOIDEA
(Solitary or nomad bees)

4. Waist between thorax and abdomen, slight or absent. > 6 mm.
Body usually flattened and soft (Fig. F5)   .   . TENTHREDINOIDEA
(Sawflies)

—With pronounced waist (pedicel) between thorax and abdomen, or
if waist indistinct, < 4 mm. Body not flattened and usually hard   . 5

5. Wingless, with a knobbly waist (pedicel). (Males and females may
have wings for a short time in their lives, but the waist is characteristic.) (Fig. F6)   .   .   .   . FORMICOIDEA
(Ants)

—Winged, with smooth waist   .   .   .   .   . PARASITICA*
(For further separation—6)

* Some ♀♀ may have long egg-laying organs (ovipositors) projecting from the abdomen.

continued

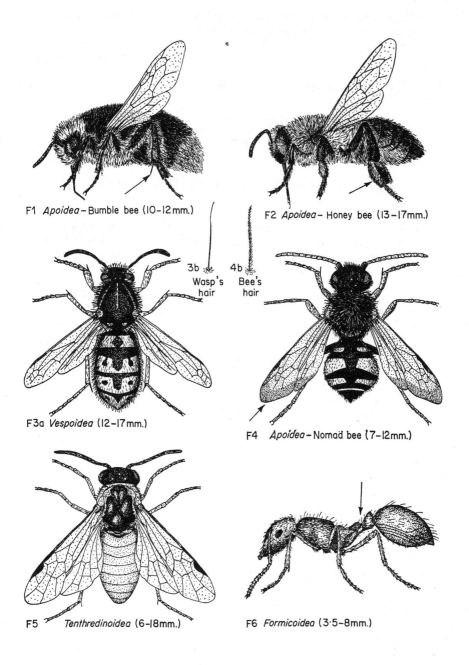

F1 *Apoidea* – Bumble bee (10-12 mm.)

F2 *Apoidea* – Honey bee (13-17mm.)

3b
Wasp's
hair

4b
Bee's
hair

F3a *Vespoidea* (12-17 mm.)

F4 *Apoidea* – Nomad bee (7-12mm.)

F5 *Tenthredinoidea* (6-18mm.)

F6 *Formicoidea* (3·5-8mm.)

6. Wings without veins, or with a vestigial vein along the leading edge. Antennae with angular bend (Fig. F7,a) or wing margins hairy (Fig. F7,b) · . . . . · . . . CHALCIDOIDEA or
PROCTOTRUPOIDEA

—Wings with veins. Antennae threadlike 7

7. Waist (pedicel) longer than the swollen part of the abdomen (gaster). Found on sandy heaths (Fig. F8) . . . . SPHECOIDEA
(Sand wasps)

—Waist shorter than the swollen part of the abdomen (gaster) . 8

8. Wings with few veins (W-shaped) concentrated near the leading edge of the forewings, which have *no* darkly-pigmented spot (pterostigma) (Fig. F9) . . . . CYNIPOIDEA
(Gall wasps)

—Wings mostly traversed with veins. Vein along leading edge of fore-wing thickened to form a darkly-pigmented spot (Fig. 10,a, b) .
. . . . ICHNEUMONOIDEA
(Ichneumon flies)

—Characters other than above . . . . . Other taxa

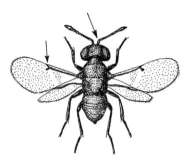

F7a *Chalcidoidea* (1–2·5mm.)

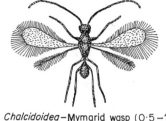

F7b *Chalcidoidea*–Mymarid wasp (0·5 –1mm.)

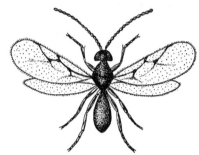

F9 *Cynipoidea* (1–5mm.)

F8 *Sphecoidea* (15–19mm.)

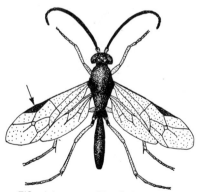

F10a *Ichneumonoidea* (1·5–23mm.)

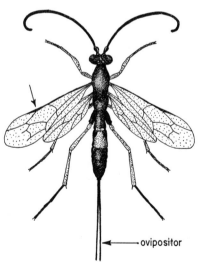

ovipositor

F10b *Ichneumonoidea* (1·5–23mm.)

Appendices A–N

APPENDIX A

Method of calculating Sums, Means, Variance and the Regression Equation

Sums, means and variance

Where there are n units or observations in a sample;

$$x_1, x_2, x_3, x_4, x_5, \ldots, x_n$$

the sum of x,

$$\sum x = x_1 + x_2 + x_3 + x_4 + x_5 + \cdots + x_n$$

the mean of x,

$$\bar{x} = \frac{\sum x}{n}$$

the sum of x^2,

$$\sum(x^2) = x_1^2 + x_2^2 + x_3^2 + x_4^2 + x_5^2 + \ldots + x_n^2$$

the square of the sum of x,

$$(\sum x)^2 = (x_1 + x_2 + x_3 + x_4 + x_5 + \cdots + x_n)^2$$

the variance of x,

$$s_x^2 = \frac{\sum(x^2) - \dfrac{(\sum x)^2}{n}}{n - 1}$$

the standard deviation of x,

$$s_x = \pm \sqrt{s_x^2}$$

the 95% Limits of x,

$$95\% \text{ Lts.}_x = \pm 2s_x \quad (\text{approx.})$$

The regression equation

Using data from Ex. 21: Given, the measured size of insects (the product of wingspan × body length in mm²) and their flight speed, in cm/sec, compute the regression equation of speed on size. The scatter diagram (Fig. III:29) shows that a double log transformation gives a linear regression, therefore, the computation must be done in logs.

Size mm²	Log. size = x	x²	Speed cm/sec	Log. speed = y	xy
25·0	1·40	1·96	91	1·96	2·74
23·0	1·36	1·85	55	1·74	2·37
16·0	1·20	1·44	76	1·88	2·26
6·6	0·82	0·67	37	1·57	1·29
2·8	0·45	0·20	46	1·66	0·75
20·0	1·30	1·69	116	2·06	2·68
41·0	1·61	2·59	85	1·93	3·11
200·0	2·30	5·29	76	1·88	4·32
520·0	2·72	7·40	435	2·64	7·18
300·0	2·48	6·15	286	2·46	6·10
126·0	2·10	4·41	192	2·28	4·79
14·0	1·15	1·32	70	1·85	2·13
72·0	1·86	3·46	305	2·48	4·61
15·0	1·18	1·39	137	2·14	2·53
6·0	0·76	0·58	76	1·88	1·43
98·0	1·99	3·96	305	2·48	4·94
512·0	2·71	7·34	650	2·81	7·62
253·0	2·40	5·76	650	2·81	6·74
340·0	2·53	6·40	167	2·22	5·62
512·0	2·71	7·34	414	2·62	7·10
—	35·03	71·20	—	43·35	80·31

The logs and squares of logs are obtained from tables, the "cross-products" (xy) from a slide rule.

$$n = 20$$

$$\sum y = 43 \cdot 35 \qquad \therefore \qquad \bar{y} = \frac{43 \cdot 35}{20} = 2 \cdot 168$$

$$\sum x = 35 \cdot 03 \qquad \therefore \qquad \bar{x} = \frac{35 \cdot 03}{20} = 1 \cdot 752$$

$$\sum (x^2) = 71 \cdot 20 \qquad (\sum x)^2 = (35 \cdot 03)^2 = 1227 \cdot 1$$

$$\sum (xy) = 80 \cdot 31.$$

$$\therefore \ s_x^2 = \frac{71 \cdot 20 - \left(\dfrac{1227 \cdot 1}{20}\right)}{19} = \frac{9 \cdot 84}{19} = 0 \cdot 518$$

$$c = \frac{\sum(xy) - \frac{\sum x \cdot \sum y}{n}}{n-1}$$

$$= \frac{80\cdot31 - \left(\frac{35\cdot03 \times 43\cdot35}{20}\right)}{19} = \frac{4\cdot38}{19} = 0\cdot231$$

Regression coefficient $\quad b = \dfrac{c}{s_x^2} \qquad\qquad = \dfrac{0\cdot231}{0\cdot518} = 0\cdot45$

Intercept $\qquad\qquad a = \bar{y} - b\bar{x} = 2\cdot168 - 0\cdot788 = 1\cdot38$

Regression equation (in logs)

$$y = 1\cdot38 + 0\cdot45x \quad \text{(see Fig. A:1)}$$

De-transformed (antilogs)*

$$y = 24x^{0\cdot45}$$

(Compare $y = 25x^{0\cdot49}$ by the graphical method, p. 83.)

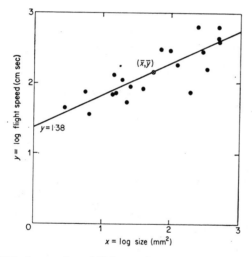

FIG. A:1. Fitted regression of flight speed on the body size of insects, in logs,
$y = 1\cdot38 + 0\cdot45x$.

* Note that, as there were no zero quantities, log N could be used instead of log $(N + 1)$; hence there was no unity to subtract when detransforming.

APPENDIX B

Table of percentages of integers up to 150

The figure in the left-hand column is given as a percentage of the quantity in the top row.

	1	2	3	4	5	6	7	8	9	10
1	100	50	33	25	20	17	14	12	11	10
2	—	100	67	50	40	33	29	25	22	20
3	—	—	100	75	60	50	43	37	33	30
4	—	—	—	100	80	67	57	50	44	40
5	—	—	—	—	100	83	71	62	56	50
6	—	—	—	—	—	100	86	75	67	60
7	—	—	—	—	—	—	100	87	78	70
8	—	—	—	—	—	—	—	100	89	80
9	—	—	—	—	—	—	—	—	100	90
10	—	—	—	—	—	—	—	—	—	100

	11	12	13	14	15	16	17	18	19	20
1	9·1	8·3	7·7	7·1	6·7	6·3	5·9	5·6	5·3	5·0
2	18	17	15	14	13	12	12	11	11	10
3	27	25	23	21	20	19	18	17	16	15
4	36	33	31	29	27	25	24	22	21	20
5	45	42	38	36	33	31	29	28	26	25
6	55	50	46	43	40	38	35	33	32	30
7	64	58	54	50	47	44	41	39	37	35
8	73	67	62	57	53	50	47	44	42	40
9	82	75	69	64	60	56	53	50	47	45
10	90·9	83	77	71	67	63	59	56	53	50
11	100	91·7	85	79	73	69	65	61	58	55
12	—	100	92·3	86	80	75	71	67	63	60
13	—	—	100	92·9	87	81	76	72	68	65
14	—	—	—	100	93·3	88	82	78	74	70
15	—	—	—	—	100	93·7	88	83	79	75
16	—	—	—	—	—	100	94·1	89	84	80
17	—	—	—	—	—	—	100	94·4	89	85
18	—	—	—	—	—	—	—	100	94·7	90
19	—	—	—	—	—	—	—	—	100	95·0
20	—	—	—	—	—	—	—	—	—	100

	21	22	23	24	25	26	27	28	29	30
1	4·8	4·5	4·3	4·2	4·0	3·8	3·7	3·6	3·4	3·3
2	9·0	9·1	8·7	8·3	8·0	7·7	7·4	7·1	6·9	6·7
3	14	14	13	12	12	12	11	11	10	10
4	19	18	17	17	16	15	15	14	14	13
5	24	23	22	21	20	19	19	18	17	17
6	29	27	26	25	24	23	22	21	21	20
7	33	32	30	29	28	27	26	25	24	23
8	38	36	35	33	32	31	30	29	28	27
9	43	41	39	38	36	35	33	32	31	30
10	48	45	43	42	40	38	37	36	34	33
11	52	50	48	46	44	42	41	39	38	37
12	57	55	52	50	48	46	44	43	41	40
13	62	59	57	54	52	50.	48	46	45	43
14	67	64	61	58	56	54	52	50	48	47
15	71	68	65	63	60	58	56	54	52	50
16	76	73	70	67	64	62	59	57	55	53
17	81	77	74	71	68	65	63	61	59	57
18	86	82	78	75	72	69	67	64	62	60
19	90·5	86	83	79	76	73	70	68	66	63
20	95·2	90·9	87	83	80	77	74	71	69	67
21	100	95·5	91·3	88	84	81	78	75	72	70
22	—	100	95·7	91·7	88	85	81	79	76	73
23	—	—	100	95·8	92·0	88	85	82	79	77
24	—	—	—	100	96·0	92·3	89	86	83	80
25	—	—	—	—	100	96·1	92·6	89	86	83
26	—	—	—	—	—	100	96·3	92·9	90	87
27	—	—	—	—	—	—	100	96·4	93·1	90
28	—	—	—	—	—	—	—	100	96·6	93·3
29	—	—	—	—	—	—	—	—	100	96·7
30	—	—	—	—	—	—	—	—	—	100

	31	32	33	34	35	36	37	38	39	40
1	3·2	3·1	3·0	2·9	2·9	2·8	2·7	2·6	2·6	2·5
2	6·5	6·2	6·1	5·9	5·7	5·6	5·4	5·3	5·1	5·0
3	9·7	9·4	9·1	8·8	8·6	8·3	8·1	7·9	7·7	7·5
4	13	13	12	12	11	11	11	11	10	10
5	16	16	15	15	14	14	14	13	13	13
6	19	19	18	18	17	17	16	16	15	15
7	23	22	21	21	20	19	19	18	18	18
8	26	25	24	24	23	22	22	21	21	20
9	29	28	27	26	26	25	24	24	23	23
10	32	31	30	29	29	28	27	26	26	25

	31	32	33	34	35	36	37	38	39	40
11	35	34	33	32	31	31	30	29	28	28
12	39	37	36	35	34	33	32	32	31	30
13	42	41	39	38	37	36	35	34	33	33
14	45	44	42	41	40	39	38	37	36	35
15	48	47	45	44	43	42	41	39	38	38
16	52	50	48	47	46	44	43	42	41	40
17	55	53	52	50	49	47	46	45	44	43
18	58	56	55	53	51	50	49	47	46	45
19	61	59	58	56	54	53	51	50	49	48
20	65	63	61	59	57	56	54	53	51	50
21	68	66	64	62	60	58	57	55	54	53
22	71	69	67	65	63	61	59	58	56	55
23	74	72	70	68	66	64	62	61	59	58
24	77	75	73	71	69	67	65	63	61	60
25	81	78	76	74	71	69	68	66	64	63
26	84	81	79	76	74	72	70	68	67	65
27	87	84	82	79	77	75	73	71	69	68
28	90·3	88	85	82	80	78	76	74	72	70
29	93·5	90·6	88	85	83	81	78	76	74	73
30	96·8	93·8	90·9	88	86	83	81	79	77	75
31	100	96·9	93·9	91·2	89	86	84	82	79	78
32	—	100	97·0	94·1	91·4	89	86	84	82	80
33	—	—	100	97·1	94·3	91·7	89	87	85	83
34	—	—	—	100	97·1	94·4	91·9	89	87	85
35	—	—	—	—	100	97·2	94·6	92·1	90·7	88
36	—	—	—	—	—	100	97·3	94·7	92·3	90
37	—	—	—	—	—	—	100	97·4	94·9	92·5
38	—	—	—	—	—	—	—	100	97·4	95·0
39	—	—	—	—	—	—	—	—	100	97·5
40	—	—	—	—	—	—	—	—	—	100

	41	42	43	44	45	46	47	48	49	50
1	2·4	2·4	2·3	2·3	2·2	2·2	2·1	2·1	2·0	2·0
2	4·9	4·8	4·7	4·5	4·4	4·3	4·3	4·2	4·1	4·0
3	7·3	7·1	7·0	6·8	6·7	6·5	6·4	6·3	6·1	6·0
4	9·8	9·5	9·3	9·1	8·9	8·7	8·5	8·3	8·2	8·0
5	12	12	12	11	11	11	11	10	10	10
6	15	14	14	14	13	13	13	13	12	12
7	17	17	16	16	16	15	15	15	14	14
8	20	19	19	18	18	17	17	17	16	16
9	22	21	21	20	20	20	19	19	18	18
10	24	24	23	23	22	22	21	21	20	20

	41	42	43	44	45	46	47	48	49	50
11	27	26	26	25	24	24	23	23	22	22
12	29	29	28	27	27	26	26	25	24	24
13	32	31	30	30	29	28	28	27	27	26
14	34	33	33	32	31	30	30	29	29	28
15	37	36	35	34	33	33	32	31	31	30
16	39	38	37	36	36	35	34	33	33	32
17	41	40	40	39	38	37	36	35	35	34
18	44	43	42	41	40	39	38	38	37	36
19	46	45	44	43	42	41	40	40	39	38
20	49	48	47	45	44	43	43	42	41	40
21	51	50	49	48	47	46	45	44	43	42
22	54	52	51	50	49	48	47	46	45	44
23	56	55	53	52	51	50	49	48	47	46
24	59	57	56	55	53	52	51	50	49	48
25	61	60	58	57	56	54	53	52	51	50
26	63	62	60	59	58	57	55	54	53	52
27	66	64	63	61	60	59	57	56	55	54
28	68	67	65	64	62	61	60	58	57	56
29	71	69	67	66	64	63	62	60	59	58
30	73	71	70	68	67	65	64	63	61	60
31	76	74	72	70	69	67	66	65	63	62
32	78	76	74	73	71	70	68	67	65	64
33	80	79	77	75	73	72	70	69	67	66
34	83	81	79	77	76	74	72	71	69	68
35	85	83	81	80	78	76	74	73	71	70
36	88	86	84	82	80	78	77	75	73	72
37	90·2	88	86	84	82	80	79	77	76	74
38	92·7	90·5	88	86	84	83	81	79	78	76
39	95·1	92·9	90·7	89	87	85	83	81	80	78
40	97·6	95·2	93·0	90·9	89	87	85	83	82	80
41	100	97·6	95·3	93·2	91·1	89·1	87	85	84	82
42	—	100	97·7	95·5	93·3	91·3	89	88	86	84
43	—	—	100	97·7	96·6	93·5	91·5	90·0	88	86
44	—	—	—	100	98·8	95·7	93·6	91·7	90·0	88
45	—	—	—	—	100	97·8	95·7	93·8	91·8	90·0
46	—	—	—	—	—	100	97·9	95·8	93·9	92·0
47	—	—	—	—	—	—	100	97·9	95·9	94·0
48	—	—	—	—	—	—	—	100	98·0	96·0
49	—	—	—	—	—	—	—	—	100	98·0
50	—	—	—	—	—	—	—	—	—	100

	51	52	53	54	55	56	57	58	59	60
1	2·0	2·0	1·9	1·9	1·9	1·8	1·8	1·7	1·7	1·7
2	4·0	4·0	3·8	3·7	3·6	3·6	3·5	3·4	3·4	3·3
3	5·9	5·8	5·7	5·6	5·5	5·4	5·3	5·1	5·1	5·0
4	7·9	7·7	7·5	7·4	7·3	7·1	7·0	6·9	6·8	6·7
5	9·9	9·6	9·4	9·3	9·1	8·9	8·8	8·6	8·5	8·3
6	12	12	11	11	11	11	11	10	10	10
7	14	13	13	13	13	13	12	12	12	12
8	16	15	15	15	15	14	14	14	14	13
9	18	17	17	17	16	16	16	16	15	15
10	20	19	19	19	18	18	18	17	17	17
11	22	21	21	20	20	20	19	19	19	18
12	24	23	23	22	22	21	21	21	20	20
13	25	25	25	24	24	23	23	22	22	22
14	27	27	26	26	25	25	25	24	24	23
15	29	29	28	28	27	27	26	25	25	25
16	31	31	30	30	29	29	28	28	27	27
17	33	33	32	31	31	30	30	29	29	28
18	35	35	34	33	33	32	32	31	31	30
19	37	37	36	35	35	34	33	33	32	32
20	39	38	38	37	36	35	35	34	34	33
21	41	40	40	39	38	38	37	36	36	35
22	43	42	42	41	40	39	39	38	37	37
23	45	44	43	43	42	41	40	40	39	38
24	47	46	45	44	44	43	42	41	41	40
25	49	48	47	46	45	45	44	43	42	42
26	51	50	49	48	47	46	46	45	44	43
27	53	52	51	50	49	48	47	47	46	45
28	55	54	53	52	51	50	49	48	47	47
29	57	56	55	54	53	52	51	50	49	48
30	59	58	57	55	55	54	53	52	51	50
31	61	60	58	57	56	55	54	53	53	52
32	63	62	60	59	58	57	56	55	54	53
33	65	63	62	61	60	59	58	57	56	55
34	67	65	64	63	62	61	60	59	58	57
35	69	67	66	64	64	63	61	60	59	58
36	71	69	68	67	65	64	63	·62	61	60
37	73	71	70	69	67	66	65	64	63	62
38	75	73	72	70	69	68	67	66	64	63
39	76	75	74	72	71	70	68	67	66	65
40	78	77	75	74	73	71	70	69	68	67
41	80	79	77	76	75	73	72	71	69	68
42	82	81	79	78	76	75	74	72	71	70
43	84	83	81	80	78	77	75	74	73	72
44	86	85	83	81	80	79	77	76	75	73
45	88	87	85	83	82	80	79	78	76	75

	51	52	53	54	55	56	57	58	59	60
46	90·2	88	88	85	84	82	81	79	78	77
47	92·2	90·4	89	87	85	84	82	81	80	78
48	94·2	92·3	90·6	89	87	86	84	83	81	80
49	96·1	94·2	92·5	90·7	89	88	86	84	83	82
50	98·0	96·2	94·3	92·6	91·0	89	88	86	85	83
51	100	98·1	96·2	94·4	92·7	91·1	89	88	86	85
52	—	100	98·1	96·3	94·5	92·9	91·2	90·6	88	87
53	—	—	100	98·1	96·4	94·6	93·0	91·4	90·8	88
54	—	—	—	100	98·2	96·4	94·8	93·1	91·5	90·0
55	—	—	—	—	100	98·2	96·5	94·8	93·2	91·6
56	—	—	—	—	—	100	98·2	96·6	94·9	93·3
57	—	—	—	—	—	—	100	98·3	96·6	95·0
58	—	—	—	—	—	—	—	100	98·3	96·7
59	—	—	—	—	—	—	—	—	100	98·3
60	—	—	—	—	—	—	—	—	—	100

	61	62	63	64	65	66	67	68	69	70
1	1·6	1·6	1·6	1·6	1·5	1·5	1·5	1·5	1·4	1·4
2	3·3	3·2	3·2	3·1	3·1	3·0	3·0	2·9	2·9	2·9
3	4·9	4·8	4·8	4·7	4·6	4·5	4·5	4·4	4·3	4·3
4	6·6	6·5	6·3	6·3	6·2	6·1	6·0	5·9	5·8	5·7
5	8·1	8·1	7·9	7·8	7·7	7·6	7·5	7·4	7·2	7·1
6	9·8	9·7	9·5	9·4	9·2	9·1	9·0	8·9	8·7	8·6
7	11	11	11	11	11	11	10	10	10	10
8	13	13	13	13	12	12	12	12	12	11
9	15	14	14	14	14	14	13	13	13	13
10	16	16	16	16	15	15	15	15	14	14
11	18	18	17	17	17	17	16	16	16	16
12	20	19	19	19	18	18	18	18	17	17
13	21	21	21	20	20	20	19	19	19	19
14	23	23	22	22	22	21	21	21	20	20
15	25	24	24	23	23	23	22	22	22	21
16	26	26	25	25	25	24	24	24	23	23
17	28	27	27	27	26	26	25	25	25	24
18	30	29	29	28	28	27	27	26	26	26
19	31	31	30	30	29	29	28	28	28	27
20	33	32	32	31	31	30	30	29	29	29
21	34	34	33	33	32	32	31	31	30	30
22	36	35	35	34	34	33	33	32	32	31
23	38	37	37	36	35	35	34	34	33	33
24	39	39	38	38	37	36	36	35	35	34
25	41	40	40	39	38	38	37	37	36	36

	61	62	63	64	65	66	67	68	69	70
26	43	42	41	41	40	39	39	38	38	37
27	44	44	43	42	42	41	40	40	39	39
28	46	45	44	44	43	42	42	41	41	40
29	48	47	46	45	45	44	43	43	42	41
30	49	48	48	47	46	45	45	44	43	43
31	51	50	49	48	48	47	46	46	45	44
32	52	52	51	50	49	48	48	47	46	46
33	54	53	52	52	51	50	49	49	48	47
34	56	55	54	53	52	52	51	50	49	49
35	57	56	56	55	54	53	52	51	51	50
36	59	58	57	56	55	55	54	53	52	51
37	61	60	59	58	57	56	55	54	54	53
38	62	61	60	59	58	58	57	56	55	54
39	64	63	62	61	60	59	58	57	57	56
40	66	65	63	63	62	61	60	59	58	57
41	67	66	65	64	63	62	61	60	59	59
42	69	68	67	66	65	64	63	62	61	60
43	70	69	68	67	66	65	64	63	62	61
44	72	71	70	69	68	67	66	65	64	63
45	74	73	71	70	69	68	67	66	65	64
46	75	74	73	72	71	70	69	68	67	66
47	77	76	75	73	72	71	70	69	68	67
48	79	77	76	75	74	73	72	71	70	69
49	80	79	78	77	75	74	73	72	71	70
50	82	81	79	78	77	76	75	74	72	71
51	84	82	81	80	78	77	76	75	74	73
52	85	84	83	81	80	79	78	76	75	74
53	87	85	84	83	82	80	79	78	77	76
54	89	87	86	84	83	82	81	79	78	77
55	90·2	89	87	86	85	83	82	81	80	79
56	91·8	90·3	89	88	86	85	84	82	81	80
57	93·4	91·9	90·5	89	88	86	85	84	83	81
58	95·1	93·5	92·1	90·6	89	88	87	85	84	83
59	96·7	95·2	93·7	92·2	90·8	89	88	87	86	84
60	98·4	96·8	95·2	93·8	92·3	90·9	89	88	87	86
61	100	98·4	96·8	95·3	93·8	92·4	91·0	90	88	87
62	—	100	98·4	96·9	95·4	93·9	92·5	91·2	90	89
63	—	—	100	98·4	96·9	95·5	94·0	92·6	91·3	90
64	—	—	—	100	98·5	97·0	95·5	94·1	92·8	91·4
65	—	—	—	—	100	98·5	97·0	95·6	94·2	92·9
66	—	—	—	—	—	100	98·5	97·1	95·7	94·3
67	—	—	—	—	—	—	100	98·5	97·1	95·7
68	—	—	—	—	—	—	—	100	98·6	97·1
69	—	—	—	—	—	—	—	—	100	98·6
70	—	—	—	—	—	—	—	—	—	100

	71	72	73	74	75	76	77	78	79	80
1	1·4	1·4	1·4	1·4	1·3	1·3	1·3	1·3	1·3	1·3
2	2·8	2·8	2·7	2·7	2·7	2·6	2·6	2·6	2·5	2·5
3	4·2	4·2	4·1	4·1	4·0	3·9	3·9	3·8	3·8	3·8
4	5·6	5·6	5·5	5·4	5·3	5·3	5·2	5·1	5·1	5·0
5	7·0	6·9	6·8	6·8	6·7	6·6	6·5	6·4	6·3	6·3
6	8·5	8·3	8·2	8·1	8·0	7·9	7·8	7·7	7·6	7·5
7	9·9	9·7	9·6	9·5	9·3	9·2	9·1	9·0	8·9	8·8
8	11	11	11	11	11	11	10	10	10	10
9	13	13	12	12	12	12	12	12	11	11
10	14	14	14	14	13	13	13	13	13	13
11	15	15	15	15	15	14	14	14	14	14
12	17	17	16	16	16	16	16	15	15	15
13	18	18	18	18	17	17	17	17	16	16
14	20	19	19	19	19	18	18	18	18	18
15	21	21	21	20	20	20	19	19	19	19
16	23	22	22	22	21	21	21	21	20	20
17	24	24	23	23	23	22	22	22	22	21
18	25	25	25	24	24	24	23	23	23	23
19	27	26	26	26	25	25	25	24	24	24
20	28	28	27	27	27	26	26	26	25	25
21	30	29	29	28	28	28	27	27	27	26
22	31	31	30	30	29	29	29	28	28	28
23	32	32	32	31	31	30	30	29	29	29
24	34	33	33	32	32	32	31	31	30	30
25	35	35	34	34	33	33	32	32	32	31
26	37	36	36	35	35	34	34	33	33	33
27	38	38	37	36	36	36	35	35	34	34
28	39	39	38	38	37	37	36	36	35	35
29	41	40	40	39	39	38	38	37	37	36
30	42	42	41	41	40	39	39	38	38	38
31	44	43	42	42	41	41	40	40	39	39
32	45	44	44	43	43	42	42	41	41	40
33	46	46	45	45	44	43	43	42	42	41
34	48	47	47	46	45	45	44	44	43	43
35	49	49	48	47	47	46	45	45	44	44
36	51	50	49	49	48	47	47	46	46	45
37	52	51	51	50	49	49	48	47	47	46
38	54	53	52	51	51	50	49	49	48	48
39	55	54	53	53	52	51	51	50	49	49
40	56	56	55	54	53	53	52	51	51	50
41	58	57	56	55	55	54	53	53	52	51
42	59	58	58	57	56	55	55	54	53	53
43	61	60	59	58	57	57	56	55	54	54
44	62	61	60	59	59	58	57	56	56	55
45	63	63	62	61	60	59	58	58	57	56

	71	72	73	74	75	76	77	78	79	80
46	65	64	63	62	61	61	60	59	58	58
47	66	65	64	64	63	62	61	60	59	59
48	68	67	66	65	64	63	62	62	61	60
49	69	68	67	66	65	65	64	63	62	61
50	70	69	68	68	67	66	65	64	63	63
51	72	71	70	69	68	67	66	65	65	64
52	73	72	71	70	69	68	68	67	66	65
53	75	74	73	72	71	70	69	68	67	66
54	76	75	74	73	72	71	70	69	68	68
55	77	76	75	74	73	72	71	71	70	69
56	79	78	77	76	75	74	73	72	71	70
57	80	79	78	77	76	75	74	73	72	71
58	82	81	79	78	77	76	75	74	73	73
59	83	82	81	80	79	78	77	76	75	74
60	85	83	82	81	80	79	78	77	76	75
61	86	85	84	82	81	80	79	78	77	76
62	87	86	85	84	83	82	81	79	78	78
63	89	88	86	85	84	83	82	81	80	79
64	90·1	89	88	86	85	84	83	82	81	80
65	91·5	90·3	89	88	87	86	84	83	82	81
66	93·0	91·7	90·4	89	88	87	86	85	84	83
67	94·4	93·1	91·8	90·5	89	88	87	86	85	84
68	95·8	94·4	93·2	91·9	90·7	89	88	87	86	85
69	97·2	95·8	94·5	93·2	92·0	90·8	90	88	87	86
70	98·6	97·2	95·9	94·6	93·3	92·1	90·9	90	89	88
71	100	98·6	97·3	95·9	94·7	93·4	92·2	91·0	90	89
72	—	100	98·6	97·3	96·0	94·7	93·5	92·3	91·1	90
73	—	—	100	98·6	97·3	96·1	94·8	93·6	92·4	91·3
74	—	—	—	100	98·7	97·4	96·1	94·9	93·7	92·5
75	—	—	—	—	100	98·7	97·4	96·2	94·9	93·8
76	—	—	—	—	—	100	98·7	97·4	96·2	95·0
77	—	—	—	—	—	—	100	98·7	97·5	96·3
78	—	—	—	—	—	—	—	100	98·7	97·5
79	—	—	—	—	—	—	—	—	100	98·8
80	—	—	—	—	—	—	—	—	—	100

	81	82	83	84	85	86	87	88	89	90
1	1·2	1·2	1·2	1·2	1·1	1·2	1·1	1·1	1·1	1·1
2	2·5	2·4	2·4	2·4	2·4	2·3	2·3	2·3	2·2	2·2
3	3·7	3·7	3·6	3·6	3·5	3·5	3·4	3·4	3·4	3·3
4	4·9	4·9	4·8	4·8	4·7	4·7	4·6	4·5	4·5	4·4
5	6·2	6·1	6·0	6·0	5·9	5·8	5·7	5·7	5·6	5·6
6	7·4	7·3	7·2	7·1	7·1	7·0	6·9	6·8	6·7	6·7
7	8·6	8·5	8·4	8·3	8·2	8·1	8·0	8·0	7·9	7·8
8	9·9	9·8	9·6	9·5	9·4	9·3	9·2	9·1	9·0	8·9
9	11	11	11	11	11	10	10	10	10	10
10	12	12	12	12	12	12	11	11	11	11
11	14	13	13	13	13	13	13	13	12	12
12	15	15	14	14	14	14	14	14	13	13
13	16	16	16	15	15	15	15	15	15	14
14	17	17	17	17	16	16	16	16	16	16
15	19	18	18	18	18	17	17	17	17	17
16	20	20	19	19	19	19	18	18	18	18
17	21	21	20	20	20	20	20	19	19	19
18	22	22	22	21	21	21	21	20	20	20
19	23	23	23	23	22	22	22	22	21	21
20	25	24	24	24	24	23	23	23	22	22
21	26	26	25	25	25	24	24	24	24	23
22	27	27	27	26	26	26	25	25	25	24
23	28	28	28	27	27	27	26	26	26	26
24	30	29	29	29	28	28	28	27	27	27
25	31	30	30	30	29	29	29	28	28	28
26	32	32	31	31	31	30	30	30	29	29
27	33	33	33	32	32	31	31	31	30	30
28	35	34	34	33	33	33	32	32	31	31
29	36	35	35	35	34	34	33	33	33	32
30	37	37	36	36	35	35	34	34	34	33
31	38	38	37	37	36	36	36	35	35	34
32	40	39	39	38	38	37	37	36	36	36
33	41	40	40	39	39	38	38	38	37	37
34	42	41	41	40	40	40	39	39	38	38
35	43	43	42	42	41	41	40	40	39	39
36	44	44	43	43	42	42	41	41	40	40
37	46	45	45	44	44	43	43	42	42	41
38	47	46	46	45	45	44	44	43	43	42
39	48	48	47	46	46	45	45	44	44	43
40	49	49	48	48	47	47	46	45	45	44
41	51	50	49	49	48	48	47	46	46	46
42	52	51	51	50	49	49	48	48	47	47
43	53	52	52	51	51	50	49	49	48	48
44	54	54	53	52	52	51	51	50	49	49
45	56	55	54	54	53	52	52	51	51	50

	81	82	83	84	85	86	87	88	89	90
46	57	56	55	55	54	53	53	52	52	51
47	58	57	57	56	55	55	54	53	53	52
48	59	59	58	57	56	56	55	55	54	53
49	60	60	59	58	58	57	56	56	55	54
50	62	61	60	60	59	58	57	57	56	56
51	63	62	61	61	60	59	59	58	57	57
52	64	63	63	62	61	60	60	59	58	58
53	65	65	64	63	62	62	61	60	60	59
54	67	66	65	64	64	63	62	61	61	60
55	68	67	66	65	65	64	63	63	62	61
56	69	68	67	67	66	65	64	64	63	62
57	70	70	69	68	67	66	66	65	64	63
58	72	71	70	69	68	67	67	66	65	64
59	73	72	71	70	69	69	68	67	66	66
60	74	73	72	71	71	70	69	68	67	67
61	75	74	73	73	72	71	70	69	69	68
62	77	76	75	74	73	72	71	70	70	69
63	78	77	76	75	74	73	72	72	71	70
64	79	78	77	76	75	74	74	73	72	71
65	80	79	78	77	76	76	75	74	73	72
66	81	80	80	79	78	77	76	75	74	73
67	83	82	81	80	79	78	77	76	75	74
68	84	83	82	81	80	79	78	77	76	76
69	85	84	83	82	81	80	79	78	78	77
70	86	85	84	83	82	81	80	80	79	78
71	88	87	86	85	84	83	82	81	80	79
72	89	88	87	86	85	84	83	82	81	80
73	90	89	88	87	86	85	84	83	82	81
74	91·4	90	89	88	87	86	85	84	83	82
75	93·6	91·5	90·4	89	88	87	86	85	84	83
76	93·8	92·7	91·6	90·5	89	88	87	86	85	84
77	95·1	93·9	92·8	91·7	90·6	90	89	88	87	86
78	96·3	95·1	94·0	92·9	91·8	90·7	90	89	88	87
79	97·5	96·3	95·2	94·0	92·9	91·9	90·8	90	89	88
80	98·8	97·6	96·4	95·2	94·1	93·0	92·0	90·9	90	89
81	100	98·8	97·6	96·4	95·3	94·2	93·1	92·0	91·0	90·0
82	—	100	98·8	97·6	96·5	95·3	94·3	93·2	92·1	91·1
83	—	—	100	98·8	97·6	96·5	95·4	94·3	93·3	92·2
84	—	—	—	100	98·8	97·7	96·6	95·5	94·4	93·3
85	—	—	—	—	100	98·8	97·7	96·6	95·5	94·4
86	—	—	—	—	—	100	98·9	97·7	96·6	95·6
87	—	—	—	—	—	—	100	98·9	97·8	96·7
88	—	—	—	—	—	—	—	100	98·9	97·8
89	—	—	—	—	—	—	—	—	100	98·9
90	—	—	—	—	—	—	—	—	—	100

	91	92	93	94	95	96	97	98	99	100
1	1·1	1·1	1·1	1·1	1·1	1·0	1·0	1·0	1·0	1·0
2	2·2	2·2	2·2	2·1	2·1	2·1	2·1	2·0	2·0	2·0
3	3·3	3·3	3·2	3·2	3·2	3·1	3·1	3·1	3·0	3·0
4	4·4	4·3	4·3	4·3	4·2	4·2	4·1	4·1	4·0	4·0
5	5·5	5·4	5·4	5·3	5·3	5·2	5·2	5·1	5·0	5·0
6	6·6	6·5	6·5	6·4	6·3	6·3	6·2	6·1	6·1	6·0
7	7·7	7·6	7·5	7·4	7·4	7·3	7·2	7·1	7·1	7·0
8	8·8	8·7	8·6	8·5	8·4	8·3	8·2	8·2	8·1	8·0
9	9·9	9·8	9·7	9·6	9·5	9·4	9·3	9·2	9·1	9·0
10	11	11	11	11	11	10	10	10	10	10
11	12	12	12	12	12	11	11	11	11	11
12	13	13	13	13	13	13	12	12	12	12
13	14	14	14	14	14	14	13	13	13	13
14	15	15	15	15	15	15	14	14	14	14
15	16	16	16	16	16	16	15	15	15	15
16	18	17	17	17	17	17	16	16	16	16
17	19	18	18	18	18	18	18	17	17	17
18	20	20	19	19	19	19	19	18	18	18
19	21	21	20	20	20	20	20	19	19	19
20	22	22	22	21	21	21	21	20	20	20
21	23	23	23	22	22	22	22	21	21	21
22	24	24	24	23	23	23	23	22	22	22
23	25	25	25	24	24	24	24	23	23	23
24	26	26	26	26	25	25	25	24	24	24
25	27	27	27	27	26	26	26	26	25	25
26	29	28	28	28	27	27	27	27	26	26
27	30	29	29	29	28	28	28	28	27	27
28	31	30	30	30	29	29	29	29	28	28
29	32	32	31	31	31	30	30	30	29	29
30	33	33	32	32	32	31	31	31	30	30
31	34	34	33	33	33	32	32	32	31	31
32	35	35	34	34	34	33	33	33	32	32
33	36	36	35	35	35	34	34	34	33	33
34	37	37	37	36	36	35	35	35	34	34
35	38	38	38	37	37	36	36	36	35	35
36	40	39	39	38	38	38	37	37	36	36
37	41	40	40	39	39	39	38	38	37	37
38	42	41	41	40	40	40	39	39	38	38
39	43	42	42	41	41	41	40	40	39	39
40	44	43	43	43	42	42	41	41	40	40
41	45	45	44	44	43	43	42	42	41	41
42	46	46	45	45	44	44	43	43	42	42
43	47	47	46	46	45	45	44	44	43	43
44	48	48	47	47	46	46	45	45	44	44
45	49	49	48	48	47	47	46	46	45	45

	91	92	93	94	95	96	97	98	99	100
46	51	50	49	49	48	48	47	47	46	46
47	52	51	51	50	49	49	48	48	47	47
48	53	52	52	51	51	50	49	49	48	48
49	54	53	53	52	52	51	51	50	49	49
50	55	54	54	53	53	52	52	51	51	50
51	56	55	55	54	54	53	53	52	52	51
52	57	56	56	55	55	54	54	53	53	52
53	58	58	57	56	56	55	55	54	54	53
54	59	59	58	57	57	56	56	55	55	54
55	60	60	59	59	58	57	57	56	56	55
56	62	61	60	60	59	58	58	57	57	56
57	63	62	61	61	60	59	59	58	58	57
58	64	63	62	62	61	60	60	59	59	58
59	65	64	63	63	62	61	61	60	60	59
60	66	65	65	64	63	63	62	61	61	60
61	67	66	66	65	64	64	63	62	62	61
62	68	67	67	66	65	65	64	63	63	62
63	69	68	68	67	66	66	65	64	64	63
64	70	70	69	68	67	67	66	65	65	64
65	71	71	70	69	68	68	67	66	66	65
66	73	72	71	70	69	69	68	67	67	66
67	74	73	72	71	71	70	69	68	68	67
68	75	74	73	72	72	71	70	69	69	68
69	76	75	74	73	73	72	71	70	70	69
70	77	76	75	74	74	73	72	71	71	70
71	78	77	76	76	75	74	73	72	72	71
72	79	78	77	77	76	75	74	73	73	72
73	80	79	78	78	77	76	75	74	74	73
74	81	80	80	79	78	77	76	76	75	74
75	82	82	81	80	79	78	77	77	76	75
76	84	83	82	81	80	79	78	78	77	76
77	85	84	83	82	81	80	79	79	78	77
78	86	85	84	83	82	81	80	80	79	78
79	87	86	85	84	83	82	81	81	80	79
80	88	87	86	85	84	83	82	82	81	80
81	89	88	87	86	85	84	84	83	82	81
82	90·1	89	88	87	86	85	85	84	83	82
83	91·2	90·2	89	88	87	86	86	85	84	83
84	92·3	91·3	90·3	89	88	88	87	86	85	84
85	93·4	92·4	91·4	90·4	89	89	88	87	86	85
86	94·5	93·5	92·5	91·5	90·5	90	89	88	87	86
87	95·6	94·6	93·5	92·6	91·6	90·6	90	89	88	87
88	96·7	95·7	94·6	93·6	92·6	91·7	90·7	90	89	88
89	97·8	96·7	95·7	94·7	93·7	92·7	91·8	90·8	90	89
90	98·9	97·8	96·8	95·7	94·7	93·7	92·8	91·8	90·9	90

	91	92	93	94	95	96	97	98	99	100
91	100	98·9	97·8	96·8	95·8	94·8	93·8	92·9	91·9	91·0
92	—	100	98·9	97·9	96·8	95·8	94·8	93·9	92·9	92·0
93	—	—	100	98·9	97·9	96·9	95·9	94·9	93·9	93·0
94	—	—	—	100	98·9	97·9	96·9	95·9	94·9	94·0
95	—	—	—	—	100	99·0	97·9	96·9	96·0	95·0
96	—	—	—	—	—	100	99·0	98·0	97·0	96·0
97	—	—	—	—	—	—	100	99·0	98·0	97·0
98	—	—	—	—	—	—	—	100	99·0	98·0
99	—	—	—	—	—	—	—	—	100	99·0
100	—	—	—	—	—	—	—	—	—	100

	101	102	103	104	105	106	107	108	109	110
1	1·0	1·0	1·0	1·0	1·0	0·9	0·9	0·9	0·9	0·9
2	2·0	2·0	1·9	1·9	1·9	1·9	1·9	1·9	1·8	1·8
3	3·0	2·9	2·9	2·9	2·9	2·8	2·8	2·8	2·8	2·7
4	4·0	3·9	3·9	3·8	3·8	3·8	3·7	3·7	3·7	3·6
5	5·0	4·9	4·9	4·8	4·8	4·7	4·7	4·6	4·5	4·5
6	5·9	5·9	5·8	5·8	5·7	5·7	5·6	5·6	5·5	5·5
7	6·9	6·9	6·8	6·7	6·7	6·6	6·5	6·5	6·4	6·4
8	7·9	7·8	7·8	7·7	7·6	7·5	7·5	7·4	7·3	7·3
9	8·9	8·8	8·7	8·7	8·6	8·5	8·4	8·3	8·3	8·2
10	9·9	9·8	9·7	9·6	9·5	9·4	9·3	9·3	9·2	9·1
11	11	11	11	11	10	10	10	10	10	10
12	12	12	12	12	11	11	11	11	11	11
13	13	13	13	13	12	12	12	12	12	12
14	14	14	14	13	13	13	13	13	13	13
15	15	15	15	14	14	14	14	14	14	14
16	16	16	16	15	15	15	15	15	15	15
17	17	17	17.	17	16	16	16	16	16	15
18	18	18	18	17	17	17	17	17	17	16
19	19	19	19	18	18	18	18	18	18	17
20	20	20	20	19	19	19	19	19	19	18
21	21	21	21	20	20	20	20	20	19	19
22	22	22	22	21	21	21	21	21	20	20
23	23	23	23	22	22	22	22	21	21	21
24	24	24	24	23	23	23	23	22	22	22
25	25	25	25	24	24	24	24	23	23	23
26	26	26	25	25	25	25	24	24	24	24
27	27	27	26	26	26	25	25	25	25	25
28	28	28	27	27	27	26	26	26	26	25
29	29	29	28	28	28	27	27	27	27	26
30	30	30	29	29	29	28	28	28	28	27

	101	102	103	104	105	106	107	108	109	110
31	31	30	30	30	30	29	29	29	28	28
32	32	31	31	31	30	30	30	30	29	29
33	33	32	32	32	31	31	31	31	30	30
34	34	33	33	33	32	32	32	31	31	31
35	35	34	34	34	33	33	33	32	32	32
36	36	35	35	35	34	34	34	33	33	33
37	37	36	36	36	35	35	35	34	34	34
38	38	37	37	37	36	36	36	35	35	35
39	39	38	38	38	37	37	36	36	36	35
40	40	39	39	38	38	38	37	37	37	36
41	41	40	40	39	39	39	38	38	38	37
42	42	41	41	40	40	40	39	39	39	38
43	43	42	42	41	41	41	40	40	40	39
44	44	43	43	42	42	42	41	41	40	40
45	45	44	44	43	43	42	42	42	41	41
46	46	45	45	44	44	43	43	43	42	42
47	47	46	46	45	45	44	44	44	43	43
48	48	47	47	46	46	45	45	44	44	44
49	49	48	48	47	47	46	46	45	45	45
50	50	49	49	48	48	47	47	46	46	45
51	50	50	50	49	49	48	48	47	47	46
52	51	51	50	50	50	49	49	48	48	47
53	52	52	51	51	50	50	50	49	49	48
54	53	53	52	52	51	51	50	50	50	49
55	54	54	53	53	52	52	51	51	50	50
56	55	55	54	54	53	53	52	52	51	51
57	56	56	55	55	54	54	53	53	52	52
58	57	57	56	56	55	55	54	54	53	53
59	58	58	57	57	56	56	55	55	54	54
60	59	59	58	58	57	57	56	56	55	55
61	60	60	59	59	58	58	57	56	56	55
62	61	61	60	60	59	58	58	57	57	56
63	62	62	61	61	60	59	59	58	58	57
64	63	63	62	62	61	60	60	59	59	58
65	64	64	63	63	62	61	61	60	60	59
66	65	65	64	63	63	62	62	61	61	60
67	66	66	65	64	64	63	63	62	61	61
68	67	67	66	65	65	64	64	63	62	62
69	68	68	67	66	66	65	64	64	63	63
70	69	69	68	67	67	66	65	65	64	64
71	70	70	69	68	68	67	66	66	65	65
72	71	71	70	69	69	68	67	67	66	65
73	72	72	71	70	70	69	68	68	67	66
74	73	73	72	71	70	70	69	69	68	67
75	74	74	73	72	71	71	70	69	69	68

	101	102	103	104	105	106	107	108	109	110
76	75	75	74	73	72	72	71	70	70	69
77	76	75	75	74	73	73	72	71	71	70
78	77	76	76	75	74	74	73	72	72	71
79	78	77	77	76	75	75	74	73	72	72
80	79	78	78	77	76	75	75	74	73	73
81	80	79	79	78	77	76	76	75	74	74
82	81	80	80	79	78	77	77	76	75	75
83	82	81	81	80	79	78	78	77	76	75
84	83	82	82	81	80	79	79	78	77	76
85	84	83	83	82	81	80	79	79	78	77
86	85	84	83	83	82	81	80	80	79	78
87	86	85	84	84	83	82	81	81	80	79
88	87	86	85	85	84	83	82	81	81	80
89	88	87	86	86	85	84	83	82	82	81
90	89	88	87	87	86	85	84	83	83	82
91	90·1	89	88	88	87	86	85	84	83	83
92	91·1	90·2	89	88	88	87	86	85	84	84
93	92·1	91·2	90·3	89	89	88	87	86	85	85
94	93·1	92·2	91·3	90·4	90	89	88	87	86	85
95	94·1	93·1	92·2	91·3	90·5	90	89	88	87	86
96	95·1	94·1	93·2	92·3	91·4	90·6	90	89	88	87
97	96·0	95·1	94·2	93·3	92·4	91·5	90·7	90	89	88
98	97·0	96·1	95·1	94·2	93·3	92·5	91·6	90·7	90	89
99	98·0	97·1	96·1	95·2	94·3	93·4	92·5	91·7	90·8	90
100	99·0	98·0	97·1	96·2	95·2	94·3	93·5	92·6	91·7	90·9
101	100	99·0	98·1	97·1	96·2	95·3	94·4	93·5	92·7	91·8
102	—	100	99·0	98·1	97·1	96·2	95·3	94·4	93·6	92·7
103	—	—	100	99·0	98·1	97·2	96·3	95·4	94·5	93·6
104	—	—	—	100	99·0	98·1	97·2	96·3	95·4	94·5
105	—	—	—	—	100	99·1	98·1	97·2	96·3	95·5
106	—	—	—	—	—	100	99·1	98·1	97·2	96·4
107	—	—	—	—	—	—	100	99·1	98·2	97·3
108	—	—	—	—	—	—	—	100	99·1	98·2
109	—	—	—	—	—	—	—	—	100	99·1
110	—	—	—	—	—	—	—	—	—	100

	111	112	113	114	115	116	117	118	119	120
1	0·9	0·9	0·9	0·9	0·9	0·9	0·9	0·8	0·8	0·8
2	1·8	1·8	1·8	1·8	1·7	1·7	1·7	1·7	1·7	1·7
3	2·7	2·7	2·7	2·6	2·6	2·6	2·6	2·5	2·5	2·5
4	3·6	3·6	3·5	3·5	3·5	3·4	3·4	3·3	3·3	3·3
5	4·5	4·5	4·4	4·4	4·3	4·3	4·2	4·2	4·2	4·2
6	5·4	5·4	5·3	5·3	5·2	5·1	5·1	5·0	5·0	5·0
7	6·3	6·3	6·2	6·1	6·0	6·0	6·0	5·9	5·9	5·8
8	7·2	7·1	7·1	7·0	7·0	6·9	6·9	6·8	6·7	6·7
9	8·1	8·0	8·0	7·9	7·8	7·8	7·7	7·6	7·6	7·5
10	9·0	9·0	8·9	8·8	8·7	8·6	8·5	8·5	8·4	8·3
11	9·9	9·8	9·7	9·6	9·6	9·5	9·4	9·3	9·2	9·2
12	11	11	11	11	10	10	10	10	10	10
13	12	12	12	11	11	11	11	11	11	11
14	13	13	12	12	12	12	12	12	12	12
15	14	13	13	13	13	13	13	13	13	13
16	14	14	14	14	14	14	14	14	13	13
17	15	15	15	15	15	15	15	14	14	14
18	16	16	16	16	16	16	15	15	15	15
19	17	17	17	17	17	16	16	16	16	16
20	18	18	18	18	17	17	17	17	17	17
21	19	19	19	18	18	18	18	18	18	18
22	20	20	19	19	19	19	19	19	18	18
23	21	21	20	20	20	20	20	19	19	19
24	22	21	21	21	21	21	21	20	20	20
25	23	22	22	22	22	22	21	21	21	21
26	23	23	23	23	23	22	22	22	22	22
27	24	24	24	24	23	23	23	23	23	23
28	25	25	25	25	24	24	24	24	24	23
29	26	26	76	25	25	25	25	25	24	24
30	27	27	27	26	26	26	26	25	25	25
31	28	28	27	27	27	27	26	26	26	26
32	29	29	28	28	28	28	27	27	27	27
33	30	29	29	29	29	28	28	28	28	28
34	31	30	30	30	30	29	29	29	29	28
35	32	31	31	31	30	30	30	30	29	29
36	32	32	32	32	31	31	31	31	30	30
37	33	33	33	32	32	32	32	31	31	31
38	34	34	34	33	33	33	32	32	32	32
39	35	35	35	34	34	34	33	33	33	33
40	36	36	35	35	35	34	34	34	34	33
41	37	37	36	36	36	35	35	35	34	34
42	38	38	37	37	37	36	36	36	35	35
43	39	38	38	38	37	37	37	36	36	36
44	40	39	39	39	38	38	38	37	37	37
45	41	40	40	39	39	39	38	38	38	38

	111	112	113	114	115	116	117	118	119	120
46	41	41	41	40	40	40	39	39	39	38
47	42	42	42	41	41	41	40	40	39	39
48	43	43	42	42	42	41	41	41	40	40
49	44	44	43	43	43	42	42	42	41	41
50	45	45	44	44	43	43	43	42	42	42
51	46	46	45	45	44	44	44	43	43	43
52	47	46	46	46	45	45	44	44	44	43
53	48	47	47	46	46	46	45	45	45	44
54	49	48	48	47	46	47	46	46	45	45
55	50	49	49	48	47	47	47	47	46	46
56	50	50	50	49	48	48	48	47	47	47
57	51	51	50	50	49	49	49	48	48	48
58	52	52	51	51	50	50	50	49	49	48
59	53	53	52	52	51	51	50	50	50	49
60	54	54	53	53	52	52	51	51	50	50
61	55	54	54	54	53	53	52	52	51	51
62	56	55	55	54	53	53	53	53	52	52
63	57	56	56	55	54	54	54	53	53	53
64	58	57	57	56	55	55	55	54	54	53
65	59	58	58	57	56	56	56	55	55	54
66	59	59	58	58	57	57	56	56	55	55
67	60	60	59	59	58	58	57	57	56	56
68	61	61	60	60	59	59	58	58	57	57
69	62	62	61	61	60	59	59	58	58	58
70	63	63	62	61	60	60	60	59	59	58
71	64	63	63	62	61	61	61	60	60	59
72	65	64	64	63	62	62	62	61	61	60
73	66	65	65	64	63	63	62	62	61	61
74	67	66	65	65	64	64	63	63	62	62
75	68	67	66	66	65	65	64	64	63	63
76	68	68	67	67	66	66	65	64	64	63
77	69	69	68	68	66	66	66	65	65	64
78	70	70	69	68	67	67	67	66	66	65
79	71	71	70	69	68	68	68	67	66	66
80	72	71	71	70	69	69	68	68	67	67
81	73	72	72	71	70	70	69	69	68	68
82	74	73	73	72	71	71	70	69	69	68
83	75	74	73	73	72	72	71	70	70	69
84	76	75	74	74	73	72	72	71	71	70
85	77	76	75	75	73	73	73	72	71	71
86	77	77	76	75	74	74	74	73	72	72
87	78	78	77	76	75	75	74	74	73	73
88	79	79	78	77	76	76	75	75	74	73
89	80	79	79	78	77	77	76	75	75	74
90	81	80	80	79	78	78	77	76	76	75

	111	112	113	114	115	116	117	118	119	120
91	82	81	81	80	79	78	78	77	76	76
92	83	82	81	81	80	79	79	78	77	77
93	84	83	82	82	80	80	79	79	78	78
94	85	84	83	82	81	81	80	80	79	78
95	86	85	84	83	82	82	81	81	80	79
96	86	86	85	84	83	83	82	81	81	80
97	87	87	86	85	84	84	83	82	82	81
98	88	88	87	86	85	84	84	83	82	82
99	89	88	88	87	86	85	85	84	83	83
100	90·1	89	88	88	86	86	85	85	84	83
101	91·0	90·2	89	89	88·8	87	86	86	85	84
102	91·9	91·1	90·3	89	89·6	88	87	86	86	85
103	92·8	92·0	91·2	90·4	90·0	89	88	87	87	86
104	93·7	92·9	92·0	91·2	90·4	90·0	89	88	87	87
105	94·6	93·8	92·9	92·1	91·3	90·5	90·0	89	88	88
106	95·5	94·6	93·8	93·0	92·2	91·4	90·6	90·0	89	88
107	96·4	95·5	94·7	93·9	93·0	92·2	91·5	90·7	90·0	89
108	97·3	96·4	95·6	94·7	93·9	93·1	92·3	91·5	90·8	90·0
109	98·2	97·3	96·5	95·6	94·8	94·0	93·2	92·4	91·6	90·8
110	99·1	98·2	97·3	96·5	95·7	94·8	94·0	93·2	92·4	91·7
111	100	99·1	98·2	97·4	96·5	95·7	94·9	94·1	93·3	92·5
112	—	100	99·1	98·2	97·4	96·6	95·7	94·9	94·1	93·3
113	—	—	100	99·1	98·3	97·4	96·6	95·8	95·0	94·2
114	—	—	—	100	99·1	98·3	97·4	96·6	95·8	95·0
115	—	—	—	—	100	99·1	98·3	97·5	96·6	95·8
116	—	—	—	—	—	100	99·1	98·3	97·5	96·7
117	—	—	—	—	—	—	100	99·2	98·3	97·5
118	—	—	—	—	—	—	—	100	99·1	98·3
119	—	—	—	—	—	—	—	—	100	99·2
120	—	—	—	—	—	—	—	—	—	100

	121	122	123	124	125	126	127	128	129	130
1	0·8	0·8	0·8	0·8	0·8	0·8	0·8	0·8	0·8	0·8
2	1·7	1·6	1·6	1·6	1·6	1·6	1·6	1·6	1·6	1·5
3	2·5	2·5	2·4	2·4	2·4	2·4	2·4	2·3	2·3	2·3
4	3·3	3·3	3·3	3·2	3·2	3·2	3·2	3·1	3·1	3·1
5	4·1	4·1	4·1	4·0	4·0	4·0	3·9	3·9	3·9	3·8
6	5·0	4·9	4·9	4·8	4·8	4·8	4·7	4·7	4·7	4·6
7	5·8	5·7	5·7	5·6	5·6	5·6	5·5	5·5	5·4	5·4
8	6·6	6·6	6·5	6·5	6·4	6·3	6·3	6·3	6·2	6·2
9	7·4	7·4	7·3	7·3	7·2	7·1	7·1	7·0	7·0	6·9
10	8·3	8·2	8·1	8·1	8·0	7·9	7·9	7·8	7·8	7·7
11	9·1	9·0	8·9	8·9	8·8	8·7	8·7	8·6	8·5	8·4
12	9·9	9·8	9·8	9·7	9·6	9·5	9·4	9·4	9·3	9·2
13	11	11	11	10	10	10	10	10	10	10
14	12	11	11	11	11	11	11	11	11	11
15	12	12	12	12	12	12	12	12	12	12
16	13	13	13	13	13	13	13	13	12	12
17	14	14	14	14	14	13	13	13	13	13
18	15	15	15	15	14	14	14	14	14	14
19	16	16	15	15	15	15	15	15	15	15
20	17	16	16	16	16	16	16	16	16	15
21	17	17	17	17	17	17	17	16	16	16
22	18	18	18	18	18	17	17	17	17	17
23	19	19	19	19	18	18	18	18	18	18
24	20	20	20	19	19	19	19	19	19	18
25	21	20	20	20	20	20	20	20	19	19
26	21	21	21	21	21	21	20	20	20	20
27	22	22	22	22	22	21	21	21	21	21
28	23	23	23	23	22	22	22	22	22	22
29	24	24	24	23	23	23	23	23	22	22
30	25	25	24	24	24	24	24	23	23	23
31	26	25	25	25	25	25	24	24	24	24
32	26	26	26	26	26	25	25	25	25	25
33	27	27	27	27	26	26	26	26	26	25
34	28	28	28	27	27	27	27	27	26	26
35	29	29	28	28	28	28	28	27	27	27
36	30	30	29	29	29	29	28	28	28	28
37	31	30	30	30	30	29	29	29	29	28
38	31	31	31	31	30	30	30	30	29	29
39	32	32	32	31	31	31	31	30	30	30
40	33	33	33	32	32	32	31	31	31	31
41	34	34	33	33	33	33	32	32	32	32
42	35	34	34	34	34	33	33	33	33	32
43	36	35	35	35	34	34	34	34	33	33
44	36	36	36	35	35	35	35	34	34	34
45	37	37	37	36	36	36	35	35	35	35

	121	122	123	124	125	126	127	128	129	130
46	38	38	37	37	37	37	36	36	36	35
47	39	39	38	38	38	37	37	37	36	36
48	40	39	39	39	38	38	38	38	37	37
49	40	40	40	40	39	39	39	38	38	38
50	41	41	40	40	40	40	39	39	39	38
51	42	42	41	41	41	40	40	40	39	39
52	43	43	42	42	42	41	41	41	40	40
53	44	43	43	43	42	42	42	41	41	41
54	45	44	44	44	43	43	43	42	42	42
55	45	45	45	44	44	44	43	43	43	42
56	46	46	46	45	45	44	44	44	43	43
57	47	47	46	46	46	45	45	45	44	44
58	48	48	47	47	46	46	46	45	45	45
59	49	48	48	48	47	47	46	46	46	45
60	50	49	49	48	48	48	47	47	47	46
61	50	50	50	49	49	48	48	48	47	47
62	51	51	50	50	50	49	49	48	48	48
63	52	52	51	51	50	50	50	49	49	48
64	53	52	52	52	51	51	50	50	50	49
65	54	53	53	52	52	52	51	51	50	50
66	55	54	54	53	53	52	52	52	51	51
67	55	55	54	54	54	53	53	52	52	52
68	56	56	55	55	54	54	54	53	53	52
69	57	57	56	56	55	55	54	54	53	53
70	58	57	57	56	56	56	55	55	54	54
71	59	58	58	57	57	56	56	55	55	55
72	60	59	59	58	58	57	57	56	56	55
73	60	60	59	59	58	58	57	57	57	56
74	61	61	60	60	59	59	58	58	57	57
75	62	61	61	60	60	60	59	59	58	58
76	63	62	62	61	61	60	60	59	59	58
77	64	63	63	62	62	61	61	60	60	59
78	64	64	63	63	52	62	61	61	60	60
79	65	65	64	64	63	63	62	62	61	61
80	66	66	65	65	64	63	63	63	62	62
81	67	66	66	65	65	64	64	63	63	62
82	68	67	67	66	66	65	65	64	64	63
83	69	68	67	67	66	66	65	65	64	64
84	69	69	68	68	67	67	66	66	65	65
85	70	70	69	69	68	67	67	66	66	65
86	71	70	70	69	69	68	68	67	67	66
87	72	71	71	70	70	69	69	68	67	67
88	73	72	72	71	70	70	69	69	68	68
89	74	73	72	72	71	71	70	70	69	68
90	74	74	73	73	72	71	71	70	70	69

	121	122	123	124	125	126	127	128	129	130
91	75	75	74	73	73	72	72	71	71	70
92	76	75	75	74	74	73	72	72	71	71
93	77	76	76	75	74	74	73	73	72	72
94	78	77	76	76	75	75	74	73	73	72
95	79	78	77	77	76	75	75	74	74	73
96	79	79	78	77	77	76	76	75	74	74
97	80	80	79	78	78	77	76	76	75	75
98	81	80	80	79	78	78	77	77	76	75
99	82	81	80	80	79	79	78	77	77	76
100	83	82	81	81	80	79	79	78	78	77
101	83	83	82	81	81	80	80	79	78	78
102	84	84	83	82	82	81	80	80	79	78
103	85	84	84	83	82	82	81	80	80	79
104	86	85	85	84	83	83	82	81	81	80
105	87	86	85	85	84	83	83	82	81	81
106	88	87	86	85	85	84	83	83	82	82
107	88	88	87	86	86	85	84	84	83	82
108	89	89	88	87	86	86	85	84	84	83
109	90·1	89	89	88	87	87	86	85	84	84
110	90·9	90·2	89	89	88	87	87	86	85	85
111	91·7	91·0	90·2	90	89	88	87	87	86	85
112	92·0	91·8	91·1	90·3	90	89	88	88	87	86
113	93·4	92·6	91·9	91·1	90·4	90	89	88	88	87
114	94·2	93·4	92·7	91·9	91·2	90·5	90	89	88	88
115	95·0	94·3	93·5	92·7	92·0	91·3	90·6	90	89	88
116	95·9	95·1	94·3	93·5	92·8	92·1	91·3	90·6	90	89
117	96·7	95·9	95·1	94·4	93·6	92·9	92·1	91·4	90·7	90
118	97·5	96·7	95·9	95·2	94·4	93·7	92·9	92·2	91·5	90·8
119	98·3	97·5	96·7	96·0	95·2	94·4	93·7	93·0	92·2	91·5
120	99·2	98·4	97·6	96·8	96·0	95·2	94·5	93·8	93·0	92·3
121	100	99·2	98·4	97·6	96·8	96·0	95·3	94·5	93·8	93·0
122	—	100	99·2	98·4	97·6	96·8	96·1	95·3	94·6	93·8
123	—	—	100	99·2	98·4	97·6	96·9	96·1	95·3	94·6
124	—	—	—	100	99·2	98·4	97·6	96·9	96·1	95·4
125	—	—	—	—	100	99·2	98·4	97·7	96·9	96·2
126	—	—	—	—	—	100	99·2	98·4	97·7	96·9
127	—	—	—	—	—	—	100	99·2	98·5	97·7
128	—	—	—	—	—	—	—	100	99·2	98·5
129	—	—	—	—	—	—	—	—	100	99·2
130	—	—	—	—	—	—	—	—	—	100

	131	132	133	134	135	136	137	138	139	140
1	0·8	0·8	0·8	0·7	0·7	0·7	0·7	0·7	0·7	0·7
2	1·5	1·5	1·5	1·5	1·5	1·5	1·5	1·5	1·4	1·4
3	2·3	2·3	2·3	2·2	2·2	2·2	2·2	2·2	2·2	2·1
4	3·0	3·0	3·0	3·0	3·0	2·9	2·9	2·9	2·9	2·9
5	3·8	3·8	3·8	3·7	3·7	3·7	3·7	3·6	3·6	3·6
6	4·6	4·5	4·5	4·5	4·4	4·4	4·4	4·3	4·3	4·3
7	5·3	5·3	5·3	5·2	5·2	5·1	5·1	5·1	5·0	5·0
8	6·1	6·0	6·0	6·0	5·9	5·9	5·8	5·8	5·8	5·7
9	6·9	6·8	6·8	6·7	6·7	6·6	6·6	6·5	6·5	6·4
10	7·6	7·6	7·5	7·5	7·4	7·4	7·3	7·2	7·2	7·1
11	8·4	8·3	8·3	8·2	8·1	8·1	8·0	8·0	7·9	7·9
12	9·2	9·1	9·0	9·0	8·9	8·8	8·8	8·7	8·6	8·6
13	9·9	9·8	9·8	9·7	9·6	9·6	9·5	9·4	9·4	9·3
14	11	11	11	10	10	10	10	10	10	10
15	11	11	11	11	11	11	11	11	11	11
16	12	12	12	12	12	12	12	12	12	11
17	13	13	13	13	13	13	12	12	12	12
18	14	14	14	13	13	13	13	13	13	13
19	15	14	14	14	14	14	14	14	14	14
20	15	15	15	15	15	15	15	14	14	14
21	16	16	16	16	16	15	15	15	15	15
22	17	17	17	16	16	16	16	16	16	16
23	18	17	17	17	17	17	17	17	17	16
24	18	18	18	18	18	18	18	17	17	17
25	19	19	19	19	19	18	18	18	18	18
26	20	20	20	19	19	19	19	19	19	19
27	21	20	20	20	20	20	20	20	19	19
28	21	21	21	21	21	21	20	20	20	20
29	22	22	22	22	21	21	21	21	21	21
30	23	23	23	22	22	22	22	22	22	21
31	24	23	23	23	23	23	23	22	22	22
32	24	24	24	24	24	24	23	23	23	23
33	25	25	25	25	24	24	24	24	24	24
34	26	26	26	25	25	25	25	25	24	24
35	27	27	26	26	26	26	26	25	25	25
36	27	27	27	27	27	26	26	26	26	26
37	28	28	28	28	27	27	27	27	27	26
38	29	29	29	28	28	28	28	28	27	27
39	30	30	29	29	29	29	28	28	28	28
40	31	30	30	30	30	29	29	29	29	29
41	31	31	31	31	30	30	30	30	29	29
42	32	32	32	31	31	31	31	30	30	30
43	33	33	32	32	32	32	31	31	31	31
44	34	33	33	33	33	32	32	32	32	31
45	34	34	34	34	33	33	33	33	32	32

	131	132	133	134	135	136	137	138	139	140
46	35	35	35	34	34	34	34	33	33	33
47	36	36	35	35	35	35	34	34	34	34
48	37	36	36	36	36	35	35	35	35	34
49	37	37	37	37	36	36	36	36	35	35
50	38	38	38	37	37	37	36	36	36	36
51	39	39	38	38	38	38	37	37	37	36
52	40	39	39	39	39	38	38	38	37	37
53	40	40	40	40	39	39	39	38	38	38
54	41	41	41	40	40	40	39	39	39	39
55	42	42	41	41	41	40	40	40	40	39
56	43	42	42	42	41	41	41	41	40	40
57	44	43	43	43	42	42	42	41	41	41
58	44	44	44	43	43	43	42	42	42	41
59	45	45	44	44	44	43	43	43	42	42
60	46	45	45	45	44	44	44	43	43	43
61	47	46	46	46	45	45	45	44	44	44
62	47	47	47	46	46	46	45	45	45	44
63	48	48	47	47	47	46	46	46	45	45
64	49	48	48	48	47	47	47	46	46	46
65	50	49	49	49	48	48	47	47	47	46
66	50	50	50	49	49	49	48	48	47	47
67	51	51	50	50	50	49	49	49	48	48
68	52	52	51	51	50	50	50	49	49	49
69	53	52	52	51	51	51	50	50	50	49
70	53	53	53	52	52	51	51	51	50	50
71	54	54	53	53	53	52	52	51	51	51
72	55	55	54	54	53	53	53	52	52	51
73	56	55	55	54	54	54	53	53	53	52
74	56	56	56	55	55	54	54	54	53	53
75	57	57	56	56	56	55	55	54	54	54
76	58	58	57	57	56	56	55	55	55	54
77	59	58	58	57	57	57	56	56	55	55
78	60	59	59	58	58	57	57	57	56	56
79	60	60	59	59	59	58	58	57	57	56
80	61	61	60	60	59	59	58	58	58	57
81	62	61	61	60	60	60	59	59	58	58
82	63	62	62	61	61	60	60	59	59	59
83	63	63	62	62	61	61	61	60	60	59
84	64	64	63	63	62	62	61	61	60	60
85	65	64	64	63	63	63	62	62	61	61
86	66	65	65	64	64	63	63	62	62	61
87	66	66	65	65	64	64	64	63	63	62
88	67	67	66	66	65	65	64	64	63	63
89	68	67	67	66	66	65	65	64	64	64
90	69	68	68	67	67	66	66	65	65	64

	131	132	133	134	135	136	137	138	139	140
91	69	69	68	68	67	67	66	66	65	65
92	70	70	69	69	68	68	67	67	66	66
93	71	70	70	69	69	68	68	67	67	66
94	72	71	71	70	70	69	69	68	68	67
95	73	72	71	71	70	70	69	69	68	68
96	73	73	72	72	71	71	70	70	69	69
97	74	73	73	72	72	71	71	70	70	69
98	75	74	74	73	73	72	72	71	71	70
99	76	75	74	74	73	73	72	72	71	71
100	76	76	75	75	74	74	73	72	72	71
101	77	77	76	75	74	74	74	73	73	72
102	78	77	77	76	75	75	74	74	73	73
103	79	78	77	77	76	76	75	75	74	74
104	79	79	78	78	77	76	76	75	75	74
105	80	80	79	78	77	77	77	76	76	75
106	81	80	80	79	78	78	77	77	76	76
107	82	81	80	80	79	79	78	78	77	76
108	82	82	81	81	80	79	79	79	78	77
109	83	83	82	81	80	80	80	79	78	78
110	84	83	83	82	81	81	80	80	79	79
111	85	84	83	83	82	82	81	80	80	79
112	85	85	84	84	82	82	82	81	81	80
113	86	86	85	84	83	83	82	82	81	81
114	87	86	86	85	84	84	83	83	82	81
115	88	87	86	86	85	85	84	83	83	82
116	89	88	87	87	85	85	85	84	83	83
117	89	89	88	87	86	86	85	85	84	84
118	90	89	89	88	87	87	86	86	85	84
119	90·9	90·2	89	89	88	88	87	86	86	85
120	91·7	90·9	90·2	90	88	88	88	87	86	86
121	92·4	91·7	91·0	90·3	89	89	88	88	87	86
122	93·1	92·4	91·7	91·0	90·3	90	89	88	88	87
123	93·9	93·2	92·5	91·8	91·1	90·4	90	89	88	88
124	94·7	93·9	93·2	92·5	91·8	91·2	90·5	90	89	89
125	95·4	94·7	94·0	93·3	92·5	91·9	91·2	90·6	90	89
126	96·2	95·5	94·7	94·0	93·3	92·6	92·0	91·3	90·6	90
127	96·9	96·2	95·5	94·8	94·0	93·4	92·7	92·0	91·4	90·7
128	97·7	97·0	96·2	95·5	94·8	94·1	93·4	92·8	92·1	91·4
129	98·5	97·7	97·0	96·3	95·5	94·9	94·2	93·5	92·8	92·1
130	99·2	98·5	97·7	97·0	96·2	95·6	94·9	94·2	93·5	92·9
131	100	99·2	98·5	97·8	97·0	96·3	95·6	94·9	94·2	93·6
132	—	100	99·2	98·5	97·7	97·1	96·4	95·7	95·0	94·3
133	—	—	100	99·3	98·5	97·8	97·1	96·4	95·7	95·0
134	—	—	—	100	99·2	98·5	97·8	97·1	96·4	95·7
135	—	—	—	—	100	99·3	98·5	97·8	97·1	96·4

	131	132	133	134	135	136	137	138	139	140
136	—	—	—	—	—	100	99·3	98·6	97·8	97·1
137	—	—	—	—	—	—	100	99·8	98·6	97·9
138	—	—	—	—	—	—	—	100	99·3	98·6
139	—	—	—	—	—	—	—	—	100	99·3
140	—	—	—	—	—	—	—	—	—	100

	141	142	143	144	145	146	147	148	149	150
1	0·7	0·7	0·7	0·7	0·7	0·7	0·7	0·7	0·7	0·7
2	1·4	1·4	1·4	1·4	1·4	1·4	1·4	1·3	1·3	1·3
3	2·1	2·1	2·1	2·1	2·1	2·1	2·0	2·0	2·0	2·0
4	2·8	2·8	2·8	2·8	2·8	2·7	2·7	2·7	2·7	2·7
5	3·5	3·5	3·5	3·5	3·4	3·4	3·4	3·4	3·4	3·3
6	4·2	4·2	4·2	4·2	4·1	4·1	4·1	4·0	4·0	4·0
7	5·0	4·9	4·9	4·9	4·8	4·8	4·8	4·7	4·7	4·7
8	5·7	5·6	5·6	5·6	5·5	5·5	5·4	5·4	5·4	5·3
9	6·4	6·3	6·3	6·3	6·2	6·2	6·1	6·1	6·0	6·0
10	7·1	7·0	7·0	6·9	6·9	6·8	6·8	6·8	6·7	6·7
11	7·8	7·7	7·7	7·6	7·6	7·5	7·5	7·4	7·4	7·3
12	8·5	8·5	8·4	8·3	8·3	8·2	8·2	8·1	8·1	8·0
13	9·2	9·2	9·1	9·0	9·0	8·9	8·8	8·8	8·7	9·6
14	9·9	9·9	9·8	9·7	9·7	9·6	9·5	9.4	9·4	9·3
15	11	11	10	10	10	10	10	10	10	10
16	11	11	11	11	11	11	11	11	11	11
17	12	12	12	12	12	12	12	11	11	11
18	13	13	13	13	12	12	12	12	12	12
19	13	13	13	13	13	13	13	13	13	13
20	14	14	14	14	14	14	14	14	13	13
21	15	15	15	15	14	14	14	14	14	14
22	16	15	15	15	15	15	15	15	15	15
23	16	16	16	16	16	16	16	15	15	15
24	17	17	17	17	17	16	16	16	16	16
25	18	18	17	17	17	17	17	17	17	17
26	18	18	18	18	18	18	18	18	17	17
27	19	19	19	19	19	18	18	18	18	18
28	20	20	20	19	19	19	19	19	19	19
29	21	20	20	20	20	20	20	20	19	19
30	21	21	21	21	21	21	20	20	20	20
31	22	22	22	22	21	21	21	21	21	21
32	23	23	22	22	22	22	22	22	21	21
33	23	23	23	23	23	23	22	22	22	22
34	24	24	24	24	23	23	23	23	23	23
35	25	25	24	24	24	24	24	24	23	23

	141	142	143	144	145	146	147	148	149	150
36	26	25	25	25	25	25	24	24	24	24
37	26	26	26	26	26	25	25	25	25	25
38	27	27	27	26	26	26	26	26	26	25
39	28	27	27	27	27	27	27	26	26	26
40	28	28	28	28	28	27	27	27	27	27
41	29	29	29	28	28	28	28	28	28	27
42	30	30	29	29	29	29	29	28	28	28
43	30	30	30	30	30	29	29	29	29	29
44	31	31	31	31	30	30	30	30	30	29
45	32	32	31	31	31	31	31	30	30	30
46	33	32	32	32	32	32	31	31	31	31
47	33	33	33	33	32	32	32	32	32	31
48	34	34	34	33	33	33	33	32	32	32
49	35	35	34	34	34	34	33	33	33	33
50	35	35	35	35	34	34	34	34	34	33
51	36	36	36	35	35	35	35	34	34	34
52	37	37	36	36	36	36	35	35	35	35
53	38	37	37	37	37	36	36	36	36	35
54	38	38	38	38	37	37	37	36	36	36
55	39	39	38	38	38	38	37	37	37	37
56	40	39	39	39	39	38	38	38	38	37
57	40	40	40	40	39	39	39	39	38	38
58	41	41	41	40	40	40	39	39	39	39
59	42	42	41	41	41	40	40	40	40	39
60	43	42	42	42	41	41	41	41	40	40
61	43	43	43	42	42	42	41	41	41	41
62	44	44	43	43	43	42	42	42	42	41
63	45	44	44	44	43	43	43	43	42	42
64	45	45	45	44	44	44	44	43	43	43
65	46	46	45	45	45	45	44	44	44	43
66	47	46	46	46	46	45	45	45	44	44
67	48	47	47	47	46	46	46	45	45	45
68	48	48	48	47	47	47	46	46	46	45
69	49	49	48	48	48	47	47	47	46	46
70	50	49	49	49	48	48	48	47	47	47
71	50	50	50	49	49	49	48	48	48	47
72	51	51	50	50	50	49	49	49	48	48
73	52	51	51	51	50	50	50	49	49	49
74	52	52	52	51	51	51	50	50	50	49
75	53	53	52	52	52	51	51	51	50	50
76	54	54	53	53	52	52	52	51	51	51
77	55	54	54	53	53	53	52	52	52	51
78	55	55	55	54	54	53	53	53	52	52
79	56	56	55	55	54	54	54	53	53	53
80	57	56	56	56	55	55	54	54	54	53

	141	142	143	144	145	146	147	148	149	150
81	57	57	57	56	56	55	55	55	54	54
82	58	58	57	57	57	56	56	55	55	55
83	59	58	58	58	57	57	56	56	56	55
84	60	59	59	58	58	58	57	57	56	56
85	60	60	59	59	59	58	58	57	57	57
86	61	61	60	60	59	59	59	58	58	57
87	62	61	61	60	60	60	59	59	58	58
88	62	62	62	61	61	60	60	59	59	59
89	63	63	62	62	61	61	61	60	60	59
90	64	63	63	63	62	62	61	61	60	60
91	65	64	64	63	63	62	62	61	61	61
92	65	65	64	64	63	63	63	62	62	61
93	66	65	65	65	64	64	63	63	62	62
94	67	66	66	65	65	64	64	64	63	63
95	67	67	66	66	66	65	65	64	64	63
96	68	68	67	67	66	66	65	65	64	64
97	69	68	68	67	67	66	66	66	65	65
98	70	69	69	68	67	67	67	66	66	65
99	70	70	69	69	68	68	67	67	66	66
100	71	70	70	69	69	68	68	68	67	67
101	72	71	71	70	70	69	69	68	68	67
102	72	72	71	71	70	70	69	69	68	68
103	73	73	72	72	71	71	70	70	69	69
104	74	73	73	72	72	71	71	70	70	69
105	74	74	73	73	72	72	71	71	70	70
106	75	75	74	74	73	73	72	72	71	71
107	76	75	75	74	74	73	73	72	72	71
108	77	76	76	75	74	74	73	73	72	72
109	77	77	76	76	75	75	74	74	73	73
110	78	77	77	76	76	75	75	74	74	73
111	79	78	78	77	77	76	76	75	74	74
112	79	79	78	78	77	77	76	65	75	75
113	80	80	79	78	78	77	77	76	76	75
114	81	80	80	79	79	78	78	77	77	76
115	82	81	80	80	79	79	78	78	77	77
116	82	82	81	81	80	79	79	78	78	77
117	83	82	82	81	81	80	80	79	79	78
118	84	83	83	82	81	81	80	80	79	79
119	84	84	83	83	82	82	81	80	80	79
120	85	85	84	83	83	82	82	81	81	80
121	86	85	85	84	83	83	82	82	81	81
122	87	86	85	85	84	84	83	82	82	81
123	87	87	86	85	85	84	84	83	83	82
124	88	87	87	86	86	85	84	84	83	83
125	89	88	87	87	86	86	85	84	84	83

	141	142	143	144	145	146	147	148	149	150
126	89	89	88	88	87	86	86	85	85	84
127	90·1	89	89	88	88	87	86	86	85	85
128	90·8	90·1	90	89	88	88	87	86	86	85
129	91·5	90·8	90·2	90	89	88	88	87	87	86
130	92·2	91·5	90·9	90·3	90	89	88	88	87	87
131	92·9	92·3	91·6	91·0	90·3	90·7	89	89	88	87
132	93·6	93·0	92·3	91·7	91·0	90·4	90	89	89	88
133	94·3	93·7	93·0	92·4	91·7	91·1	90·5	90	89	89
134	95·0	94·4	93·7	93·1	92·4	91·8	91·2	90·5	90	89
135	95·7	95·1	94·4	93·8	93·1	92·5	91·8	91·2	90·6	90
136	96·5	95·8	95·1	94·4	93·8	93·2	92·5	91·9	91·3	90·7
137	97·2	96·5	95·8	95·1	94·5	93·8	93·2	92·6	91·9	91·3
138	97·9	97·2	96·5	95·8	95·2	94·5	93·9	93·2	92·6	92·0
139	98·6	97·9	97·2	96·5	95·9	95·2	94·6	93·9	93·3	92·7
140	99·3	98·6	97·9	97·2	96·6	95·9	95·2	94·6	94·0	93·3
141	100	99·3	98·6	97·9	97·2	96·6	95·9	95·3	94·4	94·0
142	—	100	99·3	98·6	97·9	97·3	96·6	95·9	95·3	94·7
143	—	—	100	99·3	98·6	97·9	97·3	96·6	96·0	95·3
144	—	—	—	100	99·3	98·6	98·0	97·3	96·6	96·0
145	—	—	—	—	100	99·3	98·6	98·0	97·3	96·7
146	—	—	—	—	—	100	99·3	98·6	98·0	97·3
147	—	—	—	—	—	—	100	99·3	98·7	98·0
148	—	—	—	—	—	—	—	100	99·3	98·7
149	—	—	—	—	—	—	—	—	100	99·3
150	—	—	—	—	—	—	—	—	—	100

APPENDIX C

Table of Square Roots of Integers from 1 to 999

	0	1	2	3	4	5	6	7	8	9
0	0	1·00	1·41	1·73	2·00	2·23	2·45	2·64	2·83	3·00
10	3·16	3·31	3·46	3·60	3·74	3·87	4·00	4·12	4·24	4·36
20	4·47	4·58	4·69	4·79	4·90	5·00	5·10	5·19	5·29	5·38
30	5·48	5·57	5·66	5·74	5·83	5·91	6·00	6·08	6·16	6·24
40	6·32	6·40	6·48	6·56	6·63	6·71	6·78	6·85	6·93	7·00
50	7·07	7·14	7·21	7·28	7·35	7·41	7·48	7·55	7·61	7·68
60	7·74	7·81	7·87	7·94	8·00	8·06	9·12	8·18	8·24	8·30
70	8·37	8·42	8·48	8·54	8·60	8·66	8·71	8·77	8·83	8·89
80	8·94	9·00	9·05	9·11	9·16	9·21	9·27	9·32	9·38	9·43
90	9·49	9·54	9·59	9·64	9·69	9·75	9·80	9·85	9·90	9·95
100	10·00	10·05	10·10	10·15	10·20	10·25	10·29	10·34	10·39	10·44
110	10·49	10·53	10·58	10·63	10·68	10·72	10·77	10·82	10·86	10·91
120	10·95	11·00	11·04	11·09	11·14	11·18	11·22	11·27	11·31	11·36
130	11·40	11·45	11·49	11·53	11·57	11·62	11·66	11·70	11·75	11·79
140	11·83	11·87	11·91	11·96	12·00	12·04	12·08	12·12	12·17	12·21
150	12·25	12·29	12·33	12·37	12·40	12·45	12·49	12·53	12·57	12·61
160	12·65	12·69	12·73	12·77	12·80	12·84	12·88	12·92	12·96	13·00
170	13·04	13·08	13·11	13·15	13·19	13·23	13·27	13·30	13·34	13·38
180	13·41	13·45	13·49	13·53	13·56	13·60	13·64	13·67	13·71	13·75
190	13·78	13·82	13·86	13·89	13·93	13·96	14·00	14·04	14·07	14·11
200	14·14	14·18	14·21	14·25	14·28	14·32	14·35	14·39	14·42	14·46
210	14·49	14·53	14·56	14·60	14·63	14·66	14·70	14·73	14·76	14·80
220	14·83	14·87	14·90	14·93	14·97	15·00	15·03	15·07	15·10	15·13
230	15·17	15·20	15·23	15·26	15·30	15·33	15·36	15·39	15·43	15·46
240	15·49	15·52	15·56	15·59	15·62	15·65	15·68	15·71	15·75	15·78
250	15·81	15·84	15·87	15·91	15·94	15·97	16·00	16·03	16·06	16·09
260	16·12	16·16	16·19	16·22	16·25	16·28	16·31	16·34	16·37	16·40
270	16·43	16·46	16·49	16·52	16·55	16·58	16·61	16·64	16·67	16·70
280	16·73	16·76	16·79	16·82	16·85	16·88	16·91	16·94	16·97	17·00
290	17·03	17·06	17·09	17·12	17·15	17·18	17·20	17·23	17·26	17·29
300	17·32	17·35	17·38	17·41	17·44	17·46	17·49	17·52	17·55	17·58
310	17·61	17·64	17·66	17·69	17·72	17·75	17·78	17·80	17·83	17·86
320	17·89	17·92	17·94	17·97	18·00	18·03	18·06	18·08	18·11	18·14
330	18·17	18·19	18·22	18·25	18·28	18·30	18·33	18·36	18·38	18·41
340	18·44	18·47	18·49	18·52	18·55	18·57	18·60	18·63	18·66	18·68
350	18·71	18·74	18·76	18·79	18·81	18·84	18·87	18·89	18·92	18·95
360	18·97	19·00	19·03	19·05	19·08	19·10	19·13	19·16	19·18	19·21
370	19·24	19·26	19·29	19·31	19·34	19·37	19·39	19·42	19·44	19·47
380	19·49	19·52	19·54	19·57	19·60	19·62	19·65	19·67	19·70	19·72
390	19·75	19·77	19·80	19·82	19·85	19·87	19·90	19·92	19·95	19·97
400	20·00	20·02	20·05	20·07	20·10	20·12	20·15	20·17	20·20	20·22

	0	1	2	3	4	5	6	7	8	9
410	20·25	20·27	20·30	20·32	20·35	20·37	20·40	20·42	20·44	20·47
420	20·49	20·52	20·54	20·57	20·59	20·62	20·64	20·66	20·69	20·71
430	20·74	20·76	20·78	20·81	20·83	20·86	20·88	20·90	20·93	20·95
440	20·97	21·00	21·02	21·05	21·07	21·09	21·12	21·14	21·17	21·19
450	21·21	21·24	21·26	21·28	21·31	21·33	21·35	21·38	21·40	21·42
460	21·45	21·47	21·49	21·52	21·54	21·56	21·59	21·61	21·63	21·66
470	21·68	21·70	21·73	21·75	21·77	21·79	21·82	21·84	21·86	21·88
480	21·91	21·93	21·95	21·98	22·00	22·02	22·04	22·07	22·09	22·11
490	22·14	22·16	22·18	22·20	22·23	22·25	22·27	22·29	22·32	22·34
500	22·36	22·38	22·40	22·43	22·45	22·47	22·49	22·52	22·54	22·56
510	22·58	22·60	22·63	22·65	22·67	22·69	22·72	22·74	22·76	22·78
520	22·80	22·83	22·85	22·87	22·89	22·91	22·94	22·96	22·98	23·00
530	23·02	23·04	23·06	23·09	23·11	23·13	23·15	23·17	23·19	23·22
540	23·24	23·26	23·28	23·30	23·32	23·35	23·37	23·39	23·41	23·43
550	23·45	23·47	23·49	23·52	23·54	23·56	23·58	23·60	23·62	23·64
560	23·66	23·69	23·71	23·73	23·75	23·77	23·79	23·81	23·83	23·85
570	23·87	23·89	23·92	23·94	23·96	23·98	24·00	24·02	24·04	24·06
580	24·08	24·10	24·13	24·15	24·17	24·19	24·21	24·23	24·25	24·27
590	24·29	24·31	24·33	24·35	24·37	24·39	24·41	24·43	24·45	24·47
600	24·49	24·51	24·54	24·56	24·58	24·60	24·62	24·64	24·66	24·68
610	24·70	24·72	24·74	24·76	24·78	24·80	24·82	24·84	24·86	24·88
620	24·90	24·92	24·94	24·96	24·98	25·00	25·02	25·04	25·06	25·08
630	25·10	25·12	25·14	25·16	25·18	25·20	25·22	25·24	25·26	25·28
640	25·30	25·32	25·24	25·36	25·38	25·40	25·42	25·46	25·46	25·48
650	25·50	25·52	25·53	25·55	25·57	25·59	25·61	25·63	25·65	25·67
660	25·69	25·71	25·73	25·75	25·77	25·79	25·81	25·83	25·85	25·87
670	25·88	25·90	25·92	25·94	25·96	25·98	26·00	26·02	26·04	26·06
680	26·08	26·10	26·12	26·13	26·15	26·17	26·19	26·21	26·23	26·25
690	26·27	26·29	26·31	26·33	26·34	26·36	26·38	26·40	26·42	26·44
700	26·46	26·48	26·50	26·51	26·53	26·55	26·57	26·59	26·61	26·63
710	26·65	26·67	26·68	26·70	26·72	26·74	26·76	26·78	26·80	26·81
720	26·83	26·85	26·87	26·89	26·91	26·93	26·94	26·96	26·98	27·00
730	27·02	27·04	27·06	27·07	27·09	27·11	27·13	27·15	27·17	27·19
740	27·20	27·22	27·24	27·26	27·28	27·30	27·31	27·33	27·35	27·37
750	27·39	27·40	27·42	27·44	27·46	27·48	27·50	27·51	27·53	27·55
760	27·57	27·59	27·60	27·62	27·64	27·66	27·68	27·70	27·71	27·73
770	27·75	27·77	27·79	27·80	27·82	27·84	27·86	27·88	27·89	27·91
780	27·93	27·95	27·96	27·98	28·00	28·02	28·04	28·05	28·07	28·09
790	28·11	28·13	28·14	28·16	28·18	28·20	28·21	28·23	28·25	28·27
800	28·28	28·30	28·32	28·34	28·36	28·37	28·39	28·41	28·43	28·44
810	28·46	28·48	28·50	28·51	28·53	28·55	28·57	28·58	28·60	28·62
820	28·64	28·65	28·67	28·69	28·71	28·72	28·74	28·76	28·78	28·79
830	28·81	28·83	28·84	28·86	28·88	28·90	28·91	28·93	28·95	28·97
840	28·98	29·00	29·02	29·03	29·05	29·07	29·09	29·10	29·12	29·14
850	29·16	29·17	29·19	29·21	29·22	29·24	29·26	29·28	29·29	29·31
860	29·33	29·34	29·36	29·38	29·39	29·41	29·43	29·44	29·46	29·48
870	29·50	29·51	29·53	29·55	29·56	29·58	29·60	29·61	29·63	29·65
880	29·67	29·68	29·70	29·72	29·73	29·75	29·77	29·78	29·80	29·82
890	29·83	29·85	29·87	29·88	29·90	29·92	29·93	29·95	29·97	29·98
900	30.00	30·02	30·03	30·05	30·07	30·08	30·10	30·12	30·13	30·15
910	30·17	30·18	30·20	30·22	30·23	30·25	30·27	30·28	30·30	30·32
920	30·33	30·35	30·36	30·38	30·40	30·41	30·43	30·45	30·46	30·48

	0	1	2	3	4	5	6	7	8	9
930	30·50	30·51	30·53	30·55	30·56	30·58	30·59	30·61	30·63	30·64
940	30·66	30·68	30·69	30·71	30·73	30·74	30·76	30·77	30·79	30·81
950	30·82	30·84	30·85	30·87	30·89	30·90	30·92	30·94	30·95	30·97
960	30·98	31·00	31·02	31·03	31·05	31·06	31·08	31·10	31·11	31·13
970	31·15	31·16	31·18	31·19	31·21	31·23	31·24	31·26	31·27	31·29
980	31·31	31·32	31·34	31·35	31·37	31·39	31·40	31·42	31·43	31·45
990	31·46	31·48	31·50	31·51	31·53	31·54	31·56	31·58	31·59	31·61

The square root of 177 will be found in col. 7 opposite row 170.

APPENDIX D

Log$_{10}$ $(N + 1)$ to two decimal places; $N = 0$ to 1000

N	log $(N + 1)$	N	log $(N + 1)$	N	log $(N + 1)$	N	log $(N + 1)$
0	0·00	36	1·57	84–85	1·93	192–196	2·29
1	0·30	37	1·58	86–87	1·94	197–200	2·30
2	0·48	38	1·59	88–89	1·95	201–205	2·31
3	0·60	39	1 60	90–91	1·96	206–210	2·32
4	0·70	40	1·61	92–93	1·97	211–215	2·33
5	0·78	41	1·62	94–95	1·98	216–220	2·34
6	0·85	42	1·63	96–97	1·99	221–225	2 35
7	0·90	43	1·64	98–100	2·00	226–230	2·36
8	0·95	44	1·65	100–102	2·01	231–236	2·37
9	1·00	45	1·66	103–104	2·02	237–241	2·38
10	1·04	46	1·67	105–107	2·03	242–247	2·39
11	1·08	47	1·68	108–109	2·04	248–253	2·40
12	1·11	48	1·69	110–112	2·05	254–258	2·41
13	1·15	49	1·70	113–115	2·06	259–264	2·42
14	1·18	50	1·71	116–117	2·07	265–271	2·43
15	1·20	51–52	1·72	118–120	2·08	272–277	2·44
16	1·23	53	1·73	121–123	2·09	278–284	2·45
17	1·26	54	1·74	124–126	2·10	285–290	2·46
18	1·28	55	1·75	127–129	2·11	291–297	2·47
19	1·30	56–57	1·76	130–132	2·12	298–304	2·48
20	1·32	58	1·77	133–134	2·13	305–311	2·49
21	1·34	59	1·78	135–138	2·14	312–318	2·50
22	1·36	60–61	1·79	139–142	2·15	319–326	2·51
23	1·38	62	1·80	143–145	2·16	327–333	2·52
24	1·40	63–64	1·81	146–148	2·17	334–340	2·53
25	1·42	65	1·82	149–152	2·18	341–349	2·54
26	1·43	66–67	1·83	153–155	2·19	350–357	2·55
27	1·45	68	1·84	156–159	2·20	358–366	2·56
28	1·46	69–70	1·85	160–163	2·21	367–374	2·57
29	1·48	71–72	1·86	164–166	2·22	375–383	2·58
30	1·49	73	1·87	167–170	2·23	384–392	2·59
31	1·51	74–75	1·88	171–174	2·24	393–401	2·60
32	1·52	76–77	1·89	175–178	2·25	402–411	2·61
33	1·53	78–79	1·90	179–183	2·26	412–420	2·62
34	1·54	80–81	1·91	184–187	2·27	421–430	2·63
35	1·56	82–83	1·92	188–191	2·28	431–440	2·64

N	$\log (N+1)$	N	$\log (N+1)$	N	$\log (N+1)$	N	$\log (N+1)$
441–450	2·65	543–553	2·74	668–683	2·83	822–840	2·92
451–461	2·66	554–567	2·75	684–698	2·84	841–859	2·93
462–472	2·67	568–581	2·76	699–715	2·85	860–879	2·94
473–483	2·68	582–594	2·77	716–732	2·86	880–900	2·95
484–494	2·69	595–608	2·78	733–748	2·87	901–921	2·96
495–505	2·70	609–622	2·79	749–766	2·88	922–942	2·97
506–517	2·71	623–637	2·80	767–784	2·89	943–965	2·98
518–530	2·72	638–652	2·81	785–802	2·90	966–987	2·99
531–542	2·73	653–667	2·82	803–821	2·91	988–1000	3·00

Above 1000 use common logarithms, i.e. log (N).

Note: N is the number of individuals in a whole sample, in a sample unit, or, as with light trap catches, in a single night's catch.

APPENDIX E

tan β (= b, the regression coefficient)
$\beta = 0.5°$ to $90°$ and $-0.5°$ to $-90°$

Degrees	·0	·5	Degrees	·0	·5
0	0·00	0·00	40	0·83	0·85
1	0·01	0·02	41	0·86	0·88
2	0·03	0·04	42	0·90	0·91
3	0·05	0·06	43	0·93	0·94
4	0·06	0·07	44	0·96	0·98
5	0·08	0·09	45	1·00	1·01
6	0·10	0·11	46	1·03	1·05
7	0·12	0·13	47	1·07	1·09
8	0·14	0·14	48	1·11	1·13
9	0·15	0·16	49	1·15	1·17
10	0·17	0·18	50	1·19	1·21
11	0·19	0·20	51	1·23	1·25
12	0·21	0·22	52	1·27	1·30
13	0·23	0·24	53	1·32	1·35
14	0·24	0·25	54	1·37	1·40
15	0·26	0·27	55	1·42	1·45
16	0·28	0·29	56	1·48	1·51
17	0·30	0·31	57	1·53	1·56
18	0·32	0·33	58	1·60	1·63
19	0·34	0·35	59	1·66	1·69
20	0·36	0·37	60	1·73	1·76
21	0·38	0·39	61	1·80	1·84
22	0·40	0·41	62	1·88	1·92
23	0·42	0·43	63	1·96	2·00
24	0·44	0·45	64	2·05	2·09
25	0·46	0·47	65	2·14	2·19
26	0·48	0·49	66	2·24	2·29
27	0·50	0·52	67	2·35	2·41
28	0·53	0·54	68	2·47	2·53
29	0·55	0·56	69	2·60	2·67
30	0·57	0·58	70	2·74	2·82
31	0·60	0·61	71	2·90	2·98
32	0·62	0·63	72	3·07	3·17
33	0·64	0·66	73	3·27	3·37
34	0·67	0·68	74	3·48	3·60
35	0·70	0·71	75	3·73	3·86
36	0·72	0·73	76	4·01	4·16
37	0·75	0·76	77	4·33	4·51
38	0·78	0·79	78	4·70	4·91
39	0·80	0·82	79	5·14	5·39

Degrees	·0	·5	Degrees	·0	·5
80	5·67	5·97	85	11·43	12·70
81	6·31	6·69	86	14·30	16·35
82	7·11	7·59	87	19·08	22·90
83	8·14	8·77	88	28·63	38·18
84	9·51	10·38	89	57·29	114·60
			90	∞	

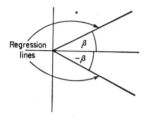

$$\tan(-\beta) = -(\tan \beta) = -b$$

APPENDIX F

The 5% level of significance for the difference between the two fractions of a divided sample

N	a	N		$a\%$
6	6	26 –	27	70·9–70
7	7	28 –	29	69·9–69
8	8	30 –	32	68·9–68
9	8+	33 –	36	67·9–67
10	9+	37 –	40	66·9–66
11	10+	41 –	45	65·9–65
12	10+	46 –	51	64·9–64
13	11+	52 –	60	63·9–63
14	12+	61 –	70	62·9–62
15	12+	71 –	83	61·9–61
16	13+	84 –	100	60·9–60
17	13+	101 –	124	59·9–59
18	14+	125 –	157	58·9–58
19	15+	158 –	205	57·9–57
20	15+	206 –	278	56·9–56
21	16+	279 –	400	55·9–55
22	17+	401 –	625	54·9–54
23	17+	626 –	1,111	53·9–53
24	18+	1,112 –	2,500	52·9–52
25	18+	2,501 –	10,000	51·9–51
		10,000 –	∞	50·9–50

N = total number (100%) in a sample divided into two fractions.
a = minimum value for the larger of the two fractions when the fractions are significantly different at the 5% level, i.e. one of the fractions must be equal to or greater than a. Above $N = 25$, a is given as a percentage of N. For example: when $N = 55$, a is between 63 and 63·9% of 55, i.e. $a = 35$.

APPENDIX G (Part I)

The Fisher–Yates test for the significance of a difference between two
proportions
(Instructions for use on p. 99)

$A+B$	B (or A)	\multicolumn{14}{c}{$C+D$}
		2	3	4	5	6	7	8	9	10	11	12	13	14	15
		\multicolumn{14}{c}{D (or C) at 5% level}
4	4	—	—	0	—	—	—	—	—	—	—	—	—	—	—
5	5	—	0	0	1	—	—	—	—	—	—	—	—	—	—
	4	—	—	—	0	—	—	—	—	—	—	—	—	—	—
6	6	—	0	0	0	1	—	—	—	—	—	—	—	—	—
	5	—	—	0	0	0	—	—	—	—	—	—	—	—	—
7	7	—	0	1	1	2	2	—	—	—	—	—	—	—	—
	6	—	—	0	0	0	1	—	—	—	—	—	—	—	—
	5	—	—	—	—	0	0	—	—	—	—	—	—	—	—
8	8	0	0	1	1	2	2	3	—	—	—	—	—	—	—
	7	—	0	0	0	1	1	2	—	—	—	—	—	—	—
	6	—	—	—	0	0	0	1	—	—	—	—	—	—	—
	5	—	—	—	—	—	0	0	—	—	—	—	—	—	—
9	9	0	0	1	1	2	3	3	4	—	—	—	—	—	—
	8	—	0	0	1	1	2	2	3	—	—	—	—	—	—
	7	—	—	0	0	0	1	1	1	—	—	—	—	—	—
	6	—	—	—	—	0	0	0	1	—	—	—	—	—	—
	5	—	—	—	—	—	—	0	0	—	—	—	—	—	—
10	10	0	0	1	2	2	3	4	4	5	—	—	—	—	—
	9	—	0	0	1	1	2	2	3	3	—	—	—	—	—
	8	—	—	0	0	1	1	1	2	2	—	—	—	—	—
	7	—	—	—	0	0	0	1	1	1	—	—	—	—	—
	6	—	—	—	—	—	0	0	0	0	—	—	—	—	—
	5	—	—	—	—	—	—	—	0	0	—	—	—	—	—
11	11	0	0	1	2	2	3	4	4	5	6	—	—	—	—
	10	—	0	0	1	1	2	3	3	4	4	—	—	—	—
	9	—	—	0	0	1	1	2	2	3	3	—	—	—	—
	8	—	—	—	0	0	1	1	1	2	2	—	—	—	—
	7	—	—	—	—	0	0	0	1	1	1	—	—	—	—
	6	—	—	—	—	—	0	0	0	0	0	—	—	—	—
	5	—	—	—	—	—	—	—	—	—	0	—	—	—	—

Column headers 2–15 are values of **C + D**; table entries are **D (or C) at 5% level**.

A + B	B (or A)	2	3	4	5	6	7	8	9	10	11	12	13.	14	15
12	12	0	0	1	2	3	3	4	5	5	6	7	—	—	—
	11	—	0	0	1	2	2	3	3	4	5	5	—	—	—
	10	—	0	0	0	1	1	2	2	3	3	4	—	—	—
	9	—	—	0	0	0	1	1	2	2	2	3	—	—	—
	8	—	—	—	0	0	0	1	1	1	1	2	—	—	—
	7	—	—	—	—	0	0	0	0	0	1	1	—	—	—
	6	—	—	—	—	—	—	0	0	0	0	0	—	—	—
	5	—	—	—	—	—	—	—	—	—	0	0	—	—	—
13	13	0	1	1	2	3	3	4	5	6	6	7	8	—	—
	12	—	0	1	1	2	2	3	4	4	5	5	6	—	—
	11	—	0	0	1	1	2	2	3	3	4	4	5	—	—
	10	—	—	0	0	1	1	1	2	2	3	3	4	—	—
	9	—	—	—	0	0	0	1	1	1	2	2	3	—	—
	8	—	—	—	—	0	0	0	1	1	1	1	2	—	—
	7	—	—	—	—	—	0	0	0	0	0	1	1	—	—
	6	—	—	—	—	—	—	—	0	0	0	0	0	—	—
	5	—	—	—	—	—	—	—	—	—	—	0	0	—	—
14	14	0	1	1	2	3	3	4	5	6	6	7	8	9	—
	13	0	0	1	1	2	2	3	4	4	5	6	6	7	—
	12	—	0	0	1	1	2	2	3	3	4	4	**5**	6	—
	11	—	—	0	0	1	1	2	2	3	3	3	4	4	—
	10	—	—	0	0	0	1	1	1	2	2	3	3	3	—
	9	—	—	—	0	0	0	0	1	1	1	2	2	2	—
	8	—	—	—	—	0	0	0	0	1	1	1	1	2	—
	7	—	—	—	—	—	—	0	0	0	0	0	1	1	—
	6	—	—	—	—	—	—	—	—	0	0	0	0	0	—
	5	—	—	—	—	—	—	—	—	—	—	—	0	0	—
15	15	0	1	1	2	3	4	4	5	6	7	7	8	9	10
	14	0	0	1	1	2	3	3	4	5	5	6	7	7	8
	13	—	0	0	1	1	2	2	3	4	4	5	5	6	6
	12	—	0	0	0	1	1	2	2	3	3	4	4	5	5
	11	—	—	0	0	0	1	1	2	2	2	3	3	4	4
	10	—	—	—	0	0	0	1	1	1	2	2	2	3	3
	9	—	—	—	—	0	0	0	1	1	1	1	2	2	2
	8	—	—	—	—	—	0	0	0	0	1	1	1	1	1
	7	—	—	—	—	—	—	—	0	0	0	0	0	1	1
	6	—	—	—	—	—	—	—	—	—	0	0	0	0	0
	5	—	—	—	—	—	—	—	—	—	—	—	—	—	—

Adapted from Finney, D. J. 1948: The Fisher–Yates test of significance in 2 × 2 contingency tables, *Biometrika* **35**, 145–156, by kind permission of the author and publishers.

APPENDIX G (PART II)

A + B	B (or A)	C + D												
		3	4	5	6	7	8	9	10	11	12	13	14	15
		D (or C) at 1% level												
5	5	—	—	0	—	—	—	—	—	—	—	—	—	—
6	6	—	0	0	0	—	—	—	—	—	—	—	—	—
7	7	—	0	0	1	1	—	—	—	—	—	—	—	—
	6	—	—	—	0	0	—	—	—	—	—	—	—	—
8	8	—	0	0	1	1	2	—	—	—	—	—	—	—
	7	—	—	0	0	0	0	—	—	—	—	—	—	—
	6	—	—	—	—	—	0	—	—	—	—	—	—	—
9	9	0	0	1	1	2	2	3	—	—	—	—	—	—
	8	—	—	0	0	0	1	1	—	—	—	—	—	—
	7	—	—	—	—	0	0	0	—	—	—	—	—	—
	6	—	—	—	—	—	—	0	—	—	—	—	—	—
10	10	0	0	1	1	2	2	3	3	—	—	—	—	—
	9	—	0	0	0	1	1	2	2	—	—	—	—	—
	8	—	—	—	0	0	0	1	1	—	—	—	—	—
	7	—	—	—	—	—	0	0	0	—	—	—	—	—
11	11	0	0	1	1	2	3	3	4	4	—	—	—	—
	10	—	0	0	0	1	1	2	2	3	—	—	—	—
	9	—	—	0	0	0	1	1	1	2	—	—	—	—
	8	—	—	—	—	0	0	0	0	1	—	—	—	—
	7	—	—	—	—	—	—	0	0	0	—	—	—	—
12	12	0	0	1	2	2	3	3	4	5	5	—	—	—
	11	—	0	0	1	1	2	2	3	3	4	—	—	—
	10	—	—	0	0	0	1	1	2	2	2	—	—	—
	9	—	—	—	0	0	0	0	1	1	1	—	—	—
	8	—	—	—	—	—	0	0	0	0	1	—	—	—
	7	—	—	—	—	—	—	0	0	0	—	—	—	—
13	13	0	0	1	2	2	3	4	4	5	5	6	—	—
	12	—	0	0	1	1	2	2	3	3	4	4	—	—
	11	—	—	0	0	1	1	1	2	2	3	3	—	—
	10	—	—	—	0	0	0	1	1	1	2	2	—	—
	9	—	—	—	—	0	0	0	0	1	1	1	—	—
	8	—	—	—	—	—	—	0	0	0	0	0	—	—
	7	—	—	—	—	—	—	—	—	0	0	0	—	—

A + B	B (or A)	C + D												
		3	4	5	6	7	8	9	10	11	12	13	14	15
		D (or C) at 1% level												
14	14	0	1	1	2	2	3	4	4	5	6	6	7	—
	13	—	0	0	1	1	2	3	3	4	4	5	5	—
	12	—	0	0	0	1	1	2	2	3	3	3	4	—
	11	—	—	0	0	0	1	1	1	2	2	2	3	—
	10	—	—	0	—	0	0	0	1	1	1	2	2	—
	9	—	—	—	—	—	0	0	0	0	1	1	1	—
	8	—	—	—	—	—	—	—	0	0	0	0	0	—
	7	—	—	—	—	—	—	—	—	—	—	0	0	—
15	15	0	1	1	2	3	3	4	5	5	6	7	7	8
	14	0	0	1	1	2	2	3	3	4	4	5	6	6
	13	—	0	0	0	1	1	2	2	3	3	4	4	5
	12	—	—	0	0	0	1	1	2	2	2	3	3	4
	11	—	—	—	0	0	0	1	1	2	2	2	2	3
	10	—	—	—	—	0	0	0	0	1	1	1	1	2
	9	—	—	—	—	—	—	0	0	0	0	0	1	1
	8	—	—	—	—	—	—	—	—	0	0	0	0	0
	7	—	—	—	—	—	—	—	—	—	—	—	0	0

APPENDIX H

Table of χ^2 for 1 degree of freedom

$P\%$	χ^2
99	0·00016
98	0·00063
95	0·0039
90	0·016
80	0·064
70	0·15
50	0·46
30	1·07
20	1·64
10	2·71
5	3·84
2	5·41
1	6·64
0·1	10·83

APPENDIX I

The 5% significance level for the number of points falling in any quadrant
of a scatter diagram
(Instructions for use on p. 80)

N	Lower	Upper	N	Lower	Upper
8–9	0	4	74–75	13	24
10–11	0	5	76–77	14	24
12–13	0	6	78–79	14	25
14–15	1	6	80–81	15	25
16–17	1	7	82–83	15	26
18–19	1	8	84–85	16	26
20–21	2	8	86–87	16	27
22–23	2	9	88–89	16	28
24–25	3	9	90–91	17	28
26–27	3	10	92–93	17	29
28–29	3	11	94–95	18	29
30–31	4	11	96–97	18	30
32–33	4	12	98–99	19	30
34–35	5	12	100–01	19	31
36–37	5	13	110–11	21	34
38–39	6	13	120–1	24	36
40–41	6	14	130–1	26	39
42–43	6	15	140–1	28	42
44–45	7	15	150–1	31	44
46–47	7	16	160–1	33	47
48–49	8	16	170–1	35	50
50–51	8	17	180–1	37	53
52–53	8	18	200–01	42	58
54–55	9	18	220–1	47	63
56–57	9	19	240–1	51	69
58–59	10	19	260–1	56	74
60–61	10	20	280–1	61	79
62–63	11	20	300–01	66	84
64–65	11	21	320–1	70	90
66–67	12	21	340–1	75	95
68–69	12	22	360–1	80	100
70–71	12	23	380–1	84	106
72–73	13	23	400–01	89	111

Reproduced from "Rapid Statistical Calculations", (M. H. Quenouille),
Charles Griffin and Company Ltd., by kind permission of author and publishers.

APPENDIX J

The use of week numbers instead of dates

The analysis of annual records, e.g. of light trap or suction trap catches, is simplified if the numbered "weeks" contain the same dates in successive years. This can easily be obtained by dropping the dates February 29 and December 31 from the calendar. The weeks then become, by number, as follows:

	Week No.	Dates		Week No.	Dates
Winter	1	Jan. 1–Jan. 7	**Summer**	27	July 2–July 8
	2	Jan. 8–Jan. 14		28	July 9–July 15
	3	Jan. 15–Jan. 21		29	July 16–July 22
	4	Jan. 22–Jan. 28		30	July 23–July 29
	5	Jan. 29–Feb. 4		31	July 30–Aug. 5
	6	Feb. 5–Feb. 11		32	Aug. 6–Aug. 12
	7	Feb. 12–Feb. 18		33	Aug. 13–Aug. 19
	8	Feb. 19–Feb. 25		34	Aug. 20–Aug. 26
	9	Feb. 26–Mar. 4		35	Aug. 27–Sept. 2
Spring	10	Mar. 5–Mar. 11	**Autumn**	36	Sept. 3–Sept. 9
	11	Mar. 12–Mar. 18		37	Sept. 10–Sept. 16
	12	Mar. 19–Mar. 25		38	Sept. 17–Sept. 23
	13	Mar. 26–Apr. 1		39	Sept. 24–Sept. 30
	14	Apr. 2–Apr. 8		40	Oct. 1–Oct. 7
	15	Apr. 9–Apr. 15		41	Oct. 8–Oct. 14
	16	Apr. 15–Apr. 22		42	Oct. 15–Oct. 21
	17	Apr. 23–Apr. 29		43	Oct. 22–Oct. 28
	18	Apr. 30–May 6		44	Oct. 29–Nov. 4
	19	May 7–May 13		45	Nov. 5–Nov. 11
	20	May 14–May 20		46	Nov. 12–Nov. 18
	21	May 21–May 27		47	Nov. 19–Nov. 25
	22	May 28–June 3		48	Nov. 26–Dec. 2
Summer	23	June 4–June 10	**Winter**	49	Dec. 3–Dec. 9
	24	June 11–June 17		50	Dec. 10–Dec. 16
	25	June 18–June 24		51	Dec. 17–Dec. 23
	26	June 25–July 1		52	Dec. 24–Dec. 30

To find α, find the intersection of the number of individuals (N) along the x axis, and the number of species (S) up the y axis and read off the value of α from the nearest of the heavy divergent lines. For example:

with 1000 moths in 130 species, α = 40

The curved broken lines indicate the accuracy of the estimate of α, such that the 95% limits are ± 2 × the given percentage of α. For example:

with 1000 moths in 43 species α is 9 ± 7·5%

i.e. the 95% limits for α are 9 ± (2 × 7·5% of 9) = 9 ± 1·35, i.e. 7·65 and 10·35

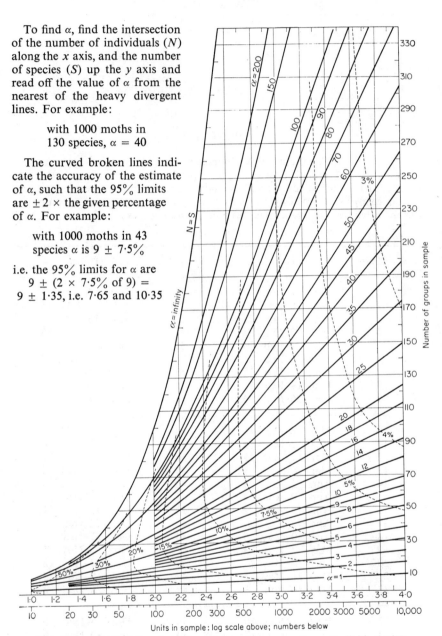

FIG. K:1. The relation between total number of units, or individuals (N), the number of groups, or species (S), and the Index of Diversity (α) in the Logarithmic Series: $S = \alpha \log_e (1 + N/\alpha)$

APPENDIX L
Pairs of random co-ordinates from 1 to 100
Select co-ordinates systematically; e.g. begin column 4, row 27 and take every 7th pair. Do not select "by eye".

x	y	x	y	x	y	x	y	x	y	x	y	x	y
98	03	91	50	48	07	33	26	12	72	56	16	42	36
56	06	03	45	71	88	05	53	56	59	31	85	96	18
98	68	89	41	08	92	98	61	65	100	78	12	66	10
96	06	13	43	38	51	85	13	34	87	98	81	88	77
09	02	71	71	51	83	04	41	70	39	95	66	67	98
54	80	19	28	78	12	03	10	48	21	03	35	95	39
40	69	56	38	68	73	54	08	09	04	72	93	90	54
100	31	39	27	95	28	68	50	71	30	80	81	22	30
96	74	73	13	82	17	39	90	56	33	85	79	47	19
51	22	81	60	13	38	56	50	97	50	32	25	73	87
94	36	05	62	26	40	59	77	40	33	08	64	69	63
07	15	62	97	48	77	25	19	17	78	97	96	33	56
15	90	31	13	43	15	23	02	39	46	80	66	58	61
04	02	97	38	80	40	55	85	90	14	26	02	78	35
39	37	32	11	96	59	68	45	60	22	03	30	58	70
29	45	81	99	32	24	69	31	35	27	98	59	34	78
28	10	45	74	18	64	37	31	37	11	64	72	47	42
23	26	11	84	43	47	66	42	100	84	98	02	33	11
75	09	14	66	89	58	33	65	12	08	76	66	97	30
46	14	40	25	61	21	76	32	60	60	97	28	86	62
22	17	44	48	55	80	43	33	60	09	53	58	54	80
86	56	41	94	30	85	28	31	67	85	14	96	68	47
91	25	07	12	41	92	97	19	62	95	32	22	13	26
46	66	64	27	62	40	82	80	48	79	24	32	22	17
29	12	80	71	13	50	03	68	88	09	30	28	19	36
29	41	27	06	78	66	65	16	12	75	04	73	16	77
43	30	54	68	51	57	24	65	61	73	42	70	78	43
66	40	02	92	66	86	02	72	48	06	83	27	03	28
96	11	83	52	19	83	79	16	71	42	24	77	93	22
31	89	38	61	51	78	04	75	85	64	82	77	78	76
64	27	01	01	79	68	40	64	48	69	33	14	23	68
34	40	21	66	73	52	06	27	14	83	04	51	15	39
84	39	32	29	63	99	62	40	09	11	50	09	58	71
76	28	04	59	86	28	100	97	54	52	60	73	57	35
61	23	38	64	97	96	50	64	50	58	93	09	48	50
62	48	48	33	93	41	38	54	35	69	91	67	61	96

x	y	x	y	x	y	x	y	x	y	x	y	x	y
09	70	82	82	40	24	46	86	38	58	49	92	36	93
81	34	63	100	06	06	90	74	72	22	67	95	18	87
58	67	94	51	97	81	66	21	04	69	54	50	88	53
39	50	60	52	65	99	87	05	68	50	56	23	09	72
36	85	98	01	16	91	46	90	16	47	11	28	12	96
62	90	26	45	62	03	11	88	20	50	77	55	85	94
77	54	59	81	92	27	11	05	39	58	35	96	38	64
32	55	10	85	45	51	33	94	92	17	02	84	53	44
12	48	08	56	100	81	24	89	15	64	49	90	40	76
85	10	36	01	05	06	15	19	46	86	75	27	02	17
90	26	78	38	12	68	05	64	48	28	92	42	95	17
78	04	32	59	07	79	57	49	58	92	34	59	35	76
60	86	60	14	52	16	77	82	52	62	71	63	26	29
96	06	87	39	16	01	24	10	55	98	61	63	77	80
28	89	60	58	89	84	50	100	44	67	32	15	46	40
30	29	06	49	51	99	44	37	46	44	68	49	37	56
95	74	01	28	08	12	90	57	30	80	50	93	61	65
01	85	58	57	69	99	50	74	89	99	92	20	93	43
10	91	76	56	88	91	44	04	62	03	21	68	21	96
05	33	72	49	59	45	32	74	17	27	13	81	95	20
04	43	68	98	84	27	75	73	19	79	25	76	97	86
05	85	03	02	65	45	76	18	93	74	83	79	69	92
84	90	04	47	48	28	100	17	17	18	61	58	04	31
28	55	04	78	93	18	54	95	42	37	48	84	61	06
89	83	36	13	18	26	69	99	35	63	07	28	85	93
73	20	71	100	45	62	25	47	26	41	46	13	21	74
10	89	88	80	26	22	47	46	98	10	32	25	15	69

Appendix L is taken from Table No. 33 of Fisher and Yates, "Statistical Tables for Biological, Agricultural and Medical Research", published by Oliver and Boyd, Ltd., Edinburgh, and by permission of the authors and publishers.

APPENDIX M

Suggested symbols

a = y-intercept, or added constant, in a regression equation
α = the index of diversity and growth rate
$b\begin{cases} = \text{regression coefficient} = \tan \beta \\ = \text{the index of aggregation and initial growth index} \end{cases}$
β = the angle of inclination of a regression line
m = the mean = $\bar{x}$ or $\bar{y}$
n = the number of sample units in a sample
N = the number of individuals in a sample or unit, e.g. moths in a moth catch
ϕ = physiographic score for a site
S = the number of groups in a mixed sample, e.g. species in a moth catch
s^2 = variance
s = standard deviation
t = time
θ = temperature
x = the independent variable
χ^2 = chi-squared
y = the dependent variable
$>$ = greater than
$<$ = less than
$\geqslant$ = greater than or equal to
$\leqslant$ = equal to or less than
$\pm$ = plus or minus
$\overset{\times}{\div}$ = times or divided by

APPENDIX N

Additional Keys and Reading

Keys to the most easily identifiable taxa of terrestrial invertebrates, more detailed than those included in this book, or more suitable for the fauna of the U.S., are listed below. The simple keys chosen usually include as many pictures as possible. There are separate, detailed, and usually difficult keys, not listed here, to most orders and to the larger families of British insects published by the Royal Entomological Society, 41 Queen's Gate, London. Books on specialized habitats, general ecology, analytical and sampling methods and museum techniques are recommended. One asterisk indicates an elementary key or introductory text; two, an advanced key or reference work. Works relevant to the British or North American fauna are marked (U.K.) and (U.S.).

Land Invertebrates (except insects)

* CLOUDSLEY-THOMPSON, J. L., and SANKEY, J. (1961). "Land Invertebrates." Methuen, London (U.K.).

Lumbricidae (Earthworms)

** GERARD, B. M. (1964). "Synopses of the British Fauna", No. 6, *Lumbricidae (Annelida)*. Linn. Soc., London (U.K. and U.S.).

Gastropoda (Snails and Slugs)

* JANUS, H. (1965). "Land and Freshwater Molluscs", Burke, London (U.K.).
* BURCH, J. B. (1962). "The Eastern Land Snails", Brown, Dubuque, Iowa (U.S.).
** PILSBRY, H. A. (1939–48). "Land Molluscs of North America (North of Mexico)", Vol. I, II. Acad. Nat. Sci., Philadelphia (U.S.).

Isopoda (Woodlice, Sow bugs)

** EDNEY, E. B. (1954). "Synopses of the British Fauna", No. 9, *British Woodlice*. Linn. Soc., London (U.K.).
** HATCH, M. H. (1947). "The Chelifera and Isopoda of Washington and Adjacent Regions", *Univ. Wash. Publ. Biol.*, **10**: 155–274 (U.S.).

Acari (Mites)

** EVANS, G. O., SHEALS, J. G. and MACFARLANE, D. (1961). "The terrestrial Acari of the British Isles", British Museum (Nat. Hist.), London (U.K.).
** BAKER, E. W. and WHARTON, G. W. (1952). "An Introduction to Acaralogy", Macmillan, New York (U.S.).

Araneida (Spiders)

** LOCKET, G. H. and MILLIDGE, A. F. (1951, 1953). "British Spiders", Vol. I, II. Ray Soc., London (U.K.).

* KASTON, B. J. (1953). "How to Know the Spiders", Brown, Dubuque, Iowa (U.S.).

** KASTON, B. J. (1948). "The Spiders of Connecticut", *State Geol. and Nat. Hist. Survey. Bull.* No. 70 (U.S.).

Chelonethi (False Scorpions)

** EVANS, G. O. and BROWNING, E. (1954). "Synopses of the British Fauna", No. 10, *Pseudoscorpiones*. Linn. Soc., London (U.K.).

Opiliones (Harvestmen)

* SANKEY, J. (1956). "How to Begin the Study of Harvestspiders", *Countryside*, **17**: 370–77 (U.K.).

** TODD, V. (1948). "Key to the Determination of British Harvestmen (*Arachnida, Opiliones*)", *Ent. mon. Mag.*, **84**: 109–113 (U.K.).

Chilopoda (Centipedes)

* EASON, E. H. (1964). "Centipedes of the British Isles", Warne, London (U.K.).

Diplopoda (Millipedes)

** BLOWER, J. G. (1958). "Synopses of the British Fauna", No. 11, *British Millipedes* (*Diplopoda*). Linn. Soc., London (U.K.).

** CHAMBERLIN, R. V. and HOFFMAN, R. L. (1958). "Check List of the Millipedes of N. America", *U.S. Nat. Museum Bull.*, **212** (U.S.).

Insects (general)

* JAQUES, H. E. (1947). "How to Know Insects", Brown, Dubuque, Iowa (U.S.).

* CHU, H. F. (1949). "How to Know the Immature Insects", Brown, Dubuque, Iowa (U.S. and U.K.).

* GATES, D. E. and PETERS, L. L. (1962). "Insects in Kansas", Rev. Edn. Extension Division, Kansas State Univ., Manhattan, Kansas (U.S.).

Lepidoptera (Moths, Butterflies)

* SOUTH, R. (1961). "The Moths of the British Isles", Series I, II. Warne, London (U.K.).

* SOUTH, R. (1941). "The Butterflies of the British Isles". Warne, London (U.K.).

* HOLLAND, W. J. (1913). "The Moth Book", Doubleday, New York (U.S.).

* HOLLAND, W. J. (1914). "The Butterfly Book", Doubleday. New York, New York (U.S.).

* EHRLICH, P. R. (1961). "How to Know the Butterflies", Brown, Dubuque, Iowa (U.S.).

Neuroptera (Lacewings)

** KILLINGTON, F. J. (1936, 1937). "A Monograph of the British Neuroptera", Vol. I, II. Ray Soc., London (U.K.).
* JAQUES, H. E. (1947). "How to Know Insects", Brown, Dubuque, Iowa (U.S.).

Trichoptera (Caddis flies)

* HICKIN, N. E. (1952). "Caddis", Methuen, London (U.K.).
** MOSELEY, M. E. (1939). "The British Caddis Flies", Routledge, London (U.K.).
* JAQUES, H. E. (1947). "How to Know Insects", Brown, Dubuque, Iowa (U.S.).

Ephemeroptera (Mayflies)

** KIMMINS, D. E. (1954). "Revised Key to the Adults of British Species of Ephemeroptera", Fresh-water Biological Assoc. Sci. Publ. No. 15 (U.K.).
* JAQUES, H. E. (1947). "How to Know Insects", Brown, Dubuque, Iowa (U.S.).

Odonata (Dragonflies)

* JAQUES, H. E. (1947). "How to Know Insects", Brown, Dubuque, Iowa (U.S.).

Thysanoptera (Thrips)

** MORISON, G. D. (1947, 48, 49). "Thysanoptera of the London Area", London Naturalist Reprint, No. 59, Buncle, Arbroath (U.K.).
** BAILEY, S. F. (1957). "The Thrips of California", Pt. I. Sub order Terebrantia. Bull. Calif. Insect Survey, 4: 143–220 (U.S.).
** STANNARD, L. J. (1957). "The Phylogeny and Classification of the North American Genera of the Sub-order Tubulifera (Thysanoptera). Illinois Biological Monographs, No. 25 (U.S.).

Hemiptera-Homoptera (Plant lice)

* VILLIERS, A. (1947). "Atlas des Hemiptères de France", Vol. II, Boubeé, Paris (In French but well illustrated) (U.K.).
* JAQUES, H. E. (1947). "How to Know Insects", Brown, Dubuque, Iowa (U.S.).

Hemiptera-Heteroptera (Bugs)

** SOUTHWOOD, T. R. E. and LESTON, D. (1954). "Land and Water Bugs of the British Isles", Warne, London (U.K.).
* JAQUES, H. E. (1947). "How to Know Insects", Brown, Dubuque, Iowa (U.S.).

Hymenoptera (Bees, Wasps, Sawflies, Parasites)

* BERLAND, L. (1958). "Atlas des Hymenoptères de France", Vol. I, II, Boubeé, Paris (In French but well illustrated) (U.K.).
* JAQUES, H. E. (1947). "How to Know Insects", Brown, Dubuque, Iowa (U.S.).

Hymenoptera (Bumble bees)

* YARROW, I. H. H. (1959). "The Identification of British Bumblebees", in FREE, J. B. and BUTLER, C. G. (1959). "Bumblebees", Collins, London (U.K.).
** MITCHELL, T. B. (1962). "Bees of the Eastern United States", Vol. II. Tech. Bull. N.C. Agric. Exp. Sta. No. 152 (U.S.).

Hymenoptera (Ants)

** COLLINGWOOD, C. A. (1964). "The Identification and Distribution of British Ants" (*Hym. Formicidae*). *Trans. Soc. Brit. Ent.*, **16**: 93–121 (U.K.).
** WHEELER, G. C. and WHEELER, J. (1963). "The Ants of North Dakota". University of North Dakota Press, Grand Forks, N. Dakota (U.S.).

Dermaptera (Earwigs)

* HELFER, J. R. (1963). "How to Know the Grasshoppers and Their Allies". Brown, Dubuque, Iowa (U.S.).

Orthoptera (Grasshoppers, Crickets, Katydids)

** RAGGE, D. R. (1965). "Grasshoppers, Crickets and Cockroaches of the British Isles". Warne, London (U.K.).
* HELFER, J. R. (1963). "How to Know the Grasshoppers and Their Allies". Brown, Dubuque, Iowa (U.S.).

Coleoptera (Beetles)

* BECHYNE, J. (1956). "Beetles". Thomas and Hudson, London (U.K.).
** JOY, N. H. (1932). "A Practical Handbook of British Beetles", Vols. I, II. Witherby, London (U.K.).
* JAQUES, H. E. (1951). "How to Know the Beetles", Brown, Dubuque, Iowa (U.S.).

Diptera (Two-winged flies)

* COLYER, C. N. and HAMMOND, C. O. (1951). "Flies of the British Isles", Warne, London (U.K.).
* JAQUES, H. E. (1947). "How to Know Insects", Brown, Dubuque, Iowa (U.S.).

Specialized Habitats

* JACKSON, R. M. and RAW, F. (1966). "Life in the Soil", Arnold, London (U.K. and U.S.).
* ENGELHARDT, W. and MERXMÜLLER, H. (1964). "Pond Life", Burke, London (U.K.).
* MACAN, T. T. (1959). "A Guide to Freshwater Invertebrate Animals", Longmans, London (U.K.).

Flora

* MCLINTOCK, D. and FITTER, R. S. R. (1965). "Pocket Guide to Wild Flowers", Collins, London (U.K.).
** CLAPHAM, A. R., TUTIN, T. G. and WARBURG, E. F. (1959). "Excursion Flora of the British Isles", Cambridge University Press, London (U.K.).
* CUTHBERT, M. J. (1948). "How to Know the Fall Flowers", Brown, Dubuque, Iowa (U.S.).
* CUTHBERT, M. J. (1949). "How to Know the Spring Flowers", Brown, Dubuque Iowa (U.S.).
* JAQUES, H. E. (1959). "How to Know the Weeds", Brown, Dubuque, Iowa (U.S.).
* POHL, R. N. (1953). "How to Know the Grasses", Brown, Dubuque, Iowa (U.S.).

General Ecology

* ODUM, E. P. (1963). "Ecology", Holt, Rinehart and Winston, New York and London (U.K. and U.S.).
** CLARK, G. L. (1965). "Elements of Ecology", Wiley, New York, London (U.K. and U.S.).

Sampling

** SOUTHWOOD, T. R. E. (1966). "Ecological Methods", Methuen, London (U.K. and U.S.).
** YATES, F. (1963). "Sampling Methods for Censuses and Surveys", 3rd Ed., Griffin, London (U.K. and U.S.).

Analysis

** ANDREWARTHA, H. G. (1961). "Introduction to the Study of Animal Populations", Methuen, London (U.K. and U.S.).
** BAILEY, N. T. J. (1964). "Statistical Methods in Biology", English Universities, London (U.K. and U.S.).
** DEFARES, J. G. and SNEDDON, I. N. (1960). "An Introduction to the Mathematics of Medicine and Biology", North Holland Publishing Co., Amsterdam (U.K. and U.S.).
** FISHER, R. A. and YATES, F. (1957). "Statistical Tables for Biological, Agricultural and Medical Research", 5th Ed., Oliver and Boyd, Edinburgh (U.K. and U.S.).
** FINNEY, D. J., LATSCHA, R., BENNETT, B. M. and HSU, P. (1964). "Tables for Testing Significance in a 2 × 2 Contingency Table". *Biometrika*, C.U.P., London (U.K. and U.S.).
** QUENOUILLE, M. H. (1959). "Rapid Statistical Calculations", Griffin, London (U.K. and U.S.).
** WILLIAMS, C. B. (1964). "Patterns in the Balance of Nature", Academic Press, London, New York (U.K. and U.S.).

The Physical Environment

** PLATT, R. B. and GRIFFITHS, J. (1964). "Environmental Measurement and Interpretation", Reinhold, New York; Chapman and Hall, London (U.K. and U.S.).
** GEIGER, R. (1965). "The Climate near the Ground", Harvard University Press, Cambridge, Mass. (U.K. and U.S.).

Collecting and Museum Methods.

* OLDROYD, H. (1958). "Collecting and Preserving Insects", Hutchinson, London (U.K. and U.S.).
* SMART, J. (1954). "Instructions for Collectors", British Museum (Nat. Hist.), London (U.K. and U.S.).
** WAGSTAFFE, R. and FIDLER, J. M. (1957). "The Preservation of Natural History Specimens", Vol. I, "Invertebrates", Witherby, London (U.K. and U.S.).

Acknowledgements

We thank the following authors of papers containing data used in exercises: Askew, R. R. (Ex. 41, 42); Arias, R. O. and Crowell, H. H. (Ex. 9); Brian, A. D. (Ex. 30); Cameron, E. (Ex. 33); Edwards, C. E. (Ex. 34, 36); Free, J. B. (Ex. 26, 28, 29); Harper, P. S. (Ex. 22); Heath, G. W. (Ex. 37); Ivy, A. C. (Ex. 1); Madge, D. (Ex. 10, 13, 14); Raw, F. (Ex. 38); Savory, T. H. (Ex. 8); Smith, R. C. (Ex. 12); Synge, A. D. (Ex. 31); Twigg, H. M. (Ex. 39) and Williams, C. B. (Ex. 32, 45). Mr. S. C. Littlewood supplied data from Burleigh School, Hatfield, England, for Exs. 2, 11, and 44.

Where figures have been redrawn, the original author is gratefully acknowledged in the legend. We thank the following publishers and authors for permission to reproduce figures: *J. Anim. Ecol.* and Askew, R. R. for Figs. IV 41:2, 41:3, 41:4, 42:1, 42:2; *J. Anim. Ecol.* and Synge, A. D. for Figs. IV:30:4, 31:1; *Bull. ent. Res.* and Cameron, E. for Fig. IV:33:2; Field Studies Council and Sinker, C. for Fig. IV:39:1a and Twigg, H. M. for Figs. IV:39:1b, 39:2; Academic Press and Williams, C. B. for Figs. IV: 32:1, 32:2, III:35, *J. exp. Biol.* and Edney, E. B. for Fig. II:4; Univ. of Chicago Press and Andrewartha, H. G. for Fig. II:12. We are indebted to the Literary Executor of the late Sir Ronald A. Fisher, F.R.S., Cambridge, to Dr. Frank Yates, F.R.S., Rothamsted and to Messrs. Oliver & Boyd Ltd., Edinburgh, for permission to reprint part of Table No. 33 from their book "Statistical Tables for Biological, Agricultural and Medical Research". We thank the following for photographs: Mr. D. J. Greathead and Anti-Locust Res. Centre, London for that of locust damage to sorghum and the Radio Times Hulton Picture Library for that of bomb damage (Plate 1); Mr. P. Webb for the picture of traps (Plate 2), Miss J. M. Thurston for Plate 3 and Mr. F. Cowland for Plate 4.

We thank the Soil Survey Unit of Great Britain for permission to reproduce categories of soil texture on p. 259; Mr. J. W. Siddorn, Imperial College, University of London, for information on the construction of thermocouples; Miss B. J. Pyrah who drew the illustrations in Chapter VI and Misses V. Kibbey, M. J. Dupuch and J. Wilson who assisted with other illustrations.

We gratefully acknowledge the help given in checking exercises and keys by the staff and many teachers, students and pupils on field courses at Flatford Mill Field Centre, Suffolk, England and by Mr. S. C. Littlewood and pupils of Burleigh School, Hatfield. We thank Drs. C. G. Johnson, Head of Entomology Department, Rothamsted and T. R. E. Southwood, Reader in Insect Ecology, Imperial College, University of London who read the manuscript and helped with many valuable suggestions, as did Mr. G. J. S. Ross, Statistics Department, Rothamsted, for Chapter III, though we alone accept full responsibility for the views expressed in the book. Dr. H. Knutson kindly made facilities available in the Entomology Department at Kansas State University, Manhattan, Kansas, U.S.A. for one of us (L. R. T.).

We thank our wives for much patient help in preparing, checking and typing the manuscript.

Subject Index

Italic numbers indicate pages on which a figure is shown.
Only the first page of a reference is listed.

A

Abcissa, 82
  see also *x*-axis
Abrostola triplasia, 173
Acalypterates, 127, 129, 131, 154, 156, 188, 189, 288
Acari, 71, 72, 133, 160, 161, 164, 208, 209, 210, 238, 239, 264, *265*, 376
  see also Mites
Achillea millefolium, 223
Activity
  and aggregation, 183
  annual cycle of, 11, 191, 197
  daytime, 7, 13, 187, 197
  effect of weather on, 8, 12, 84, 134, 141, 175
  night time, 13, 185, 187
  synchronisation with environment, 7, 196, 206
Adaptation to environment, 24, 113, 160
  morphological, 27, 153, 158
  physiological, 8, 25, 157
Adaptive colouration, 24, 151, 157
Aeolothripidae, 128, 131
Aerial
  distribution, 29, 167, 170, 175, 180
  environment, 2, 160, 166, 180
Aesculus hippocastanum, 229, 230
Aeshna grandis, 227
Aestivate, 8
Agabus bipustulatus, 226
Age-structure in population, 73, 183
Aggregation, 16, 85
  and activity, 183
  index of, 23, 89, 178, 181, 375
  in Man, 4, 180, 182
  and migration, 180
Aggression, 21, 185
  induced by Man, 186, 187
  in Man, 4, 157, Plate 1
Aglais urticae, 24, 28

Agraylea (larva), 224, 227
Agriculture, 1, 210, 212, 217
  ecology in, 1
Agriolimax reticulatus, 137, 138
Agrochola, 168
Agromyzidae, 202, 294
Agrostis, 237
  stolonifera, 223
Agrotis exclamationis, 172, 199
  ipsilon, 172
Air, sampling animals in, 126, 141, 168, 175, 197, 239, 245, 250, 253
Albatross, 166
Alcohol, 255
Alder, 226
Alderfly, larva, *225*
Aldrin, 208
Alcis repandata, 160
Aleyrodidae, 284, *285*
Alfalfa, 180
  butterfly, 145
Allometric growth, 109
Altitude, 5
Amathes c-nigrum, 173
Amblythrips ericae, 29
Amphimallon, 23
Anabolia, 228
  nervosa, 227
Anacaena limata, 227
Analysis, regression, 78, 118, 163, 247, 325
Anaplectoides prasina, 173
Ancylus lacustris, 226
Angelica sylvestris, 223
Angle Shades, 173
Anglewing, 153
Animal
  abundance, 113
  behaviour, 113
  distribution, 113
Annelida, 238, 239
  see also Earthworms

Behaviour (*contd.*)
  of populations, 2, 191
  of species, 192
  territorial, 21
  threatening, 21, 154, 185
Bell Heath, 237
Bena fagana, 172
Bent Grass, 237
Berula erecta, 223, 224
Bibionidae, 294, *295*
Bimodal distributions, 63
Biological
  control, 21, 202, 208
  equilibrium, 113
Biorrhyza pallida, 35
Birds
  and evolution of insects, 157, 160
  flight of, 2, 166
  predation by, 18, 153, 160, 206
  territorial behaviour, 21
Bird's Foot Trefoil, 237
Birthrate, 2, 5
  see also reproduction
Biston betularia, 159, 171, 174
Bithynia tentaculata, 226
Biting midges, 290
Blackflies, 284, *285*
  see also Aphididae
Blister Beetle, 31, *32*
Blood-vein Moth, 173
Blowflies, 12, 17
  see also Calliphoridae
Bluebottles, 156, 296
  see also Calliphoridae
Body temperature, 9, 36, 57, 115, 141, 144
  in Man, 36, 44, 59, 115
Bombus
  agrorum, 192, *193*
  hortorum, 192, *193*, *194*
  lucorum, 192, *193*, *194*
  pratorum, 192, *193*
Bombyliidae, 294, *295*
Booklice, 71, 257, 274
Bouin's fixative, 259
Braconid Wasp, 202
Branched Bur-reed, 226
  see also *Sparganium erectum*
Brassicae, 212
Breeding populations, 183

Bright-line Brown-eye, 173
Brimstone Moth, 174
Bristle-tails, 280, 281
Broad-bordered Yellow Underwing, 172
Broken-barred Carpet, 173
Broom Moth, 173
Buff Ermine Moth, 102, 171, 172, 174
Buff Tip Moth, 154, 172
Bugs, *see* Heteroptera; *see also* Homoptera
Bumble bees, *see* Bees, bumble
Burnished Brass Moth, 171, 173, 174
Butomus umbellatus, 221, 223

C

Cabbage
  Moth, 173
  Root-fly, 21
  White Butterfly, *see* Large White Butterfly
Cacao Thrips, 5, 6
Caddis
  flies, 257, 270, 378
  see also Trichoptera
  fly larva, *225*
Cadmium, 18
Caenis
  horaria, 228
  horaria, nymph, 226
Callimorpha jacobaeae, 172
Callimorphidae, 243
Calliphoridae, 127, 128, 131, 132, 133, 296, *297*
Calluna, 30, 237
Calopteryx
  splendens, 22
  virgo, 22
Calothysanis amata, 173
Camouflage, 24, 151, 157
Campaea margaritata, 174
Canals, 219
Cantharidae, 156, 188, 302, *303*, 312, *313*
  see also Soldier Beetle
Capsid bugs, 25, 286, *287*
  see also Mirid bugs
Carabidae, 132, 133, 155, 218, 310, *311*
  see also Ground beetles
Cardamine pratensis, 223

398 INDEX

Salad Burnet, 237
Saldidae, 155, 284, *285*
Sample unit, 95, 375
Sampling, 87, 93, 375
  aerial populations, 126, 141, 168, 175, 197, 239, 245, 250, 253
  by attraction, 97, 132
  bias, 93
  comparison of methods, 123, 125
  ground surface, 132
  by marking and recapture, 123
  by quadrats, 124, 237
  random, 96
  regular, 96
  by searching, 129
  soil, 209, 213, 218, 254
  stratified, 97
  by sweeping, 129, 187
Sand wasps, 322, *323*
Satellite Moth, 173
Saturation of soil, 138
Saturniidae, 243
Sawflies, 131, 154, 156, 274, 320, *321*, 378
Scale Insect, 27, 278, *279*
Scalloped Hazel Moth, 174
Scalops sulcipes, 15
Scarabaeidae, 316, *317*
  see also Chafer beetles
Scatopsidae, 155, 294, *295*
Scatter diagrams, 41, 67, 78, 143, 370
Sciarinae, 292, *293*
Sciomyzidae, 294
Scolytidae, 314, *315*
  see also Bark Beetle
Scolytus
  destructor, 14, 164
  multistriatus, 14, 164
Scorched Wing, 174
Scorpion flies, 257, 272, *273*
Sea
  Slater, 9, 256
  Slug, 10
Seagull, 2
Seasonal cycles, 11, 113, 197, 206
Sedge, 237
Seed weevils, *see* Apionidae
Selenia
  bilunaria, 198, 199
  lunaria, 174

Selenothrips rubrocinctus, 5, 6
Semilog paper, 73
Sepsidae, 296, *297*
Sesiidae, 243
Setaceous Hebrew Character, 172
Sex ratio, 100, 170, 184
Sexual attraction, 152
Shark Moth, 173
Sheep, 13, 16, 18, 31, 217
  Tick 16, 18
Sheeps Fescue, 237
Shield bugs, 286, *287*
Shore
  bugs, 284, *285*
  flies, 300, *301*
Shoulder-striped Wainscot, 173
Shrews, 186
Sialis
  larva, *225*
  lutaria, 226
Sigmoid curve, 51, 63
Significance
  of differences between means, 103
  of differences between two fractions, 99, 364
  of differences between two proportions, 99, 365
  levels of, 99
  meaning of, 78, 97
  χ^2 tests for, 100, 369
Significant figures, 38, 73
Silphidae, 132, 133, 306, *307*, 318, *319*
Silver
  Cloud, 173
  -Ground Carpet, 173
  Y, 171, 173, 174
Silverfish, 280, *281*
Sitaris muralis, 31, *32*
Sitodiplosis mosellana, 28
Size
  and area of territory, 22
  and competition, 164
  and environmental control, 166
  and flight speed, 167
  and rate of development, 149
Skew distributions, 48, 59, 87
Slugs, 137, 138, 139, 258, 268, *269*, 376
Small Tortoiseshell Butterfly, 24, 28
Semiothisa notata, 174

This book was first published just before the current wave of concern, especially amongst young people, for the environment. It was written in simple, non-technical language to help teachers and students appreciate the ecological principles underlying the interdependence of living things. The critical approach presented is even more necessary in the present atmosphere when there is a danger of ecology becoming a facile, emotional, panacea for all social ills.
In school teachers will appreciate the adaptability of the open-ended excercises while at higher levels students can use them as a guide in devising their own field experiments. Naturalists and student teachers may find the quantitative outlook and ideas useful for local fauna surveys and training projects, and under-graduates will find it a sound basis for more advanced studies.

"... authors and publishers to be congratulated on an excellent book ..."
Nature
"... probably the most useful, authoritative and carefully constructed guide to the concepts of the subject yet published in English ..." **Guardian**
"... a new and imaginative approach to a difficult subject ... an important reference and instructional text for beginning students ..." **Forest Science**
"... unique feature is the section of excercises in ecology ... they have all the hall-marks of excellent teaching instruments ..."
Assoc. for Science Education
"... selection is attractive ... presentation in most cases ideal ..." **Oikos**
"... well worth studying by anyone interested in ecological analysis. Many high school and elementary college courses will profit ..." **Castanea**
"... engenders a desire to get out into the field and enjoy being an ecologist ..."
Herts and Middx Trust for Nature Conservation

£2.40

_KW_

ACADEMIC PRESS
London and New York

A Subsidiary of Harcourt Brace Jovanovich, Publishers
24-28 Oval Road, London NW1, England
111 Fifth Avenue, New York, N.Y. 10003, U.S.A.

Registered office
Academic Press Inc. (London) Ltd.
24-28 Oval Road, London NW1 7DX

0.12.447156.0

Printed in England